AF212689

EXPLORATION OF THE DEEP CONTINENTAL CRUST

Edited by

H.-J. Behr, Göttingen
C.B. Raleigh, Palisades

Advisory Board

R.N. Anderson, Lamont
K. Fuchs, Karlsruhe
W.S. Fyfe, London (Canada)
H. Rischmüller, Hannover
W. Schreyer, Bochum
M. Zoback, Stanford

Deep Drilling in Crystalline Bedrock

Volume 1: The Deep Gas Drilling in the Siljan Impact Structure, Sweden and Astroblemes

Proceedings of the International Symposium
held in Mora and Orsa, September 7–10, 1987

Edited by

A. Bodén and K. G. Eriksson

With 151 Figures

Springer-Verlag
Berlin Heidelberg New York London Paris Tokyo

Lic. Eng. ANDERS BODÉN
Vattenfall
16287 Vällingby, Sweden

Professor Dr. K. GÖSTA ERIKSSON
Chalmers University of Technology
Department of Geology
41296 Göteborg, Sweden

ISBN-13:978-3-642-73454-0 e-ISBN-13:978-3-642-73452-6
DOI: 10.1007/978-3-642-73452-6

Library of Congress Cataloging-in-Publication Data. Deep drilling in crystalline bedrock / edited by A. Bodén and K. G. Eriksson. p. cm.-(Exploration of the deep continental crust) Proceedings of the Third International Symposium on Observation of the Continental Crust through Drilling, held in Mora and Orsa, Sweden, Sept. 7–10, 1987. Includes index. Contents:v.1.The deep gas drilling in the Siljan impact structure,Sweden and astroblemes.ISBN-13:978-3-642-73454-0(U.S.:v.1) 1. Earth–Crust–Congresses. 2. Rocks, Crystalline–Congresses. 3. Boring–Congresses. I. Bodén, A. (Anders), 1946- . II. Eriksson, K. G. (K. Gösta), 1922- . III. International Symposium on Observation of the Continental Crust through Drilling (3rd : 1987 : Mora, Sweden and Orsa, Sweden) IV. Series. QE511.D29 1988 551.1'3–dc 19 88-12158 CIP

This work is subject to copyright. All rights are reserved, whether the whole or part of the material is concerned, specifically the rights of translation, reprinting, re-use of illustrations, recitation, broadcasting, reproduction on microfilms or in other ways, and storage in data banks. Duplication of this publication or parts thereof is only permitted under the provisions of the German Copyright Law of September 9, 1965, in its version of June 24, 1985, and a copyright fee must always be paid. Violations fall under the prosecution act of the German Copyright Law.

© Springer-Verlag Berlin Heidelberg 1988
Softcover reprint of the hardcover 1st edition 1988

The use of registered names, trademarks, etc. in this publication does not imply, even in the absence of a specific statement, that such names are exempt from the relevant protective laws and regulations and therefore free for general use.

2132/3130-543210

Preface Volume 1

The rapid evolution of deep-drilling technique during recent years has, together with an increased scientific and economic interest for the origin and composition of deeper parts of the Earth's crust, motivated many countries to develop programs for investigation and, hopefully, also exploration of deep-situated natural resources within their territories. This fact was the motive for Lamont-Doherty Geological Observatory of Columbia University and the Inter-Union Commission on the Lithosphere to organize the first symposium dealing with "Observation of the Continental Crust through Drilling", in Tarrytown, USA, 1984. Presentation of ongoing or planned, deep-drilling programs and advances in drilling and logging technology were the two main themes of this meeting. The second symposium took place in Seeheim, Federal Republic of Germany, 1985, and was dominated by a review of national programs and a broad presentation of technique and coring problems in deep drilling. The third country as host in this sequence of symposia was Sweden, 1987. The meeting was arranged in Mora and Orsa, in the Siljan Ring area. The main subject during this meeting was the progress of the Deep Gas Project in the Siljan Ring in Precambrian rocks. Of great interest were also, among others, the presentations of the KTB-project in northern Bavaria, FRG, the Cajon Pass project along the San Andreas fault, California, USA, and the plans to deepen the super-deep well on the Kola Peninsula, Sovjet-Union, down to 15 kilometers.

In search of alternative energy resources in Sweden, Vattenfall (The Swedish State Power Board), initiated by professor T. Gold, USA, had for several years investigated the possibilities of a natural gas reservoir in the Siljan Ring area. This work led to a decision to drill a deep well to 7.5 km with spud in on July 1st, 1986.

The drilling project was still in progress during the symposium and had, at that time, reached a vertical depth in the crystalline bedrock of about 6300 meters. The drilling was furloughed in September 1987 but will according to plans commence again in April 1988. The drilling was initially planned as purely commercial, but as it also developed interesting scientific ideas it was motivated to invite interested scientists and engineers to discuss related issues at a Third International Symposium on Observation of the Continental Crust through Drilling. The symposium plans were approved by the Inter-Union Commission on the Lithosphere in Singapore, 1986.

About 250 participants took part in the symposium, representing 20 countries. During the fourth day of the symposium a one-day excursion took place in the Siljan Ring area.

Vattenfall together with Chalmers University of Technology and the Geological Survey of Sweden were in charge of the organization of the symposium.

The large number of attending scientists and engineers from many parts of the world and their active participation in the symposium is good evidence of the increasing interest in the research of deeper parts of the earth's crust. The proposals for future cooperation between different countries gives good hope for continued support to future deep drilling projects which are necessary for an expanding knowledge of the earth's crust in order to provide sufficient basis for exploration and utilization of natural resources.

The papers presented at this symposium have, for practical reasons, been printed in two volumes. This volume (I) contains mainly the papers dealing with the Deep Gas Project in the Siljan Ring. Volume II contains papers describing deep-drilling projects and technology from other areas of the world. Together they give a good reflection of the state of the art in the deep-drilling R&D-front in 1987.

For the steering group

K.Gösta Eriksson   T.Lindbo   A.Wesslén

A.Bodén
Secretary-General

The Third International Symposium on Observation of the Continental Crust Through Drilling was Organized by the Following Staff

Steering Committee

K Gösta Eriksson, Professor, Chalmers University of Technology
Tord Lindbo, Project Director of the Deep Gas Drilling, Vattenfall
Arne Wesslén, Director General, Geological Survey of Sweden

Organizing Committee

Per-Olov Karlsson, Vattenfall (Chairman)
Bengt Collini, University of Uppsala
K Gösta Eriksson, Chalmers University of Technology
Gunnar Gustafson, Chalmers University of Technology
Ulf Lindblom, Chalmers University of Technology
Elisabeth Rosén, Chalmers University of Technology

Sven-August Hultin, Vattenfall (Chairman of the Sponsor Committee)

Coordinators

Anders Bodén, Vattenfall (Secretary-General)
Ingvar Rhen, Viak AB
Mats Öberg, Hagconsult AB

Scientific Reviewers

John Beswick, United Kingdom; Laust Börsting Pedersen, Sweden; Bengt Collini, Sweden; Gunnar Gustafson, Sweden; Ulf Lindblom, Sweden; Thomas Lundqvist, Sweden.

Contents

Volume 1: The Deep Gas Drilling in the Siljan Impact Structure, Sweden and Astroblemes

Proceedings of the Third International Symposium on Observation of the Continental Crust through Drilling held in Mora and Orsa, Sweden, September 7th-10th, 1987

General Papers

Technical Papers

Scientific Papers

Contributors

You will find the addresses at the beginning of the respective contribution

General Papers

Siljan Drilling Experiment: Participation of the Gas Research Institute

F. FISH and A.W. GORODY[1]

Introduction

This paper will present the history and reasons for the participation of the Gas Research Institute (GRI) in the Gravberg-1 drillhole, what benefits GRI expected to receive from participation, what elements comprised GRI's participation, and what results were obtained. GRI's course in this endeavor has been guided by its strategy of collecting fundamental information on deep natural gas resources which will lead to the future recovery of these resources.

History

The first research funded by GRI in the geosiences area was a University Grant to Cornell University with Dr. Thomas Gold as principal investigator. The objective was the "elucidate the role of abiogenic methane in the outgassing of the Earth with the ultimate aim of determining whether there is a large enough supply of deep abiogenic methane to contribute significantly to our energy needs and, if so, whether it is possible to tap the supply." GRI is still supporting Dr. Gold's research on the basis that the consequences of his view that large amounts of the methane presently found near the Earth's surface are of deep abiogenic origin is so important that research is warranted even though the probability of his being correct is perceived by most geoscientists as being very small. Even if only 10 or 20% of the methane is abiogenic, it constitutes a large resource, a resource which could be overlooked in conventional exploration efforts.

Vattenfall representatives had visited GRI in the summer of 1983 to discuss a proposed drill hole in the Siljan Ring in Sweden. Vattenfall approached GRI in late 1984 with a proposal that GRI make a financial contribution to the drilling. A contract was signed in November 1985 which provided Vattenfall with $4.32 million to partially support the drilling. GRI, according to the contract terms, was: (1) to have access to all data and to receive copies of all data upon request; (2) to collect gas samples, cuttings samples and core samples. No direct participation in the project was specified in the contract.

---

1 Gas Researche Institute 8600 West Bryn Mawr Avenue Chicago, IL 60631, USA

However, as will be discussed below, GRI's participation evolved into a significant direct participation in the scientific sampling, analysis and interpretation aspect of the project.

Reasons for GRI Participation

GRI had two reasons for participation in the project: (1) it was a possible means of providing support for Dr. Gold's abiogenic methane hypothesis; (2) GRI would have a part in an important scientific and engineering experiment for less than 15% of its total costs.

The abiogenic hypothesis as developed by Dr. Gold consists of four essential ingredients (Gold 1987): (1) hydrocarbons were incorporated in the upper mantle of the Earth as the Earth accreted during the formation of the solar system; (2) the hydrocarbons have outgassed in the most stable form, methane, more or less at a constant rate over geologic time; (3) this methane has formed a large part of all hydrocarbone deposits; (4) the outgassing has taken place not only along faults and rifts but also through solid rock, including crystalline rocks, in which it collects in "stacked domains" starting at a depth between 3 and 10 km, depending on the strength of the rock. These domains consist of porous, gas-containing zones where the porosity is maintained by the high pressure of the gas. The domains are separated by "critical layers" in which no porosity exists because of the weight of the overlying rock.

The Siljan crater seemed an excellent place to test the hypothesis for several reasons (Gold 1987, p. 171). The drilling would take place in igneous rock which had been crushed by an ancient meteorite impact, which gave hope that porosity would be present. There was evidence at the surface that the crushed granite had been cemented by minerals deposited by circulating fluids. Therefore, all the ingredients of a gas deposit were possibly present: an abiogenic gas source; a reservoir of crushed rock; and a caprock of either the cemented rock or the top critical layer. The virtual absence of sedimentary rocks all but removed uncertainty about the source of any hydrocarbons which might be found.

This wildcat drilling venture was also an opportunity for GRI to add emphasis to the scientific aspects of the program and to collect unusual and important scientific information related to natural gas at a unique drilling site. This information included:

1. Molecular and isotopic composition of gases present in the rocks to deduce the origin of the gases.

2. Nature of the porosity in igneous rocks which is important for determining migration routes and potential reservoir characteristics.

3. Nature of the cap rock (if any).

4. Evaluation of how conventional surface exploration methods and well logging methods apply to igneous rocks.

GRI's Participation

Once the contract with Vattenfall was signed, GRI arranged for other contractors to analyze and interpret the gas and rock samples from the drillhole. A contract was awarded to Global Geochemistry Corporation to analyze and interpret gas samples. Ongoing contracts at the University of Wisconsin and Argonne National Laboratory were redirected to analyze and interpret the rock cuttings and rock core samples. Dr. Gold, at Cornell University, did overall interpretations of the data. Description of the results from these investigations are given elsewhere in this volume.

Early in 1986, GRI formed a Scientific Advisory Committee (SAC) to advise GRI on the conduct of the scientific aspects of the drilling program. The SAC was composed of Dr. Gordon MacDonald (chairman), Mitre Corporation; Dr. Charles Maxwell, University of Texas; Dr. Theodore Weisman, Duquesne University; Dr. Charles Mankin, Oklahoma Geological Survey; and Dr. George Claypool, United States Geological Survey (now at Mobil Research and Development center). Seven meetings of the SAC, between May 1986 and September 1987 were attended by the SAC members, by three to five representatives from Vattenfall, GRI's other contractors working on Siljan, GRI staff, and visitors such as representatives from DOSECC.

The SAC provided GRI with excellent advice, most of which was quickly implemented thanks to the attendance of Vattenfall representatives at the meetings and their concurrence in proposed actions to be taken. The SAC recommendations which were instituted included improved gas sample handling procedures, vertical seismic profiling, additional tests, sidewall cores acquisition and a recommendation that a geologist/geochemist, familiar with drilling for hydrocarbons but also with laboratory science, be hired to act as chief scientific advisor at the drill site. In a consultation between Vattenfall and GRI, it was decided that this person should be a GRI representative. John Castano, newly retired from Shell Oil Company, was hired as a consultant. This expanded GRI's participation in the science on a day-to-day basis through having an experienced person as our representative at the drill site. John has also made many valuable contributions to the scientific program.

The SAC later recommended that we retain a consultant with broad experience in well logging to give a second opinion on the well logging records. The team of Robert Sneider and J. Dennis Loren as well as Lloyd Fons was retained, and they supplied GRI and Vattenfall with their interpretations of the logs. Sneider visited the drill site and instituted a procedure for inspection of the cuttings integrated with interpretation of the drilling parameters, to identify fracture zones as drilling progressed. The diagnosis of fracturing made by this procedure correlated very well with subsequent well logs.

Results Obtained

Only a brief description of the chief scientific results are given here; other papers in this volume contain more detailed reporting. The results are given with a personal opinion of their significance.

1. Gravberg was chosen as the drilling site chiefly for two reasons: the presence of a negative (-20 mgal) gravity anomaly which means that less dense rock must lie below the surface, and the presence of four seismic reflectors at 1440, 2800, 4740 and 7400 m. Any of these reflectors might represent a nonporous cap rock above a porous reservoir rock.

   a) The cause of the gravity anomaly has not been found. There is no evidence of either a less dense type of rock or a more porous rock formation beneath the surface covered by the gravity anomaly. This is one of the most important unanswered questions of the drilling program.

   b) The first three seismic reflectors have been penetrated and are caused by dolerite sills with perhaps some contribution from adjacent fractured zones in the granite.

2. Vertical seismic profiling was very useful in confirming the presence of reflectors indicated by surface seismic reflection methods. Depths at which dolorite sills were intersected with the drill bit agreed very well with predicted VSP depths. Weak, diffuse reflecting intervals correspond to intervals of relatively intense fracturing.

3. All evidence gathered, which includes age dating of dolerites exposed at the surface, age dating of dolerites drilled, and textural petrographic evidence, indicates that the dolerite intrusion preceded the impact event by as much as 1/2 billion years.

4. High, near-lithostatic pressures were never encountered. Instead, all the rocks encountered have pore pressures near hydrostatic values.

5. Numerous large, highly eccentric, wellbore breakouts developed regularly as a result of high horizontal stresses in the area. These were most pronounced while drilling unfractured rock intervals. Breakouts caused severe drilling problems requiring sidetracking of the well, made complete core recovery nearly impossible and caused severe problems in running and interpreting wireline logs.

6. Detailed methods for examination of cuttings were developed on site, and made it possible to identify fractured intervals while drilling. Fractured intervals were also identified on the basis of drilling parameters and electrical logging data.

7. The granite is extensively fractured. Petrographic evidence demonstrates that more than one mechanism led to fracturing;

the meteor impact event, and faulting both before and after the impact event. Fractures appear to have provided the most important conduit for circulating fluids. Repeated sidetrack drilling and careful analysis of cuttings had demonstrated that this conduits have limited areal extent at depth and cannot be correlated beyond distances of tens of meters.

8. Calcite occurs as the principal cement in fractured granites down to a depth of 3.2 km and lesser amounts are found as deep as 5 km. Stable carbon and oxygen isotope analyses and fluid inclusion data indicate that these carbonates primarily precipitated from reactions involving surface waters. Siljan granite has therefore been exposed to the effects of extremely deep weathering. This is further corroborated by oxidation profiles of iron-bearing minerals. No clear evidence has been found which would indicate that calcite is derived from the oxidation of mantle methane at this site.

9. There has been extensive circulation of fluids through these rocks in the past. As a result, most fractures have been cemented and original minerals have been altered along the most permeable fractured pathways. Based on the variety of cement compositions examined, and on the wide range of fluid inclusion compositions encountered in cements, the fluid circulation history of the area has been extremely complex.

10. Very little porosity was found, with calculated values ranging from 1 to 5% on the basis of electric logs and core analysis. Much of the porosity results from alteration of feldspars to secondary minerals, and remaining porosity is due to partially healed fractures and intercrystalline voids. Most porosity is not extensively interconnected, and nowhere is porosity sufficient for commercial exploitation.

11. Permeability, as reported from special core analysis, drill stem tests, and leak-off tests, was very low and in the microdarcy range. All gas shows were discontinuous, suggesting limited permeability across a small volume of rock. Based on detected quantities of hydrogen and helium, these rocks are permeable to these gases over geologic time. At 6047 m in the original hole, a highly fractured zone was encountered, circulation was lost, and 110 m^3 of 13.7 pounds/gal drilling fluid invaded the formation. Initial reports suggested that permeability had been encountered; subsequent analysis suggests that lost circulation was caused by the heavy mud hydrofacturing weak rock. Once mud weights were reduced, fluid loss to the formation was negligible.

12. Gases were collected during the drilling program in a variety of ways. None of the collection methods allows direct quantification of gas volumes encountered, because all gases were mixed to some unquantifiable degree with air and

with drilling mud. In general, the highest volumes of hydrocarbon gases were detected in trace quantities, an order of magnitude below quantities normally detected as background values in most oil and gas provinces.

a) Methane, detected at all levels and in all rock types, was the most abundant hydrocarbon collected. Several different drilling fluids were used during the drilling operations, and detected hydrocarbons had differing retention times and solubilities in each. In spite of measurement uncertainties created by changing drilling fluids, methane concentrations appeared to increase with depth. Methane quantities encountered while drilling dolerite sills were an order of magnitude higher than those encountered while drilling granite. Hydrocarbon compositions were also distinctly different in both rock types. The dolerites contain a "dry" gas with methane carbon isotope compositions similar to those encountered at mid-ocean ridges. Although these gases appear to be of abiogenic origin, their origin is still in question. Hydrocarbon gases in the granite are "wetter", have lighter stable carbon isotopic values similar to thermogenic gases associated with mature source rocks, and have relatively high proportions of unsaturated gases. The origin of these hydrocarbon gases is uncertain: they may be indigenous to the granite perhaps resulting from a Fischer-Tropsch type reaction, they may be derived from overlying sedimentary rocks, or they may be a product of drilling.

b) Hydrogen has been found at most levels. Its origin has been controversial: there is still no consensus of opinion as to how much of the detected hydrogen was generated in some way by the drilling process, or how much it had accumulated in the rock over geologic time. Stable deuterium isotope data have not been helpful in distinguishing between the two potential sources because H_2 and aqueous drilling fluids rapidly re-equilibrate on the way to the surface. Below 6000 m, hydrogen was detected in large quantities approaching 20% by volume of gas sample. Because increases in helium abundances partially correlated with increased hydrogen abundances (Pearson's correlation coefficient of 0.7), it is likely that at this depth, the hydrogen gas is indigenous.

c) Helium was detected intermittently in small quantities, often exceeding the concentration of hydrocarbon gases. The largest quantities were detected along with hydrogen below 6000 m in the first sidetracked hole. He^3/He^4 isotopic data are consistent with a radiogenic, crustal origin and not with a mantle origin for most of the helium. Strong correlations were not observed between increasing concentrations of hydrocarbon gases and increasing concentrations of helium or hydrogen.

13. The drilling has been stopped at 6600 m because of lack of funds. If permission is obtained and money can be raised,

the hole will be competed to its target depth of 7500 m. The hole will then penetrate the strongest reflector of all at about 7400 m. It seems likely that this reflector will also prove to be a dolerite sill, but it could be be the top of a zone of greater porosity.

Conclusions

Gravberg-1 has not yet been a successful wildcat hole. None of the carefully gathered data provide compelling evidence confirming either that expected mantle-sourced hydrocarbons have been found, or that they have accumulated in commercial quantities at this site. However, a maximum amount of scientific information, consonant with the commercial aim of this project, has been extracted from the drilling program. From this point of view, GRI's participation has been a success. That this happened is due to the excellent spirit of cooperation between Vattenfall and GRI.

References

Gold Thomas, **Power from the Earth**, J M Dent & Sons Ltd London 1987

Preparatory Investigations – an Overview

P.-O. Karlsson[1]

Introduction

In view of the fact that the Deep Gas Project is based on theories which as yet, at least from a commercial point of view, are not confirmed, and which deviate from the conventional theories regarding the origin of hydrocarbons, the preparatory stages of the work took the form of a widely based scientific study on the application of the deep gas hypothesis in the Siljan Ring structure, with field investigations, theoretical analyses and examination of the results by a large number of experts.

The questions to which answers were sought were whether there are connections with the inner regions of the earth as a source of the gas, whether there is a reservoir and whether or not there is a caprock, i.e., an impermeable rock formation over the reservoir to retain possible deposits of gas.

The theoretical analyses showed that the crystalline rock can serve both as a caprock and as a reservoir, with a successively greater proportion of gas appearing in the pores with depth. A distinct caprock of the type that can be found in sedimentary rock is less likely. A reservoir is believed to have low permeability, at the same time as the thickness of the reservoir is potentially very great

Field Investigations

Relatively extensive geophysical and geochemical field investigations of different types, as well as drilling of a number of shallow core boreholes, were carried out to verify further the caprock and reservoir conditions and to provide basic data for locating the deep borehole on the best site:

The following investigations gave especially valuable information for the well location:
- Seismic investigations
- Magneto-telluric measurements
- Gravimetric measurements
- Geochemical analyses of soil samples for determination of possible methane-derived carbonates and variations in trace elements.

---

[1] Vattenfall, S-162 87 Vällingby, Sweden

Reflection Seismic Investigations

During February to March 1985, an approximately 50-km-long reflection seismic investigation was carried out. The investigations were arranged in the form of a cross with an east to west (E-W) profile and a north to south (N-S) profile. The E-W profile extended beyond the directly visible area affected by the impact with the intention of documenting the effects of the impact in the rock mass outside the "impact centre". As a complement to these two lines, two more profiles were carried out in August; a total of about 15 km. One of the lines constituted an extension of the earlier so-called N-S line to the north past Lake Skattungen. The other line was arranged in an east-west alignment in the form of a cross over the N-S line on a level with Gravberg in the north-eastern part of the structure.

The interpretation of the reflection seismic investigation verified the assumption of a central uplift area with its centre located slightly south-west of the geographical centre of the Siljan Ring.

Within the Gravberg area, the seismic investigations indicated the presence of more distinct horizons than within any other

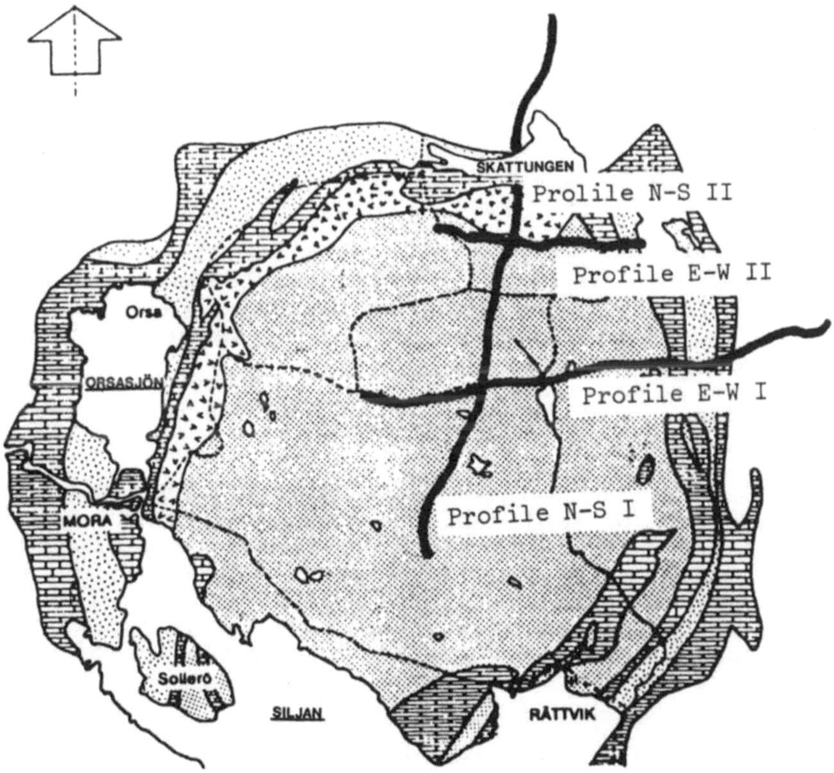

Fig. 1. Extension of the seismic lines over the Siljan Ring area

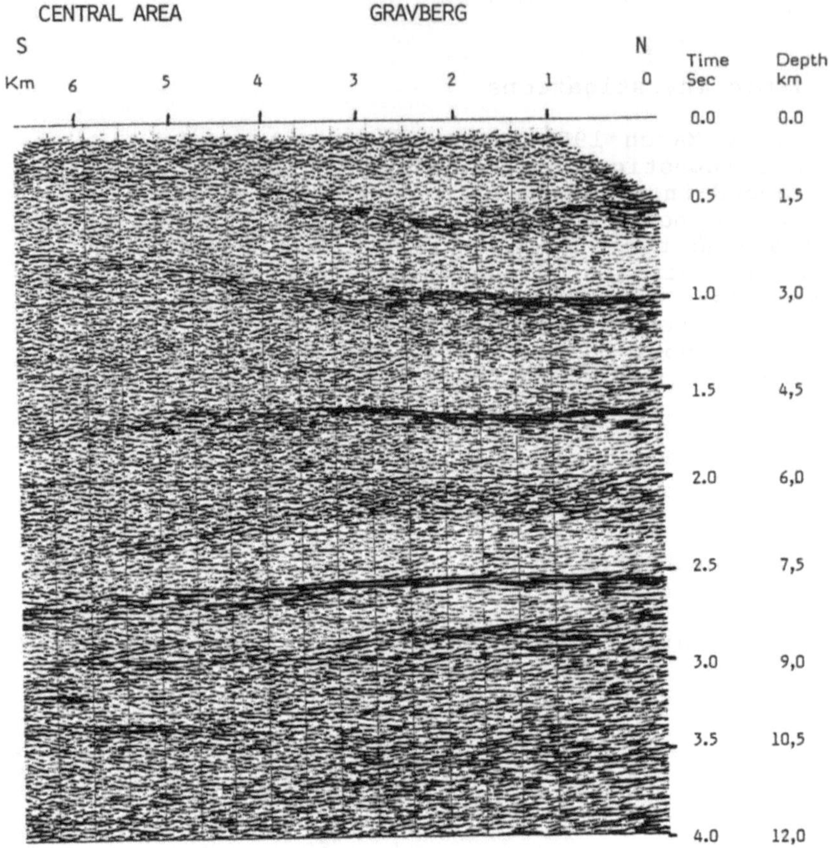

Fig. 2. Reflection seismic profile from the northern part of the north-south line

part of the Siljan Ring area studied. These extremely distinct horizons were interpreted as possible indicators of a reservoir. Especially the signs of inverted velocities between the third and fourth horizon indicated the presence of favourable reservoir conditions.

The seismic profiles showed no continuous reflecting horizons within the central area.

The seismic investigations also confirmed that there are a large number of faults or fracture zones, the majority of which within the central part can be deemed to be impact-generated. These zones were registered down to a depth of about 18 km, which constitutes the lower measuring limit for data collection. The confirmation of these zones was important, since they are possible migration routes for the seepage of deep gas up to a reservoir.

Magneto-Telluric Measurements

The magneto-telluric measurements within the Ring showed in general a high conductivity, which could be interpreted as fractured liquid- and gas-filled rock. The magneto-telluric measurements were not sufficiently sensitive for detailed selection of a drilling site, but primarily served to show that the Siljan Ring, in a regional comparison, deviated from the surrounding area. The results supported the model suggesting the presence of one large deposit. Local positive results which indicated shallower depths to conductive levels were found in Gravberg and in the central area.

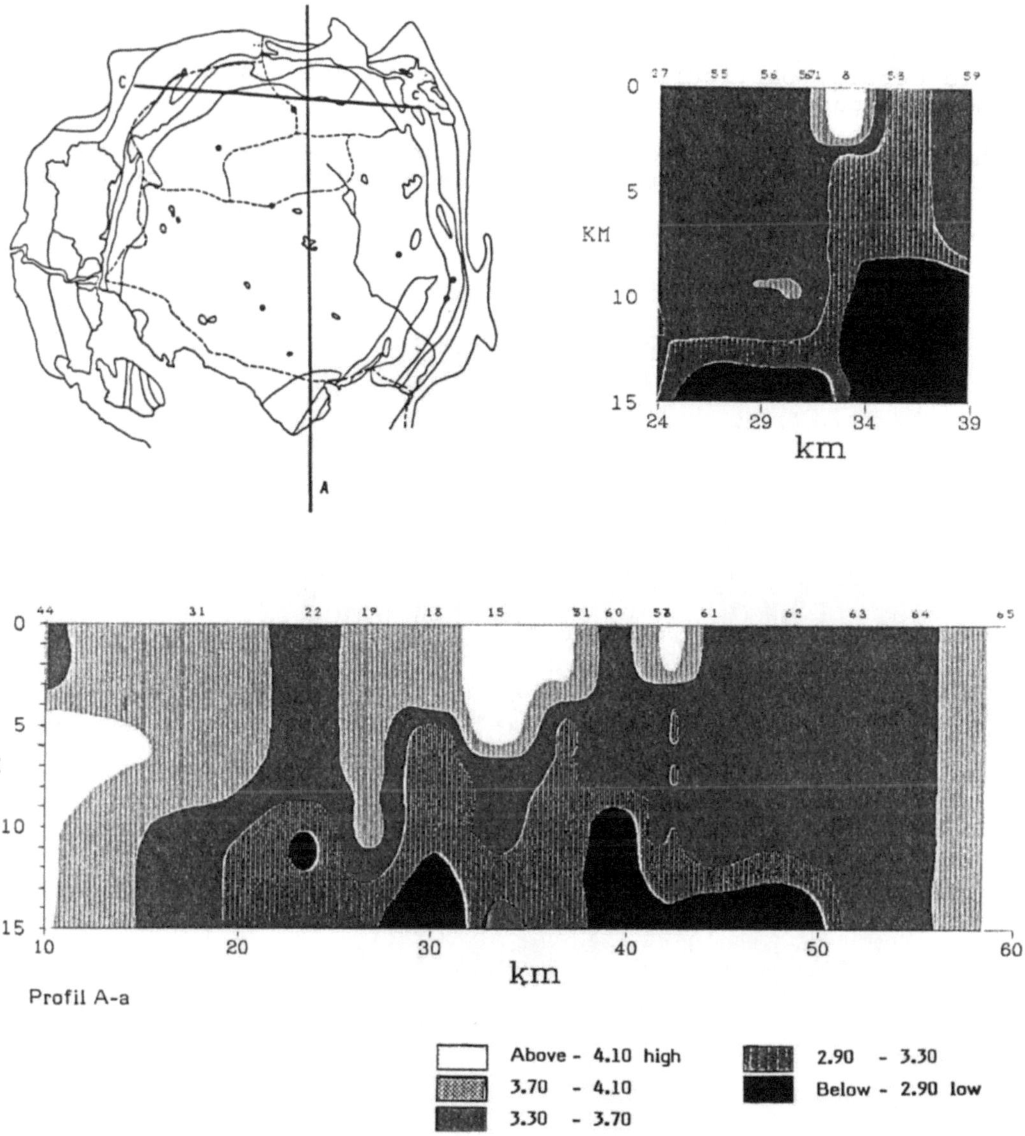

Fig. 3. Summary of the magneto-telluric investigations showing depth down to layers with different conductivity

Gravimetric Investigations

In 1985, an extended gravimetric measurement was conducted in the Siljan Ring area as a complement to measurements carried out earlier.

The gravity chart, Fig. 4, showed a remarkably well-accentuated negative anomaly located mainly in the northern part of the Siljan Ring area. The recorded negative anomalies were interpreted to be at least in part connected with the lighter types of sedimentary rock in the actual Ring which surround the granitic cupola, particularly in the eastern and southern parts of the structure.

The marked negative anomaly in the north, however, could not be totally explained by the near-surface sedimentary rock types,

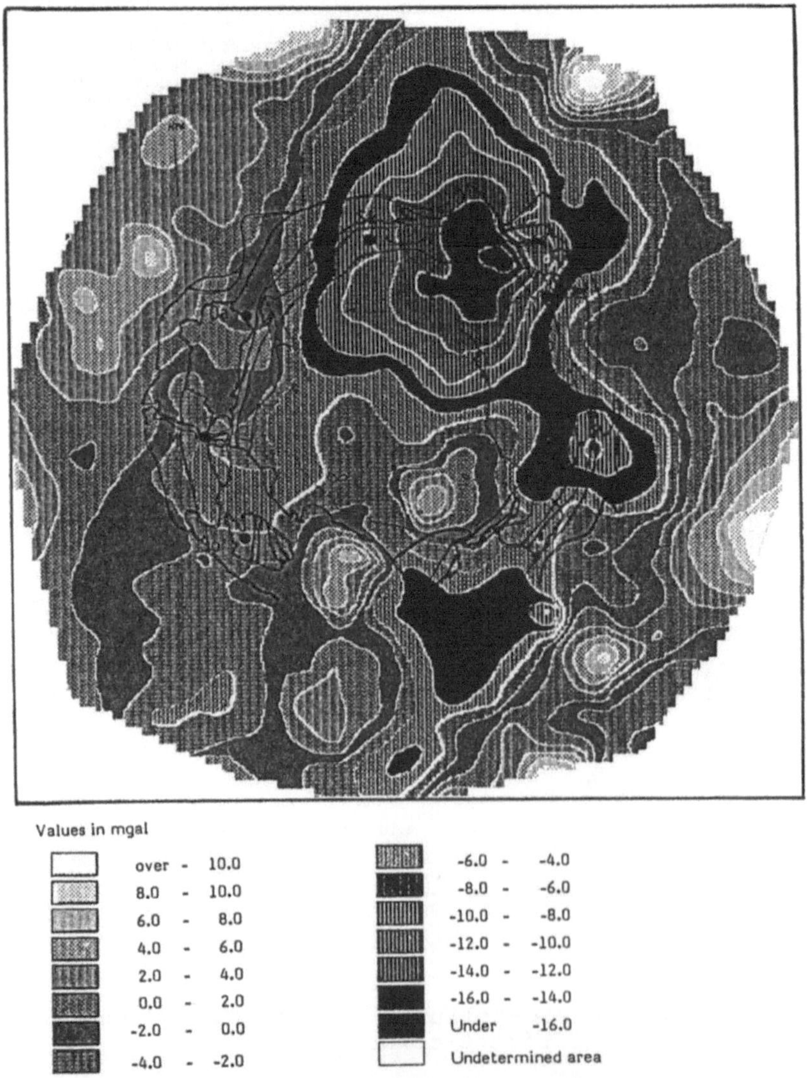

Fig. 4. Gravity chart showing variations in the gravity field

but was interpreted as depending on underlying causes, indicating a porosity maximum. Studies of satellite photos showed that this part of the area also displays the highest level of tectonic influence.

Analyses of Soil Samples

An extensive soil sampling programme was performed in three different stages. A total of over 250 soil samples were taken from both within and outside the Siljan Ring area and analysed.

The geochemical analyses displayed clear traces of methane and heavy hydrocarbons. The ΔC-interpretation of soil samples resulted in a distinct halo in the area. The halo coincided with analysed trace elements, for example vanadium and nickel. The sedimentary rocks surrounding the Siljan Ring area only partially coincided with the irregular ΔC-halo. Particularly in the north-western corner of the area, higher ΔC-values were measured essentially outside the known northern limit for the distribution of the sedimentary rocks. The high ΔC-values appeared to coincide with valley dikes, whereas the raised areas displayed low ΔC-values. The question of increased ΔC-values caused by sediment transported by ice was ruled out, since the ice movement was from the north-west and the increases were north of the sedimentary ring. One explanation for the phenomenon was that there has been a long-term leakage of hydrocarbon in the large dikes, which has caused a rise in the content of methane-derived carbonate. The possibility that the sediment itself had generated the high ΔC-values was contradicted by the fact that the majority of the biological matter in the sediment shales has a low level of maturity, but also by the fact that the sediment only coincided to a certain degree with high ΔC-values within the area.

The indication of greatest gas leakage in the north-north-western corner agreed in a significant way with methane analyses on lake waters. One explanation for the appearance of more leakage in that area was that the tectonic pattern is considerably more complex there. This was seen from a number of independent analyses, including satellite photo analysis.

Within the central area there was no clear evidence of slow gas leakage. That coincided with the conventional assumption of a reservoir and caprock behaviour with "tight" centre and leakage around the edges of the reservoir.

The fact that the Siljan Ring has been subjected to glaciation has led to a number of questions of the origins of ΔC and trace metal anomalies in the area.

With the intention of reaching agreement on the subject, additional processing of the values obtained from the geochemical analyses was undertaken. The processing was carried out with the aid of advanced statistical and analytical prospecting methods. With the aid of data calculations, a valuation can be made of each analysis point in relation to all the other

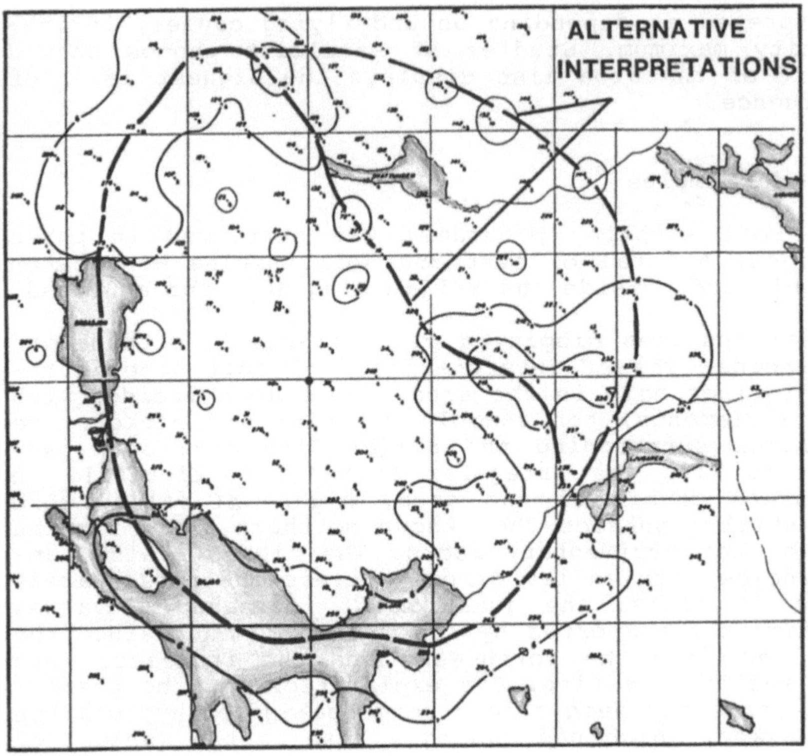

Fig. 5. ΔC-interpretation of soil samples

analysis points. With this process it was possible to establish and eliminate the natural background value for the area.

This method reinforced the theory that gas leakage from below is responsible for the halo pattern.

Well Site Location

To summarize, it was concluded that an exceptionally high proportion of results had been obtained supporting the possibility that the Siljan Ring structure constitutes a reservoir for deep gas.

On the basis of the investigations carried out, two different parts of the Siljan Ring area were judged to be of high priority in connection with the choice of the site for a deep borehole; the central uplift area in the middle and the Gravberg area in the north eastern part of the structure. It was considered that within both these areas favourable caprock and reservoir conditions may exist.

The overall assessment made by various experts led to a recommendation to locate the first borehole in Gravberg. It was ge-

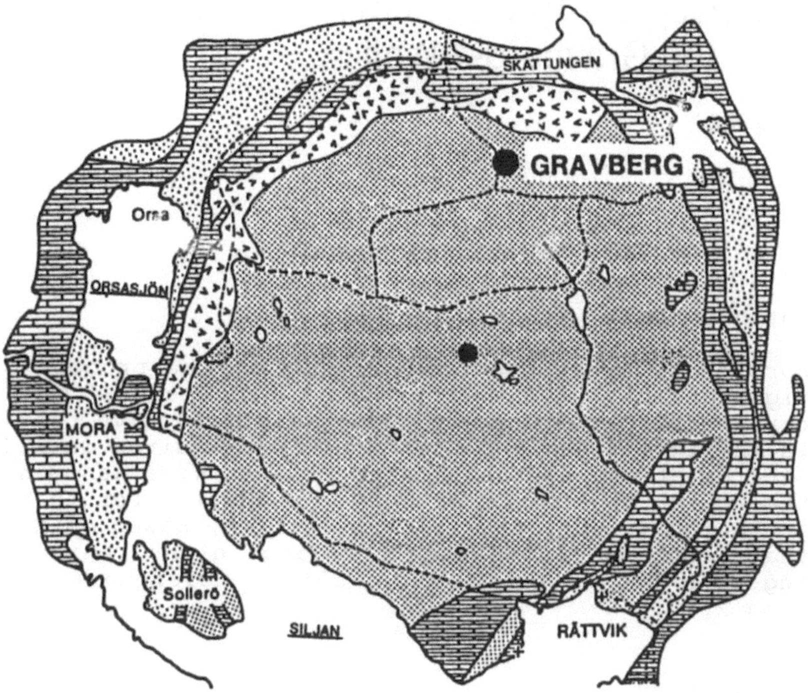

Fig. 6. Alternative locations of drill site. The Gravberg and the central area

nerally agreed that the investigation results were extremely interesting. The relatively unanimous recommendation was based primarily on the results of the seismic investigations, which could indicate a connection with hydrocarbon deposites.

The Deep Earth Gas Theory with Respect to the Results from the Gravberg-1 Well

T. GOLD[1]

Introduction

For nearly every observation made in the process of the investigations of the Siljan prospect and the drilling of the Gravberg well, two different interpretations have been offered. One is the interpretation that would be required if all hydrocarbon manifestations and associated effects were assumed to be due to the action of surface biology. We shall refer to this as interpretation 1. The other interpretation, based on the supposition that reducing fluids are ascending in the region and originating from very deep levels, will be referred to as interpretation 2.

Observation

In the early phase of the investigations it was known that various seepages of hydrocarbons existed in the Siljan Ring area. In particular, oil seepages were known in the several limestone quarries on the eastern edge of the ring and methane seepages from water wells were known in different parts, but mainly in the area of the ring rather than in the interior.

Interpretation 1. The oil seeps were considered to be the result of organic sediments deposited in the limestone that filled the ring-shaped depression, known to be about 300 m deep. The impact event or subsequent volcanic intrusives supplied sufficient heat at one time to convert the organic content of these sediments into petroleum, and this petroleum is now seeping up.

The methane seeps were attributed to organic surface material as currently produced in the woodland soil, washed by surface water into the rock below. Gas was then produced and released from it.

No explanation was offered why the Siljan Ring area was significantly favoured with hydrocarbon seeps compared with other areas containing similar depths of deposits or similar woodland soils.

---

1 38 Madingley Road, Cambridge CB3 OEX, United Kingdom

Interpretation 2. The impact area contains rock shattered to a great depth, presumably through the entire crust, and this provides easier pathways for the escape of deep fluids than the surrounding more solid rock. The entire Scandinavian shield area is underlaid by an area of hydrocarbon-rich mantle, which shows itself in all the productive hydrocarbon areas surrounding it and the numerous tar occurrences in the Swedish bedrock. Methane and hydrogen are seeping up and are chanelled into the most permeable pathways, which at shallow levels are on the interface between the sediments of the Ring and the metamorphic interior. These gas streams carry up heavy hydrocarbon molecules of petroleum which are shed and deposited at shallow depths in porous rock. Both the oil seeps of the area and the methane seeps are derived from this common source of hydrocarbons at depth.

Observation

In the early phase of the investigations five holes were drilled, distributed over the interior of the ring to depths varying between 200 and 600 m. Continuous cores were obtained. All cores showed fractured granite, the pieces being a few centimetres in size, deformed so that the junctions between them were generally less than 1 mm wide. These junction lines were clearly visible and were mostly filled with a cementing mineral found to be a mixture of calcite and clay. The volume proportions of calcite cements in the rock were estimated as between 1 and 1/10th%.

On investigation, the calcite proved to be isotopically anamolous. The carbon isotope distribution was significantly lighter (less carbon 13) than the ordinary marine carbonate sediments. The range of variation of the isotope value of these carbonates (on the conventional PDB scale) was from 0 to -25 per mil, while normal marine carbonates rarely fall outside the range of ±5. Such calcite cements are known in other areas where they overlie gas or oil fields and are thought to derive from the oxidation of methane and the combination of the resulting CO_2 with local calcium oxide.

If an estimate is made of the proportion of the anomalous calcite in the rock on to a depth of 500 m and it is taken as 1/2%, and if it is further assumed that the five holes distributed over the Siljan impact area are representative of the 44-km diameter circle, then the astonishingly large quantity of 9.5×10^9 tons of carbonate will be involved and 2.1×10^{12} cubic meters of methane (STP) would have been necessary to create this.

In the deep Gravberg well the same calcite was seen, but with a concentration that decreased with depth. Even so, if one assumed a concentration in the rock of only five parts in 10,000 between 500 m and 6.1 km, the additional amount of methane that would have been necessary to create this over the entire area would amount to 10^{10} tons or 2.3×10^{12} cubic metres.

Interpretation 1. The explanation for the sedimentary portion of the ring is in terms of upwelling hydrocarbons from sedimentary biodeposits. However, for the occurrences of this material in non-sedimentary rock, including the interior of the Siljan Ring area, an origin from the downward transportation of surface-derived biological material has been suggested, even when the depth at which it was observed was as great as 6 km. The isotopic spread is considered to imply that two different sources of carbon were involved in producing the carbonates.

Interpretation 2. Such anomalous carbonate cements are known in many other locations as characteristically overlying oil and gas fields and infusing a vertical column irrespective of the nature of the formation or its stratigraphy. For this reason it is believed that these carbonates derive from methane which is bacterially oxidized with oxygen available in the formation and the CO_2 so formed then combined with locally available calcium oxide. The carbon isotope value of the product is then dependent on that of the originating methane and the fractionation process that takes place in its upward migration and in the process of oxidation. This fractionation is a progressive process, the methane remaining in the upward stream becoming progressively lighter as the heavy isotope is preferentially taken away in the oxidation. The carbonates so formed then show the corresponding istopic effect. It is characteristic of these carbonates that they show a very much larger variability in the carbon isotope value than any other form of terrestrial carbon, with the exception of methane. Such cements are recognised not only in sediments overlying oil and gas fields, but also in numerous other areas of basement rock, such as other parts of the Scandinavian shield, the Canadian shield, the Kola Peninsula and Siberia.

It is presumed that methane and other hydrocarbons are percolating through the crust in many locations though at locally quite variable rates. Any depth range in which the temperatures are sufficiently low for bacterial oxidation to take place may then show these carbonates, and they may also form at depths where the temperature is sufficiently high for direct oxidation to take place. There is thus no difference in the origin between the basement and the sedimentary deposits of calcite cements. The regionally very uneven permeability of crystalline rock compared with the more uniform permeability of most sediments tends to make the isotopic fractionation effect very uneven there and for this reason very different isotope values are recorded in crystalline rock even from neighbouring locations. It is not necessary to assume that at each level more than one source of carbon made a significant contribution.

Observation

The shallow test holes that were drilled initially all yielded hydrogen as the principal gas. Some methane and heavier hydrocarbons were also seen in all holes, the proportion of heavier hydrocarbons being highest in the granite below one of the major surface oil seeps. The hydrogen appeared to come in bursts during drilling, suggesting that pockets of it were penetrated by the drill. On some occasions, values as high as 2% hydrogen were measured in the discharge pipe, the rest being mainly air entrained with the drilling water.

Interpretation 1. The hydrogen was considered to be a product of the chemical reactions of the drilling water with the iron in the formation or in the drilling machinery. The hydrocarbons were considered to be due to downward seepage from surface bio-products, even though the highest C2 to C5 values were recorded not in the sediments of the ring but in the granite below it.

Interpretation 2. Hydrogen is often seen accompanying hydrocarbons and is presumed to be the product of hydrocarbon dissociation. This could have taken place at a deeper level where the temperature is high enough for partial dissociation to occur, or at shallow levels by bacterial action. If any substantial reservoir exists that is filled with hydrocarbons usually containing a small proportion of hydrogen, then the overlying denser rocks may be preferentially infused with hydrogen, just because hydrogen is the most mobile component and can best penetrate the dense rock. It is this combination of circumstances that is seen as the reason for hydrogen being found so frequently in petroliferous areas. Also this frequent availability of free hydrogen is taken as the explanation why hydrocarbons are so largely hydrogen-saturated, a property that would be unexpected for a biogenically derived product.

The hydrocarbons recorded are considered as a component of the upwelling gases and the high values of C2 to C5 observed underlying the oil seeps would be in accord with the common observation that a high ratio of these to methane is an indicator of an oil deposit rather than of a dry gas deposit. The oil in the overlying sediments would then be regarded as a consequence of upwelling heavy hydrocarbons in this particular sector.

Observation

The Gravberg well then showed that the effects observed in the shallow holes all continued to deep levels. The calcite cements continued to be present and of anomalous and greatly variable isotopic composition, although the general abundance of these cements decreased with depth. Hydrogen has consistently been the principal gas in the rock, although methane and helium have also been seen and all gas values have increased with depth. Helium was unfortunately not measured at the shallower levels so no comparison is possible.

At each of the three seismic reflecting levels that have been penetrated, some volcanic intrusives were encountered and associated with them more highly fractured zones of granite. The hydrocarbon levels were greatly enhanced in each case, both in the dolorite intrusives and in the surrounding fractured granite.

The correlations of the gas abundancies with depth and with the nature of the formation confirm that the major amounts of both hydrogen and hydrocarbons that have been seen derive indeed from the formation and not from some contaminants introduced by the drilling. Whether minor additions of these gases did derive from artifacts has been much discussed, but it is difficult and perhaps not very important to settle this question so long as the major gas quantities observed are clearly products of the formation.

Helium measurements have given very definitive and very remarkable results. Firstly, the quantities of helium observed were very large compared with such observations in wells in other parts of the world. Secondly, the helium levels showed a strong correlation with the levels of hydrogen, therefore providing again clear evidence that the hydrogen was derived from the formation. Both the hydrogen and helium values that were seen were higher in the sidestepped hole drilled with water as the drilling fluid than they were in the original hole drilled with a heavier mud. The highest values were observed on occasions when the downhole pressure was decreased even further during tripping. It thus appears that these gases come into the well from the formation, being there at a pressure not far from the hydrostatic value at each depth, so that small variations around this value in the well bore had a strong effect on the rate of entry.

Carbon isotope measurements of the methane observed below 4 km have given values as heavy as -20 PDB (while surface biogenic methane is usually in the range of -40 to -80).

Interpretation 1. The helium and at least a major part of the hydrogen is derived from the formation; the helium from the local radioactivity, the associated hydrogen from unknown sources. The quantities of the hydrocarbons are too small to be taken seriously. They could be derived from some transportation by groundwater of bio-products or from traces of carbon in the crystalline rock. Also carbon in drilling steel and carbide bits have been mentioned as possibly responsible. The association of hydrocarbons with details of the formation is unexplained.

Interpretation 2. The quantitative aspects of the helium deserve attention. The accelerated entry of hydrogen and helium into the well at a pressure lower than the hydrostatic value suggests that the fluid pressure is close to hydrostatic but that the permeability of the rock that was encountered was low so that flow-rates were small. The hydrostatic pressure at 6000 m is 600 atm. Hydrogen is the principal gas there and helium is frequently at a level of 2% of the hydrogen. The partial pressure of helium is thus of the order of 2% of 600 atm or 12 atm. This is a quite unusually high partial pressure of helium, possibly higher than any that has been seen elsewhere in wells. The fact that helium is so largely correlated with hydrocarbons globally has meant that the enhanced presence of helium has been regarded as a favourable indicator for the presence of hydrocarbons and it must be so regarded in this case.

To convert the partial pressure of helium into a concentration by mass in the rock, one needs to have some idea of the porosity in which the gases reside. Porosity levels for the small molecules of hydrogen or the atom of helium are concerned with smaller spaces than the porosity levels that would be measured with air or other gases, and effective porosities in rock for the helium are rarely less than 2% in any rock whose density has not been much enhanced above the uncompressed density (2% porosity for helium in granite would not imply any significant value of the permeability). The density of helium at 400 K and 12 atm is 1.46×10^{-3} g/cm^3. The mass fraction of helium in a rock of 2% porosity and density 2.6 is hence 1.1×10^{-5}. An average mix of uranium and thorium would produce approximately 1α particle (helium atom) in the last 360 million years. The mass of the helium atom is approximately 1/50th of that of the uranium and thorium. These elements would therefore have to be present in the rock in a concentration by mass of 5.5×10^{-4} if they were to be responsible for the observed helium, even on the assumption that no helium at all had escaped in 360 million years (a most unlikely assumption and certainly incompatible with any suggestions of groundwater seepage). In reality the mass fraction of uranium plus thorium is much less, probably nearer 5×10^{-6} and most probably the major part of the helium that has been generated has since escaped. We are therefore seeing here much more helium than could have been produced in the local rock.

The partial pressure of helium could not have been increased beyond that at its source in any process of diffusion or bulk flow down a pressure gradient. The helium must thus have been produced at a much deeper level where the rock is compressed to a density substantially greater than the uncompressed density, and where the effective helium porosity is much less. The helium pressures reached there may be very large, comparable with the local lithostatic pressure, and the bulk flow from there to shallow levels, following approximately the lithostatic pressure gradient, can then result in helium pressures up to the local lithostatic value. The space concentration of the helium would then reach much higher values than could be produced by the mean radioactivity of the local rock, having been supplied from a much larger volume of rock at deeper levels. Equilibration favours the accumulation of gases where the porosity is larger, but of course the process could never go beyond an equalisation of partial pressures.

The helium measurements have thus given information that gases are present that have migrated up from deeper levels. Diffusion through the rock cannot provide sufficient flow even for helium and for times that are as long as the age of the Earth. A bulk flow is required to account for substantial transportation. For this a carrier gas must have been available, since a local abundance of a gas is required that is much higher than that which helium could achieve, in order to hold open pathways for such a flow. It is therefore most probable that the gas which is now found together with the helium has acted as its carrier gas. In the present case, this seems to have been chiefly hydrogen, and both hydrogen and methane have commonly been seen as the carrier gases for helium in other parts of the world.

These considerations will account for the observation that high helium values were observed generally in conjunction with high hydrogen values.

Hydrogen thus appears to be coming up from deep levels: but whether hydrogen or methane is the primary gas is very difficult to establish, for there can be much interchange between the two. Methane can be dissociated to produce hydrogen and leave the carbon behind; hydrogen can be transformed into methane with locally available carbon in the rock. The principal fact to note is that hydrogen is streaming up; whether at any level it is combined with carbon or not is of local significance only.

Observation

A region of substantial porosity was penetrated by the drill on February 23rd 1987 and some 900 barrels of drilling fluid (density 1.5) was lost in a little more than a half hour. Previous to this occurrence, very little drilling fluid had been lost, and leak-off tests were all negative. It is clear that there was a sudden transition to a different regime, also demonstrated by a fivefold increase of the drilling rate of progress. Porosity values of 5 to 8% were subsequently measured.

Whether the pore spaces were filled with hydrocarbon gases or liquids or with water could unfortunately not be observed, since the drilling fluid pressure evidently dominated over the pore fluid pressure. It is thought unlikely, however, that a domain filled with water only would have allowed as rapid a drainage of drilling fluid as that which occurred.

The gravity survey of the impact area, and specifically the north-north-eastern sector had demonstrated a substantial gravity-low. Underlying porosity was regarded as a possible cause of this. Similar gravity deficiencies have been noted in numerous other large impact structures. In the drilling down to 6 km no significant porosity was encountered, nor was any sufficiently large change in rock density observed to account for the gravity data.

Interpretation 1. The porosity that was observed was thought to be the consequence of fracturing under stresses that exist in the area. The detailed causes of these stresses are not known but stresses have been measured in the course of the drilling operation. What stress patterns could create a domain of fracture porosity at depth and leave the overlying rock dense is not clear.

Also an interpretation of local gravity deficit in terms of porosity would require a very large volume of porous rock to underlie the region and it may be difficult to attribute that to a stress-created effect. Alternatively, the apparent gravity low may be the result of an erroneous reduction of the data or the presence in the region of a rock of lesser density.

Interpretation 2. High pressure fluids penetrating the rock from below will create porosity domains. The shallowest of them has to be at a depth sufficient for the overlying rock to be compressed and made tight. Gases penetrating to levels shallower than this will escape rapidly and will thus not build up significant porous volumes. The circumstances seen here of a tight rock above and a sudden transition to a porous rock below is thus the expected one. A depth of 6 km is a reasonable one at which the overburden weight could cause the rock to close in a long period of time. The fact that no gas at lithostatic pressure was observed may be due to a comparatively recent leakage of some of the content of this particular domain. The origin of the stress patterns of the area may lie in shifting fluid contents.

On the basis of a fluid-induced porosity, one may well expect that the entire gravity deficit seen over the area can be so explained. In that case, a large pore-space domain must underlie the region. As this would cause a substantial seismic reflection, we assume that the strong reflector R4 indicates the beginning of this region, there being no other strongly reflecting level in evidence.

The fact that numerous other large impact structures show similar gravity deficits suggests a common explanation. Porosity of fractured rock would provide such an explanation, while other interpretations involve individual coincidental effects.

Observation

In the entire body of rock surveyed, both by the shallow test holes and the deep well, we see an acidic, oxidizing rock, granite, apparently infused everywhere by reducing gases, chiefly hydrogen. Moreover, we see the granite change from a pink colour at shallow depths to a grey at deeper levels, which presumably implies a more reduced state of the iron there.

Interpretation 1. The origin of the hydrogen is not known. The oxidation state of the granite, and its variation with depth may have been dictated by the percolation of water and its mineral content.

Interpretation 2. There have been many calculations of chemical oxidation equilibrium between rocks of various levels of oxygen fugacity and reducing fluids like hydrogen, methane and other hydrocarbons. On the basis of such equilibrium considerations, the dominant presence of reducing gases in granite would not have been expected. The fact that this is clearly observed here - as indeed also in the deep well in Kola - suggests that such equilibrium considerations do not apply here, and probably not in many other instances.

It was on the basis of chemical equilibrium theory that many investigators considered it impossible for hydrocarbons to exist at deeper levels. We believe these considerations to be inapplicable, when a sufficiently large quantity of a reducing fluid makes its way through fractures or pore spaces of solid rock. In that case the initial value of the oxygen fugacity of the rock soon becomes irrelevant, once the available oxygen on the crack surfaces is used up, and after that reducing fluids in the pore spaces will not be oxidized.

The consideration that it is not molecular diffusion through the rock but only bulk flow in cracks that can transport fluids up from deep levels means that in regions where such flow occurs the oxidation level is completely dominated by the fluids. It is this situation we see in the Siljan holes.

We believe also that the more reduced state of the iron in the granite at deeper levels, as indicated by the colour change with depth, is the result of the stronger reducing action of the fluids at the higher concentrations there.

Observation

A black sticky substance was found to be blocking the drill pipe at the bottom end when circulation had been stopped for several days at a depth of 5520 m. The blockage was so complete that circulation could not be resumed. On retrieval, the substance was black, sticky when wet, powdery when dry, showed no fluorescence, and was strongly magnetic. The extract that was obtained with dichloromethane was analysed by gas chromatography. The report by Geolab Nor states "the gas chromatogram of the material showed two main features, a series of peaks corresponding to the saturated hydrocarbons from C11 to C20, peak abundance at C13 and a large but poorly resolved peak centred on C22 with wings between C21 and C26." "None of the drilling additives supplied by the client showed chromatographic patterns which corresponded with the material."

Recent information shows that the material contains virtually no ordinary rock minerals, no granite fines that would be expected in the drilling fluid, but is made up almost entirely of submicroscopic particles of magnetite, held together by oils.

Interpretation 1. The material was derived from drilling additives and lubricants that got into the hole. It is not known why the gas chromatogram is dissimilar from those. We also do not know why the hydrocarbons present only in very dilute concentration in the drilling fluid, would concentrate to form a plug at a particular level around 5520 m.

Interpretation 2. The smooth sequence of alkanes in the carbon number range from C11 to C20, certainly looks like a distillation sequence. It is interesting to note that very similar distillation sequences have been recorded at a depth around 10 km in the super deep well in Kola. Hydrocarbons coming up from deeper levels may well be expected to suffer distillation processing, and indeed such sequences are not rare. The particular position of the carbon number peak will vary, of course, between one case and another as the conditions of temperature and pressure of the distillation process no doubt vary.

We believe that a detailed analysis of the chemical and isotopic composition of this substance is warranted, from which it should be possible to conclude definitely whether or not the material is a natural product from the formation. If it is a formation product it may have entered the drill pipe from below and congealed there.

If such an analysis shows this substance to be derived from the formation, it may contain the most important clues about the origin and the types of hydrocarbons that exist there. An extremely thorough study of this material would then be indicated.

The presence of light alkanes, heavier than C5, may have gone unnoticed in the past, and a periodic analysis of the drilling fluid for those is indicated.

Major Questions About Deep Continental Structures

G.J. MacDonald[1]

Introduction

The combination of increasingly sophisticated geophysical exploration tools and expanded drilling for hydrocarbons has produced a wealth of knowledge about limited regions of the Earth's continental crust, the sedimentary basins. We have a much more limited understanding of the crystalline basement that underlies areas of sediments and appears at the surface over a significant fraction of the Earth's land area. The requirements of society, coupled with growing scientific interests, suggest that during the next few decades, efforts to explore continental depths may be comparable to those of the last three decades in investigating the oceanic crust. Ocean crust studies provided the data that led to the plate tectonic "revolution" that has transformed solid earth geophysics.

Increased knowledge of the Earth's crust will contribute significantly to societies' growing needs. The requirement for new sources of energy has led to the exploration of buried hot rocks that, through stimulation, could emulate natural heat sources, such as the geothermal areas at Wairakei in the Taupo volcanic region of New Zealand. The search for new, large deposits of hydrocarbons, particularly natural gas, was the principal motivation for the deep drilling in Siljan, Sweden. Rare and valuable mineral deposits are found in exposed Precambrian crystalline shield areas around the world. Wide regions that may contain similar deposits are covered by thin sedimentary sheets that overlie unexplored crystalline basement. Shield areas are often regarded as highly stable and unchanging over time. Yet historically, some of the largest earthquakes have devastated shield areas that are lightly covered with younger sediments, one example being New Madrid, Missouri, in the early 19th century. Understanding the stress conditions in such regions, and in the deep crust elsewhere, may hasten reliable prediction of earthquakes. The presumed tectonic stability of the basement has prompted the suggestion that it be used as the site of repositories for the high-level waste generated by nuclear reactors. Without safe repositories, the nuclear contribution to the world's energy dilemma will vanish. In order to confidently construct such a repository, the state of stress

---

1 The MITRE Corporation, 7525 Colshire Drive, McLean, VA 22102, USA

within the crust must be ensured, as well as the permeability of the constituent rocks and their long-term tectonic stability. Such assurances can come only from a detailed understanding of crustal conditions and how they change.

In addition to such practical considerations, there are many deep scientific issues whose resolution requires the study of continental crust. Under current concepts, the oceanic crust is, at most, some 200 million years old, with a mean age of 100 million years, and is being continually regenerated and removed. The continental crust represents a much longer period of Earth's history, containing structures up to 40 times the age of those preserved in the oceanic crust. Evidence of the processes that have taken place within the outer regions of the Earth over its full history is hidden in the continental crust. The variation of the intensity and character of sedimentation, faulting, folding, metamorphism, volcanism and igneous intrusion over geologic time can be derived only from the study of the crust. Determination of the degree to which the mantle and crust have interchanged fluids requires an examination of the crust. The determination as to whether the crust is made up of material derived from the mantle or recycled from older crust can only be answered by the study of crustal rocks and fluids. Further, the crust may provide clues as to the nature of the underlying upper mantle.

While plate tectonics has had widely acclaimed success in explaining features of the oceanic crust, the theory has labored under a conspicuous deficiency. Plate tectonics fails to account, in detail, for most of the geology of continents - the very place where most observations are made. The past decade has seen numerous attempts to develop further the theory in order to account for continental geology. The complexity of basement geology has hindered all but the most ad hoc attempts to patchup plate tectonics theory with continental observations. In situ exploration of continental geology is required in order to make further progress in developing a theory of continental tectonics.

A description of the crust can be separated into several components. The chemical state of the solid fraction is defined by mineralogy, thermal conditions and geometrical distribution. In the past, the state of stress within the crust was generally regarded as one in which matter was under hydrostatic pressure. Recently derived evidence shows that anisotropic stress conditions previal in many regions. Crustal mechanical properties critically depend on the fluid phases present in pore spaces, fractures and microcracks. Fluid-filled cracks in the Earth's crystalline basement have generally been thought to be nonexistent or of minor importance. New evidence, to be reviewed shortly, suggests the opposite. Even in ancient shield areas, fluids play a significant role in fixing the properties of the deep crust.

A description of the crust's fluid phase requires a determination of the location of the fluid within the rock, the rock's permeability and the chemical composition of the fluid. Of particular importance is the oxidation state of the fluid phase relative to that of the solids. Such information determines whether the fluid was in equilibrium with the solid phases at the time of crystallization. The overall time history of the crust, grouped into periods of initial formation and later modification, is a critical component in crustal description. The mineralogy of fracture fillings and the chemistry and geometrical distribution of fluid inclusions are particularly important in describing the history of fluids at depth.

Large sections of the basement are exposed at the Earth's surface. While the evidence derived from such exposures is of great importance, it cannot yield satisfactory information as to the internal state of stress or of the nature of fluids at depth. The raising of crustal blocks to the surface destroys evidence of deep rock mechanisms as fluids and stresses are released. Near-surface positioning allows surface and groundwaters to add still another epoch of modification to what may be an already complicated history. Only deep drilling can provide data with respect to in-place stresses and fluids. The drilling process itself brings about local changes in the fluid-stress condition, but these are relatively minor compared with changes in crustal material brought to the surface by tectonic processes.

In order to highlight the relevance of the deep drilling at Kola and Siljan to major questions concerning the crust, the discussion will begin with the interpretation of a significant feature noted in seismic reflection studies of shield areas - strong, nearly horizontal reflectors in the upper crust. The interrelated issues of stress within the crust, the presence of fluids and crustal permeability are taken up next. Then follows a consideration of the interaction of fluids with their host rock, in which the relative oxidation state of solids and fluids plays a key role. The discovery of abundant hydrogen in the deep Siljan and Kola wells sheds light on the role of microcracks in deep basement rocks. The discussion concludes with a few cautionary notes on the use of conventional logging techniques in the exploration of crystalline basement rocks.

Dolerite Sills and Seismic Structure

Reflection seismic surveys providing data down to depths of 40 to 60 km have produced a wealth of information regarding the geometric disposition of materials that have differing seismic velocities. Many thousands of kilometers of profiles have been obtained by the COCORP program in the United States (Hauser and Oliver 1987), LITHOPROBE in Canada (Percival and Berry 1987), BIRPS in the United Kingdom (Matthews and Cheadle 1986) DEKORP in West Germany and ECORS in France (Bois et al. 1986, Percival and Berry 1987). A puzzling and surprising feature of many of the profiles, particularly those taken in shield areas, is the presence of parallel, but not always horizontal, layered structures having a limited geographical extent (Hauser and Oliver 1987, Smythe et al. 1982, Oliver et al. 1983, Allmendinger et al. 1987). Figure 1 schematically illustrates these

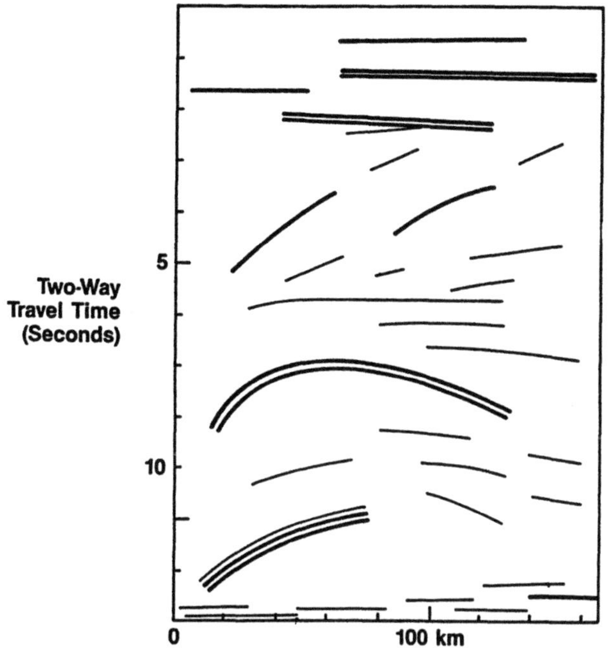

Fig. 1. Schematic illustration of continental seismic reflectors. Both near the surface and near the Moho, strong, horizontal reflectors are observed. At middle depths, there are fewer strong reflectors and these show a variety of geometrical forms

findings. In the upper crust, strong, nearly horizontal reflectors are found at 0-10 km depths in regions where no other strong reflectors are noted. At somewhat greater depths, the strong reflectors may be horizontal but more commonly display an arch or depressed basin-like structure. Just above the discontinuity that is usually identified with the Moho, discontinuous but horizontally extensive structures reappear.

The horizontal, upper crust layers have been variously interpreted as thrust fault structures, buried sediments and igneous intrusions. In one case, a layered structure in the crystalline basement southeast of Phoenix, Arizona was incorrectly interpreted as an extension of the Overthrust Belt of the Western United States and led to a flurry of prospecting excitement and a 28,000 foot dry hole (Keith 1980).

In preparation for the Siljan hole, reflection profiles were taken, which showed strong reflectors at 0.6, 0.9, 1.6 and 2.6 s, two-way travel times (see Fig. 2). Indeed, because such reflectors were judged to be of potential interest for hydrocarbon exploration, their presence influenced the choice of location of the drill hole. Subsequent drilling has shown that the upper three reflectors, at depths of 1460, 2540 and 4740 m, are dolerite sills, with thicknesses of 30, 100 and 60 m, respectively. As of September 1st, 1987, the fourth reflector had

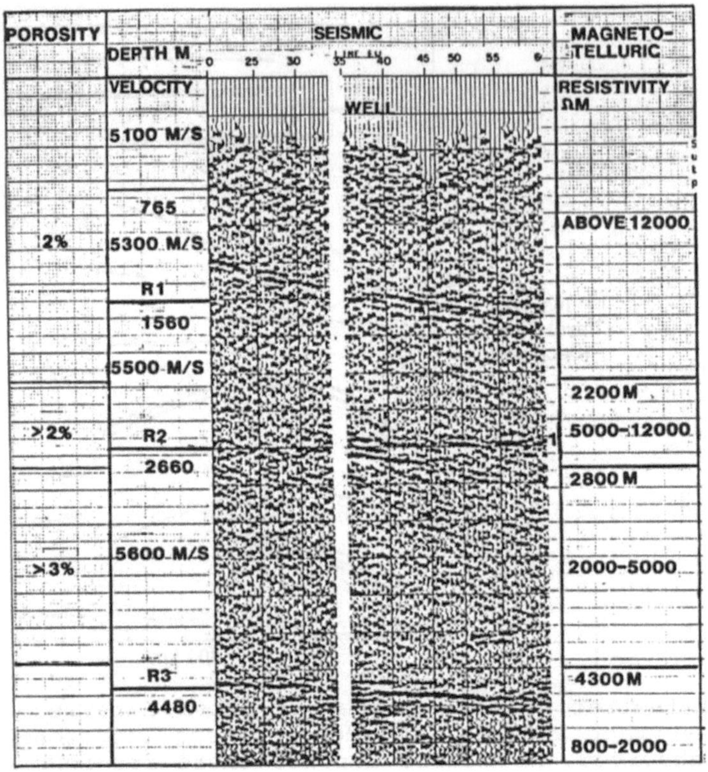

Fig. 2. North-south seismic reflection section through Gravberg, Sweden drill site

not been penetrated. Associated with all three sills are regions of healed fractures in the adjacent granite. The presence of sills in an otherwise homogeneous granite basement came as a surprise, since dolerite sills are usually found in sedimentary sequences.

Stratiform sills of dolerite, thousands of square kilometers in extent and sometimes thousands of cubic kilometers in volume, are present in every continent. Typically, these sills are found intruding young, unfolded sediments that overlie a stable crystalline basin; for example, the Win Sill in England and the Triassic Sill in the Eastern United States. Surface expressions of sills in crystalline basins are rare, in part because of the scarcity of large, vertical, exposed sections of basement rock. However, explorers in the Antarctic report horizontal sills intruding metamorphic and granitic rocks that are some miles across (Bradley 1965, Gould 1935, David and Priestley 1914, Ferrar 1925). A major sill, a few kilometers in extent, occurs at the contact of an upper red granite and a lower grey granite. Webb and McKelvey (1959) describe the sill as cutting across the grey-red granite contact and passing through a sandstone lying unconformably on the granite. The sill domes upward in apparent response to the overlying topography.

The physics of sill intrusion is well understood (Anderson 1951, Pollard 1973). In response to hydrostatic pressure forces, a silicate melt will rise through overlying rock, provided its density is less than that of the host rock. Opening a crack in the rock by hydraulic fracture, the melt will be propelled upward by means of the hydrostatic head, which is fixed by the difference in density between melt and host rock, and by the depth of the melt source. Since hydraulic fractures operate on the tensile strength of the rock, this strength can be quite low and is highly dependent on the fluid pressure within the pores of the host. If the pore fluid pressure equals the lithostatic pressure of the host rock, the tensile strength approaches its zero pressure value. As the melt rises through a channel, it cools and increases in density. At some point, the liquid becomes denser than the host rock. Under hydrostatic conditions, the heavier melt would spread horizontally, lifting the less-dense overlying rock.

At Siljan, the density of the in-place granite averages about 2620 kg/m^3. The density of a dolerite melt will depend on the temperature, total iron content and state of oxidation of iron in the melt, as is illustrated in Fig. 3 (Bottinga and Weill 1970). For intrusion in the upper crust at depths less than 10 km, the effect of pressure can be neglected. For the Siljan case, the dolerite would be denser than the granite at melt temperatures on the order of 1300°C or lower.

From dimensional arguments, a sill of thickness Δz will spread a horizontal distance L in a time T,

$$L = \frac{(\rho_r - \rho_m)\, g(\Delta z)^2}{a\eta}\, T,$$

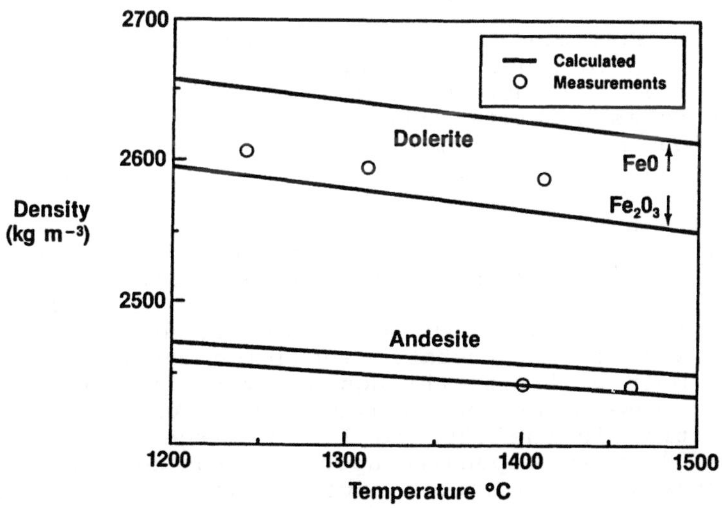

Fig. 3. Calculated and observed density at atmospheric pressure of rock melts. (After Bottinga and Weill 1970)

where η is the dynamic viscosity and a is a dimensionless constant determined by the detailed geometry - for a long thin crack a = 12, provided it is driven by a constant hydrostatic head proportional to the difference in density of the host rock ρ_r and the melt ρ_m. For T measured in years and Δz measured in meters, the time required to cool to one-half the temperature difference between the host rock and the melt is

$$T_{\frac{1}{2}} = 1.33 \times 10^{-2} (\Delta z)^2$$

(Jaeger 1964). The thermal time constant for a 100 m sill is 130 years; that of a 1 m sill is about 4 days. These cooling times must be taken as maximum estimates, since a smaller cooling than half the initial temperature difference would drop the temperature below the solidus, and the melt would become immobile. The calculation does not take crystallization heat into account, nor the heat carried off by volatilized fluids in the host rock. However, these latter affects can be expected to be minor.

A thick sill can spread over large distances in a relatively short time because of the fourth power dependence of the spreading distance L on thickness. Further, the viscosity of basaltic melts is low - approximately 10 kg/m/s at 1300°C, increasing to 4×10^3 kg/m/s at 1200°C (Kushiro 1980). This low viscosity is well illustrated in the rapid advances of basaltic lavas at the surface, where the interaction with the soil and atmosphere can remove heat far more rapidly than can solid granite at depth. Even for small density differences - 10 kg/m^3 - a dolerite could easily move 100 km in a year, provided the vertical channel remained unimpeded. Indeed, from the expression for L, it is clear that the vertical transport of melt through a narrow channel limits the extent of horizontal sills, rather than viscous resistance to the horizontal spreading.

The similarity of the seismic reflection profiles taken at Siljan to those from the upper crust in shield areas around the world, together with the identification of the Siljan reflectors as dolerite sills, strongly suggests that dolerite intrusions may be a common feature of crustal structures. Since most sills have been encountered in flat-lying sedimentary sequences, it apparently has been assumed that such structures were unimportant in the deeper crust. The association of sills with sediments led to the commonly held notion that magma would rise until it encountered weak bedding planes, at which point the melt would wedge its way into the sediments. The observation of sills in the Antarctic and in the Siljan area suggests that hydrostatic mechanics may control sill emplacement. A melt will rise through a crystalline basement if its density is less than the surrounding rock. Melts of basaltic composition in the 1200 to 1300°C range have densities of 2600 to 2700 kg/m^3, depending upon the iron content and oxidation state. Crystalline granites have densities in the 2600 to 2660 kg/m^3 range, while metamorphic sequences generally have higher densities. Basaltic melt would continue rising through heavier basement until reaching the surface of less dense rocks. Indeed, the observed distribution of sills in sediments shows a clear pattern. In regions with few sills, those sills tend to be intruded at the base of the covering sediments, or in the lower reaches of the sediments. For example, the Karoo dolerites are concentrated in sediments near the crystalline basement and become less numerous higher up the section; in the underlying basement, they seem to be absent (DuToit 1920, 1939).

On the basis of these observations, we would expect dolerite sills in those basement regions having a density less than about 2700 kg/m^3. At these densities, upflowing melts of basaltic concentration would tend to move laterally in the upper crust. The strong, near-surface, horizontal reflections noted in seismic surveys in shield areas are thus likely to be dolerite sills. Further, if the host rock has a low tensile strength, the intruding sill will mimic the overlying topography, although the relative amplitude is magnified by the ratio $\rho_r/(\rho_r - \rho_m)$, since under hydrostatic conditions fluid surfaces of equal pressure will be surfaces of equipotential. Certain of the dome-like structures noted in reflection surveys in the upper and middle crust could represent older dolerite intrusions that conformed to irregular surface topography at the time of their emplacement.

The potential importance of sills in driving vertical motions within the continental crust has been noted by McKenzie (1984). The introduction of sills from melts derived from partial melting of the mantle could provide the uplift associated with epeirogenic movements of the crust.

Dolerite Sills - Methane

Two further points should be noted about the Siljan dolerites. The granitic rocks within a few meters of the granite-dolerite contact appear more highly fractured than granite at some distance from the contact. The fractures have been healed with assemblages of secondary minerals including chlorite, epidote and quartz. Local fracturing probably resulted from the vaporization of water contained within pore spaces in the granite. The change in volume of the fluid upon heating and vaporizing led to pore pressures in excess of the lithostatic pressure, with consequent local fracturing. The dolerite is relatively unaltered and appears not to have been affected by the local release of volatiles.

At a depth of 4.7 km, the Siljan dolerite sill has been found to contain methane in a dry gas, with small concentrations of ethane, propane, etc. The dry gas associated with these dolerites is characteristically heavy in carbon isotope concentrations with $\delta^{13}C = -25$ to $-18‰$ (Jeffrey personal communication 1987). While methane is not routinely reported from basalts and dolerites, it has been observed in basalts not exposed to the atmosphere or where appropriate measurements have been made. Byers et al. (1983) report reduced gas species - principally CH_4 and CO, but including C_2H_4 and C_2H_6 - trapped in basaltic glass solidified at ocean depth and associated with the Galapagos Spreading Center. The volcanic material released the gases when heated to 800°C.

In the Kola Peninsula super deep well, Soviet investigators have found methane with a heavy carbon isotope ratio, $\delta^{13}C$ = -10‰, in intrusive gabbros (Karus et al. 1987). The regional setting, coupled with the heavy carbon isotope ratio of the Siljan dolerite and Kola gabbros, suggests that these rocks were mantle-derived. The trapped methane is consistent with a wustite-iron oxygene fugacity - lower than that of the quartz-fayalite-magnetite assemblage at the same temperature and the one usually assumed to be pertinent to basalts. Adjacent to the Siljan dolerite is a red granite with hematite-chlorite-epidote accessories in the crack-filling materials, indicating that they were deposited by a highly oxidizing fluid. Some fraction of the original methane emplaced with the Siljan dolerite has thus been preserved over the same 900-million-year period since intrusion. (Åberg 1987).

Fluids in Crystalline Basement Rocks

A wide variety of rock properties, such as electrical conductivity, compressive strength and seismic velocities, critically depend upon the presence or absence of a separate fluid phase. The significance of a fluid phase is well known to those engaged in hydrocarbon exploration in sedimentary rocks. When first deposited, and after burial to a moderate depth of 2 or 3 km, a sediment is saturated with water and is permeable. The hydrostatic pressure is given by $p_f = \rho_f gz$, where ρ_f is the density of the fluid and z is the depth. In considering the mechanical properties of the two-phase material, the stresses normal to any surface are reduced by the fluid pressure p_f from the value the normal stress would have in a single-phase elastic material; the fluid carries part of the overall stress, and the reduction in normal stress is purely a physical phenomenon. As a sediment is buried, and the fluid has a clear channel to the surface, pore pressure is maintained at its hydrostatic value. The fluid is partially expelled from the rock by compaction. Beyond a certain depth, the fluid becomes isolated from the surface; the pore pressure builds up and can approach the local lithostatic pressure, or exceed it by a small amount. The mechanical properties of the rock are basically determined by the ratio of the fluid pressure to the lithostatic pressure, p_r. At depths of 10 to 20 km, the combined effects of temperature and pressure are such that stress differences lead to plastic deformation, provided the ratio p_f/p_r <1 (Rutter 1974, Fyfe et al. 1978). As the ratio p_f/p_r approaches unity, fluid-assisted fracturing takes place at conditions representative of any depth. Similarly, the electrical properties of the material are dominated by the fluid phase. Since solid minerals are insulators at low to moderate temperatures, currents can only be carried by the fluid, which in many cases will contain dissolved salts.

Water in pores and cracks in sediments will most likely be retained to some degree, even under strong compression. Pressures build up, and local fracturing may take place as the channels to the surface are blocked, processes which equalize the local fluid and lithostatic pressure. Only under unusual circumstances will compaction and heating lead to elimination of a fluid phase; laboratory workers know the difficulty of drying even a small sample. But what of metamorphic and granitic rocks that constitute the bulk of the exposed crystalline basin? Do these rocks contain a separate fluid phase that significantly affects their physical characteristics? Some workers - for example, Fyfe et al. (1978) and Brace (1980) - argue that liquid-filled cracks exist in crystalline rocks; others believe compaction and heating completely remove any fluid phase. Evidence relating to fluids in crystalline rocks at depths of a few kilometers is scarce. As noted earlier, surface observation provides only disturbed and altered samples. A small fraction of fluid phase, 0.01 to 0.1% by volume, does not measurably alter the propagation velocities of compressional or shear waves, from which the bulk of subsurface data has been derived. The failure of rocks by fracture at depths greater than 10 to 20 km, as demonstrated by deep-lying earthquakes, can be argued to require the presence of a fluid with pressure approximately equal to that of lithostatic (Fyfe et al. 1978).

The foregoing conclusion is supported by analysis of observation of electrical conductivity in the crust. The large-scale conductivity obtained - for example, in magnetotelluric observation, which corresponds to depths down to 50 km - cannot be explained in terms of observation of rock samples in the laboratory (Garland 1981). The inferred conductivities at depths of 5 to 15 km range from 10^{-3} to 10^{-2}/Sm, or in terms of resistivity, 10^3 to 10^2 Ωm (Shankland and Ander 1983). The conductivity of dry granite is about 10^5/Sm, so there is a discrepancy of two to three orders of magnitude between inferred deep electrical conductivity and of that measured on dry rock samples. The observed conductivity could be explained if the basement rock contained connected porosity filled with water at volume concentrations on the order of 0.01 to 0.1%. A variety of conducting materials other than water has been proposed over the years, including hydrated minerals, graphite and sulfides (Stanley et al. 1977, Jiracek et al. 1983, Connerney et al. 1980). While continuous fracture fillings of hydrated minerals may locally increase conductivity, alternatives to water generally fail to provide the required electrical properties. Other fluids, such as carbon dioxide and methane, could not yield the high conductivity associated with water that contains dissolved salts.

The supposition that microcracks filled with water exist in the basement receives support from Kola well observations (Borevsky et al. 1987). Free water flowed into the borehole at depths of 6.3 km and deeper, well below zones in which aquifers have been identified. The basement rocks in the Kola region are predominantly metamorphic. The released water, interpreted as having been retained for over 10^9 years and taken to be of meteoric origin, was trapped in sediments and kept within the body of the rock during metamorphism.

The presence of fluid inclusions within basement rocks provides evidence of ancient fractures. Whether a fracture is open or closed at any depth depends on the relative efficiencies of crack growth under differential stresses, and on crack healing brought about by chemical solution and deposition. Chemical processes seal cracks; imposed stresses maintain open fractures or create new ones. The healing of cracks proceeds as open, nonsolid regions are reduced, leaving trapped tubular regions of fluid, which may survive as fluid inclusions. Further healing will transform the trapped bubbles into near-spheres distributed along the plane coincident with the original crack (Smith and Evans 1984). In some regions, the fluid does not proceed to the bubble stage; even with healing and fracture filling, there can remain a fluid phase occupying microfractures within the solid matrix. This fluid, with an internal pressure equal to lithostatic pressure, provides an explanation for the observed electrical conductivity and mechanical properties of the basement.

While the discussion has emphasized the presence of fluid as a separate phase, there is the possibility, discussed by Etheridge et al. (1983), of a grain boundary "phase". Such a phase is partly or wholly bound to the solid phase and is capable of altering the bulk properties of the rock. Initially, framework silicates, quartz or feldspar, are fractured. A large number of open silicon bonds can be formed, which can attract hydroxyls or deformed water molecules. A thin layer of water, several molecules thick, can then become attached to the solid phase. This intermediate surface "phase" will have markedly different transport properties than that of a fluid phase. The combined solid- and surface-phase system could possess electrical and mechanical properties that differ from those of the dry rock. This possibility has not been extensively explored.

Permeability of Basement Rocks

The flow rate of a fluid through a rock is determined by the product of the permeability[2] and the pressure gradient driving the flow. Permeability is determined by porosity and the geometric arrangement of the pore spaces, their connectivity in particular. In general, it is difficult to calculate a value for permeability from pore space geometry, but there is a vast literature on the subject (Dullen 1979).

---

[2] The literature on permeability contains a chaotic jumble of units and dimensions. For our purposes, the unit is one that depends only on properties of the rock (m^2). The historical unit, darcy, equals $10^{-12}/m^2$. The hydraulic conductivity or coefficient of permeability depends on the properties of the flowing fluid; for water under crustal conditions, a permeability of $10^{-12}/m^2$ is equivalent to a hydraulic conductivity of $10^{-5}/ms$.

Brace (1980) has reviewed measurements of rock permeability, both in the laboratory and from field observations such as leak-off and drill stem tests. The laboratory measurements of crystalline rocks fall in the range of 10^{-17} to 10^{-21} m^2. As in the case for electrical conductivity, field measurements on crystalline rocks at shallow depths of 3 km or less exhibit permeabilities several orders of magnitude greater than the laboratory values. In situ values of permeability of crystalline rocks range from 10^{-18} to 10^{-13} m^2, with most test holes showing regions of permeability as high as 10^{-15} to 10^{-13} m^2.

Laboratory observations presumably represent measurements of fluid flow through microcracks produced by the sampling process or through the uplift of deep rocks. The measurements also sample only small scale permeability, while in-hole measurements sample larger scale permeability. Observations secured at Kola at a depth of 6.37 km showed a permeability of 10^{-18} to 10^{-19} m^2 (Borevsky et al. 1987). Leak-off tests in the Siljan hole at depths between 4160 and 5020 m also gave values of about $10^{-17}/m^2$ (Lindbo 1987). Given the difficulty of making the measurements in a broken-out hole, no great significance should be attached to the obtained numerical values. The importance of these observations is that they demonstrate a large-scale permeability in crystalline metamorphic and granitic rocks at depths of 4-6 km.

The observed in situ deep permeabilities at Kola and Siljan lie at the upper range of observed laboratory values. In-place permeabilities can be interpreted in terms of the effect of pore pressure. In laboratory experiments, Brace et al. (1968) find that permeability rapidly increases as the ratio of pore pressure to lithostatic pressure approaches unity. The relatively high values of permeability in the dense, low porosity rocks of Kola and Siljan thus may be the result of fluids present at low concentrations, but at a pressure equal to that of the overlying rock. The observed permeabilities are low enough to be consistent with the observed retention of fluid at depth. Significantly higher permeabilities would allow the fluid to escape to the surface.

A further complication is that permeability is generally a tensor quantity, thus there is no a priori reason to expect the large-scale vertical permeability to equal the large-scale horizontal permeability. Dolerite sills or extensive horizontal regions of fracture filling may both reduce vertical permeabilities relative to horizontal permeabilities. In situ observations presently measure bulk permeability without attempting to observe anisotropic rock properties.

Oxidation State of Fluids at Depth

Estimates of the chemical composition of the fluids that have flowed through basement rocks can be obtained from analysis of the included fluids and of the mineralogy of fracture fillings and alteration rocks. While the oxidation state of the Siljan dolerite is reduced toward the iron-wustite buffer, the bulk of fluids that have permeated the basin at Siljan has been highly oxidizing. The red coloring of the granite and the presence of mineral assemblages, such as hematite, chlorite and epidote, provide strong evidence for the oxidizing nature of the pore fluids.

Observations from the Kola well bore also provide evidence for the flow of oxidizing fluids through the basement. Fractured intervals at 3 km and 4.5 km have been identified as aquifers extending to the near-surface (Borevsky et al. 1987). In Siljan, a red granite gives way to grey granite at depths greater than about 5 km, but fracture-filling minerals remain highly oxidized at all depths (Lindbo 1987). The deep circulation of meteoric waters suggested by observations at Kola and Siljan is consistent with a number of isotopic studies. Taylor (1977), and Norton and Taylor (1979), argue that isotopic data taken in Western United States batholiths support the occurrence of meteoric circulation to depths of 10 to 20 km.

Hydrogen and Methane in Deep Basement Rocks

One of the unexpected discoveries in the drilling of the Kola well was the large quantity of hydrogen. For example, the mud flowing out of the hole is described as "boiling" with hydrogen (Borevsky et al. 1987, p. 277). Below a depth of 2.8 to 3 km, hydrogen became the predominant gas in the drilling mud. Helium also became more abundant in the mud shows at these depths, and was found in concentrations comparable to that of hydrogen at depths greater than 6 km. As Borevsky, Vartanyan and Kulikov note, helium is an inert gas and therefore provides the best evidence of the flow of water and hydrogen into the well from the surrounding rock, although these investigators recognize the possible contributions of hydrogen produced by drilling.

In analyzing the gases contained in rocks recovered from the Kola well, Karus et al. (1987) found a high concentration of nitrogen - 20 cm^3/kg - in the metasediments, as well as varying concentrations of methane. The correlation of the nitrogen and methane with metamorphosed sediments, and the methane's light carbon isotopic concentration (^{13}C = -47‰), led Karus et al. (1987) to conclude that these gases are of biogenic origin. These investigators do not comment on the origin of the hydrogen but note its general association with helium.

In the Siljan hole, the concentration of hydrogen found in the drilling muds generally increases with depth (Lindbo 1987). At times, peaks in the hydrogen content of the outgoing mud coincide with peaks in helium concentration, although there are long periods of drilling in which hydrogen is found without any detectable helium. Methane concentrations are higher than hydrogen at shallow depths but are lower at depths greater than 5 km. The hydrocarbon gases associated with the granite are distinctively different from those associated with the dolerite sills. The hydrocarbon gases contained in the granite are wet, having high proportions of ethane, butane, etc., and the carbon isotopic composition is light, $\delta^{13}C$ = -40 to -30 (Jeffrey personal communication 1987).

The presence of hydrogen in both the outgoing muds and the core samples at Kola and Siljan can be interpreted in terms of the aqueous chemistry of framework silicates. Laboratory investigators working with vacuum in glass systems are well aware that moisture in these systems leads to the production of hydrogen. Fracturing of framework silicates, including quartz and feldspars, as well as glass, breaks their structure, so that silicon-oxygen-silicon bonds are severed. Fracturing of chains, sheet, and ortho and ring silicates tends to take place along cleavage planes, breaking metal-oxygen bonds. Because the silicon-oxygen bond is largely covalent, breakage can lead to the formation of radicals such as Si· and SiO· (Cotton and Wilkinson 1980, Schrader et al. 1969 a, b, Soda 1961, Arends et al. 1963). Using electron-spin-resonance techniques, Arends et al. (1963) identified the radicals and found that they were stable up to temperatures of about 360°C. Schrader et al. (1969 b) studied the reaction of radicals with steam, which forms the SiOH group. Hydrogen radicals formed during the interaction of water with silicon-based radicals react to form hydrogen gas.

Kita et al. (1982) investigated the generation of hydrogen through the reaction of crushed quartz with water. They found hydrogen generation in the 25 to 270°C temperature range, with an increase in the rate of hydrogen formation at temperatures up to about 200°C, and a decrease at higher temperatures. The decrease at high temperatures is attributed to the decreasing stability of the SiO radical; the remaining Si· and SiO· radicals form only SiOH group. Laboratory observations of hydrogen generation from quartz and feldspar have also been made by Sugisaki et al. (1983). Laboratory observation of hydrogen, generated by crushed framework silicates reacting with water, is complimented by the common observation that hydrogen is found in soils and groundwaters in active fault zones (Ware et al. 1984/85, Sugisaki 1984/85, Sato et al. 1984/85, Satake et al. 1984/85, Sato et al. 1986). The concentration of hydrogen in tectonically active regions can be interpreted in terms of the formation of silicon-based radicals that are created by the crushing of framework silicates, with the consequent formation of hydrogen gas.

Hydrogen occurrences in Kola and Siljan can be interpreted as the result of silicon-based radical formation in water saturated microcracks. The existence of microcracks is inferred from the observed inflow of helium and water into the well at Kola, and from the suspected inflow of helium at Siljan. The temperature range over which silicon radicals are stable is qualitatively consistent with observation in the Kola hole, where hydrogen increases with depth to about 9 km, or to a temperature of about 160°C. The hydrogen would react with dissolved carbon compounds in the water to form methane and higher carbon compounds.

It is not possible to obtain an estimate of the amount of hydrogen generated in microfractures, since hydrogen will be generated by the drilling process. The pulverizing of the quartz-feldspar-rich granite at Siljan presents an ideal situation for generation of hydrogen through the process discussed above. Indeed, a case could be made that all the hydrogen is of this origin. However, the association of high helium values with high hydrogen, as seen at Kola and suspected at Siljan, argues for an inflow of hydrogen into the well. It should also be noted that the reduction of water by iron during drilling will also generate hydrogen.

Cautionary Notes on Geophysical Logging in Crystalline Basement

Geophysical well logging is a highly sophisticated technique developed by the oil industry. Logging forms the basis for the evaluation of potential oil and gas deposits in a wide variety of sedimentary rocks. The techniques for interpretation have evolved over many years of experiment and experience. Logs are compared with core samples and with the actual behavior of rocks in the reservoir under development. Much less expertise is available on the use of classical logging techniques in crystalline rocks. The experience gained thus far in logging granites and metamorphics suggests that classical geophysical logs should be interpreted with caution.

Marine (1979) made a number of geophysical logs from seven drill holes in crystalline rocks at the Savannah River site in South Carolina. The logs included sonic, gamma, neutron, temperature and caliper readings. Because the holes had also undergone hydraulic pumping and slug and packer tests (Marine 1967), it was possible to compare zones that were hydraulically conductive with the geophysical logs. While sonic logs appeared to give the best results, none of the logs gave consistent correlations. A similar lack of correlation has been reported at the Stripa mine in Central Sweden (Nelson et al. 1980). In contrast, Soonawala (1983) reported good correlation between geophysical logs and television viewing of boreholes in the Loc du Bonnet granites of Manitoba. However, hydraulic tests were not conducted. Katsube and Hume (1987) obtained good qualitative correspondence between focused-electrode logs and hydraulic measurements.

A promising approach to determination of fracture zone permeability is the use of tube waves (Paillet 1983). Tube waves are guided wave modes associated with Stoneley wave propagation and are peculiar to a fluid-filled hole in an elastic solid. An explosion some distance from the hole leads to an excitation of tube waves at fractures by conversion of compressional waves. There appears to be a high degree of correlation between tube wave amplitude and the magnitude of hydraulic conductivity, as determined by packer tests.

Part of the difficulty of applying classical logging techniques to crystalline rocks is the nature of the fractures in such rocks. In sediments, the permeability resulting from flow through joints, faults or bedding planes is comparable in magnitude to the flow through small samples of the whole rock. In crystalline rock, these permeabilities can differ by orders of magnitude. A well log measures properties of rocks where stress conditions have been irreversibly altered through the drilling process. Near-hole shifts in stress conditions can open or close microfractures out to a distance of a meter or more. The log will measure the altered rock. In most sediments, microfractures contribute little to permeability, and changes in the stress field caused by drilling produce only minor alterations in bulk rock properties.

Summary

Outside of sedimentary basins, our knowledge of the continental crust is largely based on examinations of surface-exposed basement rock and on seismic observation. Application of reflection seismology - developed for hydrocarbon exploration - has greatly enriched the fundamental data base with respect to the nature of the crystalline basement. The interpretation of strong reflectors and diffractors in seismic reflection sections is limited by the lack of ground truth. Horizontal structures in basement areas are unlikely to be exposed at the Earth's surface, yet they are frequently noted in reflection sections of ancient shields. At Siljan, four such reflectors have been observed in the upper 8 km. Drilling has penetrated the first three of these reflectors and established that they are dolerite sills. This observation, together with an elementary consideration of intrusion dynamics, suggests that dolerite sills may be found in low density (less than 2650 to 2700 kg/m^3) homogeneous basement. The emplacement of such horizontal structures may have been responsible for vertical uplift in an otherwise stable crust.

Of critical importance in understanding the mechanical, electrical and thermal properties of the basement is whether a light, inviscid fluid permeates the rocks. Four critical questions concerning the fluid phase require answers, if the properties of the crust are to be understood: To what depths does a separate fluid phase exist? What is the chemical nature of the fluid phase? What is the value of the fluid pressure at the depths at which a separate fluid phase exists? What is the permeability at these depths? In sedimentary sections with large porosity and permeabilities, these questions can be readily answered. The task is far more difficult in basement terrains.

Shallow drilling (3 km or less) has established the existence of fracture zones that act as aquifers, permitting circulation of meteoritic waters to depth. In the Kola well, similar fracture regions extend to 6 km. At greater depths, down to 11 km, fracture regions containing water exist. In Siljan, the principal evidence of a fluid phase at depths greater than 5 km comes from observation of helium inflow. Water flow into the Siljan well is minor, if present at all. The dominant fluid phase at Kola is water at all depths. A case can be made for a similar situation at Siljan, given the nature of observed alteration products and fluid inclusions.

Data on fluid pressures at depth is unsatisfactory. Kozlovsky (1987) and his colleagues make a case for fluid pressure at or near lithostatic pressure at Kola, but the evidence is not clear. Difficulties in carrying out down-hole permeability tests on both the Siljan and Kola wells lead to the conclusion that the permeability at depths below 6 km is greater than 10^{-17} to 10^{-18} m^2. If the conclusion at Kola is correct - that helium has remained trapped for 10^9 years in spite of the presence of water, which is capable of removing it - the permeability is 10^{-21} m^2 or less.

Surface observations of electrical conductivity in deep crustal rocks are strongly suggestive of a separate fluid phase. Recent measurements of polarized splitting of shear wave seismic propagation can also be interpreted in terms of fluid-filled aligned microcracks (Crampin 1987). In situ observations at Kola and Siljan provide sketchy support for these interpretations.

References

Åberg G (1987) K-Ar Analysis from the deep earth gas project at Siljan, Southern Central Sweden. Unpubl Tech Rep

Allmendinger R, Nelson K, Potter C, Barazangi M, Brown L, Oliver J (1987) Deep seismic reflection characteristics of the continental crust. Geology 15:304-310

Anderson E (1951) The dynamics of faulting and dyke formation with applications to Britain. Oliver and Boyd, Edinburgh

Arends J, Dekker A, Perdok W (1963) Color centers in quartz produced by crushing: Phys Status Solidi 3:2275-2279

Bois C, Cazes M, Damotte B, Galdéano A, Hirn A, Mascle A, Matte P, Raoult J, Torreilles G (1986) Deep seismic profiling of the crust in Northern France: The ECORS project. In: Reflection seismology: a global perspective. Barazangi M, Brown L, (eds): Geodyn Ser, vol: 13. Am Geophys Un, Washington D C

Borevsky L, Vartanyan G, Kulikov T (1987) Hydrogeological essay. In: Kozlovsky Ye (ed) The superdeep well of the Kola Peninsula, Springer, Berlin Heidelberg New York Tokyo

Bottinga Y, Weill D (1970) Densities of liquid silicate systems calculated from partial molar volumes of oxide components, Am J Sci 269:169-182

Brace W (1980) Permeability of crystalline and argillaceous rocks. Int J Rock Mech Min Sci Geomech 17:241-251

Brace W, Walsh J, Frangos W (1968) Permeability of granite under high pressure. J Geophys Res 73:2225-2236

Bradley J (1965) Intrusion of major dolerite sills. Trans R Soc N Z 3:27-55

Byers C, Muenow D, Garcia M (1983) Volatiles in basalts and andesites from the Galapagos spreading center, 85° to 86° W. Geochim Cosmochim Acta 47:1551-1558

Connerney J, Nekut A, Kuckes A (1980) Deep crustal electrical conductivity in Adirondacks. J Geophys Res 85:2603-2614

Cotton A, Wilkinson G (1980) Advanced inorganic chemistry. Wiley, New York

Crampin S (1987) Geological and industrial applications of extensive-dilatancy anisotropy. Nature (London) 328:491-496

David E, Priestley R (1914) Geology of South Victoria Land. In: Br Antarct Exped Rep 1907-1909. Heinemann, London

Dullen F (1979) Porous media: fluid transport and pore structure. Academic Press, London New York

DuToit A (1920) The Karov dolerites. Trans Geol Soc S Afr 33: 1-42

DuToit A (1939) Geology of South Africa. Oliver and Boyd, Edinburgh

Etheridge M, Wall V, Vernon R (1983) The role of the fluid phase during regional metamorphism and deformation. J Metamorph Geol 1:205-226

Ferrar H (1925) The geologic history of the Ross Dependency. N Z J Sci Technol 7:354-361

Fyfe W, Price N, Thompson A (1978) Fluids in the Earth's crust. Elsevier, Amsterdam

Garland G (1981) The significance of terrestrial electrical conductivity variations. Annu Rev Earth Planet Sci 9:147-174

Gould L (1935) Structure of the Queen Maud Mountains, Antarctica. Geol Soc Am Bull 46:354-361

Hauser E, Oliver J (1987) A new era in understanding the continental basement: The impact of seismic reflection profiling. In: Fuchs K, Froidevaux C (eds) Structure and dynamics of the lithosphere-asthenosphere system. Geodyn Ser vol 16. Am Geophys Un, Washington D C

Jaeger J (1964) Thermal effects of intrusions. Rev Geophys 2: 443-466

Jiracek G, Gustafson E, Mitchell P (1983) Magnetotelluric results opposing magma origin of crustal conductors in the Rio Grande Rift. Tectonophysics 94:299-326

Karus E, Nartikoyev V, Bartashevich O, Gigashvili G, Ikorsky S, Pavlova M, Petersilje I, Pisarnitskaya T (1987) Gases and organic matter. In: Kozlovsky Ye (ed) The superdeep well of the Kola Peninsula. Springer, Berlin Heidelberg New York Tokyo

Katsube T, Hume J (1987) Permeability determination in crystalline rocks by standard geophysical logs. Geophysics 52:342-352

Keith S (1980) The Great Southwestern Arizona overthrust oil and gas play. Az Bur Mines Min Tech Field Not 10:1-9

Kita I, Matsuo S, Wakita H (1982) H_2 Generaion by reaction between H_2O and crushed rock: an experimental study on degassing from an active fault zone. J Geophys Res 87:10,789-10,795

Kozlovsky Ye (1987) The superdeep well of the Kola Peninsula. Springer, Berlin Heidelberg New York

Kushiro I (1980) Viscosity, density, and structure of silicate melts at high pressures and their petrologic applications. In: Hargraves R (ed) Physics of magmatic processes. Princeton Univ Press, Princeton

Lindbo T (1987) Deep gas project, Gravberg Well status report 0-5.5 km. Vattenfall, January 1987

Marine I (1967) The permeability of fractured crystalline rock at the Savannah River Plant near Aiken, South Carolina. US Surv Prof Pap 575-B:203-211

Marine I (1979) Hydrology of buried crystalline rock at Savannah River Plant near Aiken, South Carolina. Dep Energy Rep DOE/SR-WM-79-2, Washington D C

Matthews D, Cheadle M (1986) Deep reflections from the Caledonides and Varescides west of Britain and comparison with the Himalayas. In: Barazangi M, Brown L (eds) Reflection seismology: a global perspective. Geodyn Ser, vol 13. Am Geophys Un, Washington D C

McKenzie D (1984) A possible mechanism for epeirogenic uplift, Nature (London) 307:616-618

Nelson P, Magnusson K, Rachiele R (1980) Application of borehole geophysics at an experimental waste storage site. Lawrence Berkely Lab Rep LBL-11982

Norton D, Taylor H (1979) Quantitative simulation of the hydrothermal systems of crystallizing magmas on the basis of transport theory and oxygen isotope data: an analysis of the Skaergaard Intrusion. J Petrol 20:421-486

Oliver J, Cook F, Brown L (1983) COCORP and the continental crust. J Geophys Res 88:3329-3347

Paillet F (1983) Acoustic characterization of fracture permeability at Chalk River, Ontario. Can Geotech J 20:468-476

Percival J, Berry M (1987) The lower crust in continents. In: Fuchs K, Froidevaux C (eds) Composition, structure and dynamics of the lithosphere-asthenosphere system. Geodyn Ser, vol 16. Am Geophys Un, Washington D C

Pollard D (1973) Derivation and evaluation of a mechanical model for sheet intrusions. Tectonophysics 19:233-269

Rutter E (1974) The influence of temperature, strain rate and interstitial water in the experimental deformation of calcite rocks. Tectonophysics 22:311-334

Satake H, Ohashi M, Hayashi Y (1984/85) Discharge of H_2 from the Atotsugawa and Ushikubi Faults, Japan, and its relation to earthquakes. PAGEOPH 122:185-193

Sato M, Sutton A, McGee K (1984/85) Anomalous hydrogen emissions from the San Andreas Fault observed at the Cienega Winery, Central California. PAGEOPH 122:376-391

Sato M, Sutton A, McGee K, Russell-Robinson S (1986) Monitoring of hydrogen along the San Andreas and Calaveras Faults in Central California in 1980-1984. J Geophys Res 91:12,315-12,326

Schrader R, Glock G, Köhnke K (1969a) Mechanochemie des Systems Quarz-Ruß. Z Chem 9:156-157

Schrader R, Wissing R, Kubsch H (1969b) Zur Oberflächenchemie von Mechanisch Aktiviertem Quarz. Z Anorg Allg Chem 365: 191-198

Shankland T, Ander M (1983) Electrical conductivity, temperatures, and fluids in the lower crust. J Geophys Res 88: 9475-9484

Smith D, Evans B (1984) Diffusional crack healing in quartz. J Geophys Res 89:4125-4135

Smythe D, Dobinson A, McQuillin R, Brewer J, Matthews D, Blundell D, Kelk B (1982) Deep structure of the Scottish Caledonides revealed by the MOIST reflection profile. Nature (London) 299:338-340

Soda R (1961) Infrared absorbtion spectra of quartz and some other silica modifications. Bull Chem Soc Jpn 34:1491-1495

Soonawala N (1983) Geophysical logging in granites. Geoexploration 21:221-230

Stanley W, Boehl J, Bostick F, Smith H (1977) Geothermal significance of magnetotelluric sounding in the Eastern Snake River Plain-Yellowstone Region. J Geophys Res 82:2501-2514

Sugiasik R (1984/85) Relation between hydrogen emission and seismic activities. PAGEOPH 122:175-184

Sugisaki R, Ido M, Takeda H, Isobe Y, Hayashi Y, Nakamura N, Satake H, Mizutani Y (1983) Origin of hydrogen and carbon dioxide in fault gases and its relation to fault activity. J Geol 91:239-258

Taylor H (1977) Water/rock interaction and the origin of H_2O in granite batholiths. J Geol Soc London 140:509-558

Ware R, Roecken C, Wyss M (1984/85) The detection and interpretation of hydrogen in fault gases. PAGEOPH 122:392-402

Webb P, McKelvey B (1959) Geologic investigations in South Victoria Land, Antarctica. N Z J Geol Geogr 2:120-136

Conclusions – Exploratory Work

T. Lindbo[1]

Introduction

This paper presents preliminary compressed status conclusions of the drilling to 6337 m TVD before the furlough beginning on Sept 1st, 1987.

After spud-in on July 1st, 1986, the first 4 km were drilled very fast and easily with more and more reduced cost. Thus it appeared possible for the investors to change TD to 7.5 km in order to investigate the 4th reflector where the US deep gas experts saw the best possibility of finding a productive gas reservoir. However, between 4 and 6.5 km serious problems developed and we were forced to sidetrack and redrill twice. Thus the funds were exhausted so that the drilling had to be stopped and new funding efforts are needed for the drilling to be able to continue to TD 7.5 km.

Pores and Fracture Systems

The results from wireline logs including borehole seismics and gravity log, analysis of cuttings and gas shows, together with drilling parameters, mainly the rate of penetration, are all important when determining the properties of the rock and its capacity to constitute a gas reservoir. In general, all results point to a very heterogenous rock and because of that the rate of penetration was highly variable. The rock can be described as consisting of zones with matrix porosity and fractures which are mostly healed but sometimes open. Between these zones homogeneous rock is present. All observations imply that the rock, even at 6 km, may contain open fractures which could act as a gas reservoir.

Pore Pressure

A large loss of the drilling fluid occurred at 6047 m in the original hole, indicating the presence of a pore pressure less than that of the heavy water-based mud. In addition, the build-up of mudcake in the old hole with sticking of the drillstring indicates a low pore pressure. When drilling sidetrack I with

[1] Swedish State Power Board, 162 87 Vällingby, Sweden

only lubricated freshwater and tripping, thus slightly reducing the pressure in the hole, interesting gas shows of hydrogen and helium were seen. The conclusion of that is that the pore pressure even between 6 and 6.5 km is somewhat below the hydrostatic of freshwater. The rock mechanics studies indicate a stress field with anisotropic horizontal stresses. The maximum stress is of the same order as the vertical stress, and the minimum may be around 60 to 70% of the vertical. The maximum pore pressure that can be expected in the rock is on the order of the minimum horizontal stress, that is 50% to 60% above the hydrostatic of freshwater. At what depth a transition occurs from pore pressure near the hydrostatic of freshwater to that of the minimum horizontal stress is unknown. However, it is below this transition zone that an overpressured reservoir can first be expected to be encountered.

Gases

Two types of hydrocarbon gases have been detected. The highest concentration has been found in connection with the dolerites and thus the seismic reflectors. This gas is a dry gas consisting of more than 90% of C_1 (methane) and isotopic analysis points to a probable mantle origin. The other gas is found in the granite and is a wet gas with more than 10% C_2-C_5 and of lighter isotopic character. No gas inclusions have been detected in the quartz crystals that have been looked at. Thus the gas may exist only in the fractures and pores, which is consistent with the theory of gas coming from below after the impact.

Below 4 km the gas shows of hydrogen increased and deeper down even helium, with a maximum of both gases in the "seismic band" between 5.9 and 6.4 km. When tripping, the hole gave repeatable high shows of hydrogen, indicating a possible reservoir. Because of low pressure and comparatively small hydrocarbon gas shows, probably all hydrocarbon gases are dissolved in water. However, hydrogen and helium with low and no solubility, respectively, may exist as free gas.

Gas readings from the mud when drilling in the "band" shows a clear correlation with depth between hydrogen, helium and even methane. The conclusion here is that all these gases come from the formation and may have migrated upwards from below from a level where they exist in a certain ratio.

Drilling

The breakouts are presently interpreted as caused by anisotropic horizontal stresses with the minimum stress component parallel to the large axis of the oval hole. Breakout free zones normally show extensive natural fracturing and stresses may not be anisotropic in these zones. The hole deviation is interpreted as a function of the structure of the rock layers, fracture systems and the breakouts. Below 5.5 km, the deviation was stabilized at 35 to 45°, and this deviation is expected to continue down to TD. The technical problems when drilling with different types of water-based mud seem to have been solved by using oil-based mud.

Gravity and Seismics

One of the main reasons for choosing the Gravberg site was a relatively large gravity anomaly measured from the surface. Regardless of the anomaly, a possible reservoir porosity below 7 km would be difficult to detect from the surface. Densities from a recent borehole gravity survey cannot explain the gravity anomaly by the rock penetrated up to now.

Reflector R4 is most likely associated with a dolerite. The intensity of it points both to a thicker dolerite and more fractured rock around it. Although it is realized that one is close to the resolution limit, recent velocity analysis indicates a velocity inversion between R4 and R5, also indicating more fractured rock. Using the amplitude of the seismic reflection to predict if the fractures are open and possibly overpressured is not possible, since the effects of dolerite cannot be taken into account in a quantitative manner. Since the previous dolerites have shown clear evidence of the presence of dry gas in the rock, one would expect to encounter additional dry gas at the R4 reflector.

Summary

The data obtained so far shows that the granite has the potential to act as a reservoir. Of great importance is that the rock already drilled through has contained open fractures at great depth and there is no reason for this not to be the case below 6.3 km. The dry hydrocarbon gas found in association with the dolerite is probably of nonbiological origin, but is in all certainty coming from the formation. The wet gas detected when drilling in granite also comes from the formation.

The increase in hydrogen and helium below 5 km and the repeatability of the shows in the trip gas, indicating gas influx from the formation, is extremely interesting. Hydrogen and helium have much better capacity than hydrocarbon gases to penetrate fine fissures in the rock connecting open fractures. The premise that hydrogen should be found above a methane reservoir makes it important to drill deeper.

Analyses of the data up to now indicate two sections of the borehole which have a chance to produce gas. The first section is between 4.6 km and 5.3 km around the third reflector. If this section is productive, it could serve as a supplement to a more productive reservoir deeper down. The other section between 5.9 and 6.4 km could be interpreted as a hydrogen reservoir with traces of hydrocarbon.

It must be clearly stated that all these reservoir possibilities must be tested, with gas flow to the surface and volumes calculated, before anything is confirmed. The question of testing when drilling through interesting sections or before the furlough in drilling has been discussed. The risk of destroying the borehole by doing so is imminent and, therefore, all tests are planned for TD at 7500 m. In order to evaluate, the gas potential of low porosity and permeability reservoirs, we must rely on the experts opinion as to what the chances are of producing gas from these formations as a basis for future funding. It should be kept in mind that when drilling in such reservoirs in an overbalanced situation there will be no influx of gas into the wellbore and very little gas will be detected at the surface.

The motive to drill to 7.5 km depth was explained already 1 year before spud-in by involved American experts. Based mainly on the seismics, they said that of course there was a possibility of finding a gas formation around the third reflector, but reflector R4 near 7.5 km and the surrounding rock has the best potential for gas production. The presence of a potential minor gas reservoir in association with the R3 reflector makes it imperative to continue drilling to R4.

When predicting the R4 formation based upon R3 and the "band", one may assume that it could be formed by a thick dolerite sill surrounded with open fracture systems. The whole structure may be a magnitude bigger than R3. Even if R4 most likely is underpressured, it can have good commercial reservoir possibilities. If it is overpressured, even better.

Technical Papers

Gravberg-1: Well Design and Drilling Preparations

A. J. Beswick[1]

Introduction

Following the encouraging results from the various preliminary studies, a deep exploration borehole was proposed as part of the Swedish Deep Gas Project. This paper briefly describes the well design and drilling preparation phases of the project up to the commencement of drilling on 1st July 1986.

First, it is relevant to review the overall project objectives. The principal goal was commercial exploration, not science. However, in a pioneering endeavour of this kind, a scientific approach is important and so from the outset, the project was in part a scientific investigation.

Following the preliminary studies, the planning of a deep exploration well in the Siljan Ring structure commenced in early 1985. There were many aspects to consider, not least where to drill and the logistics of implementing a major drilling campaign in central Sweden.

Drilling Objectives

The drilling objectives were first and foremost satety. Secondly, if the well encountered gas in commercial quantities, there must be some scope to complete the well as a producer. Thirdly, the conditions must be such as to allow the evaluation of the well both commercially and scientifically. Last, but not least, the operation must be at an acceptable cost.

Logistics

The location in central Sweden, some distance from the main oilfield supply bases scattered around Europe, presented obvious logistics implications. The project location is shown on the map on Fig. 1. Most of the equipment and supplies had to be imported into Sweden. A bonded store area was established at the drilling site, which greatly simplified the import and re-export problems with considerable cost savings. This arrange-

---

1 Kenting Drilling Services Limited, Castle Donington, Derby DE7 2NP, England

Fig. 1. Location map

ment also allowed certain goods to be maintained on site on a consignment basis.

The location north of latitude 60° in northern Europe also implied severe winter conditions.

Site Selection

The well location was obviously the subject of considerable debate. Two sites were shortlisted, one in the centre of the structure and one to the north east near the hamlet of Gravberg. The two sites are shown on the map of the Siljan area on Fig. 2. The central location involved more logistics problems with respect to power supply, road access and daily travel to and from the drilling site, particularly in winter, but otherwise there was no real concern from a drilling organisational standpoint. There were many facets of the geological setting of the two sites which coloured the perception of what would be found during drilling and how the well design should be approached, but these were somewhat secondary in the general debate about the geological prognosis. After reviewing all the information from the preliminary studies, and much discussion, it was decided to drill the deep exploration well at the Gravberg site.

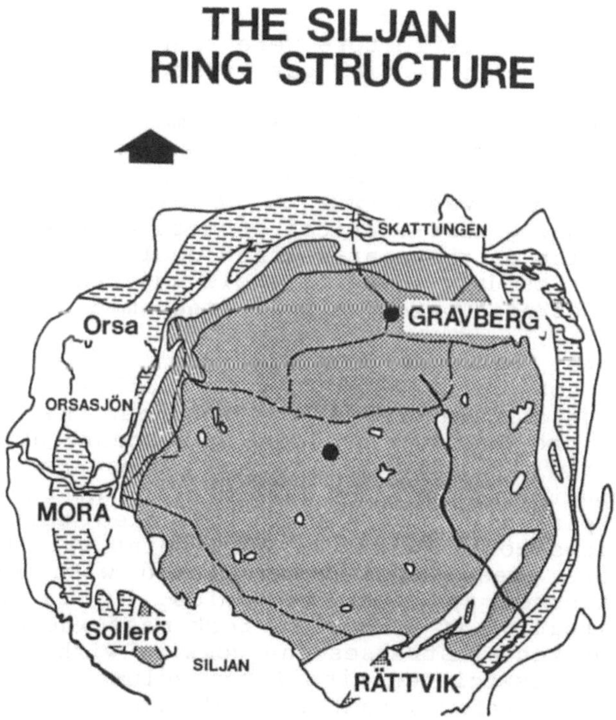

Fig. 2. Location of two shortlisted sites

The terrain in the area is heavily wooded with timber haulage tracks affording access across the roughly circular area of the impact structure inside the public highways around the perimeter. Noteworthy is the fact that the tree cover and glacial overburden obscure much of the geology, except for a few scattered outcrops in river channels and around the perimeter of the structure.

Information Available for the Well Design

The well design was based on previous knowledge and experience from other deep drilling in crystalline rock elsewhere together with the geological prognosis for the particular site selected which was based on the results of the preliminary studies and the regional geological setting. The main aspects of what was known about the conditions likely to be encountered at Gravberg are summarised below.

Geology. It was known that the rock was likely to comprise a series of granites virtually from the surface, although the levels and types at depth were speculative. Strong seismic reflectors were a feature of the reflection seismic profile in this area. Four reflectors were indicated in the first 7500 m.

The seismic profile in the vicinity of the selected site together with the deep well location are shown on Fig. 3. Interpretation of these reflectors gave rise to much discussion. In part, the reflectors were seen as a possible manifestation of a series of potential gas "reservoirs".

The Gravberg location was almost at the centre of a low gravity anomaly, indicating the presence of a relatively high porosity rock mass. The presence of fractured zones was anticipated and the site was known to be close to a north west-south east-trending near-vertical structural feature. More detailed information on the near-surface geology was provided by a 600 m deep continuously cored borehole adjacent to the Gravberg site which penetrated predominantly fresh medium-grained granite.

Pore Pressures and Fracture Gradient. One of the important aspects with respect to the design of the well was the prediction of pore pressures throughout the depth to be investigated. At one extreme, the deep gas theory applied to the Siljan scenario suggested that pore pressures under suitable caprock conditions could be lithostatic. At the other extreme, previous experience in granites indicated that sub-hydrostatic conditions could exist. At this point the whole subject of credible pore pressures becomes linked with the question of in situ stresses and rock mechanics. It is difficult to envisage a situation in a granite rock mass where significant anisotropic horizontal stresses can coexist with abnormally high pore pressure associated with large volumes of fluids. It was considered, therefore, that it would be realistic to modify the maximum pore pressure at 5000 m to 100 MPa. Notwithstanding this reduction in highest perceived pore pressures at 5000 m, some thought had to be given to the actual eventuality of encounter

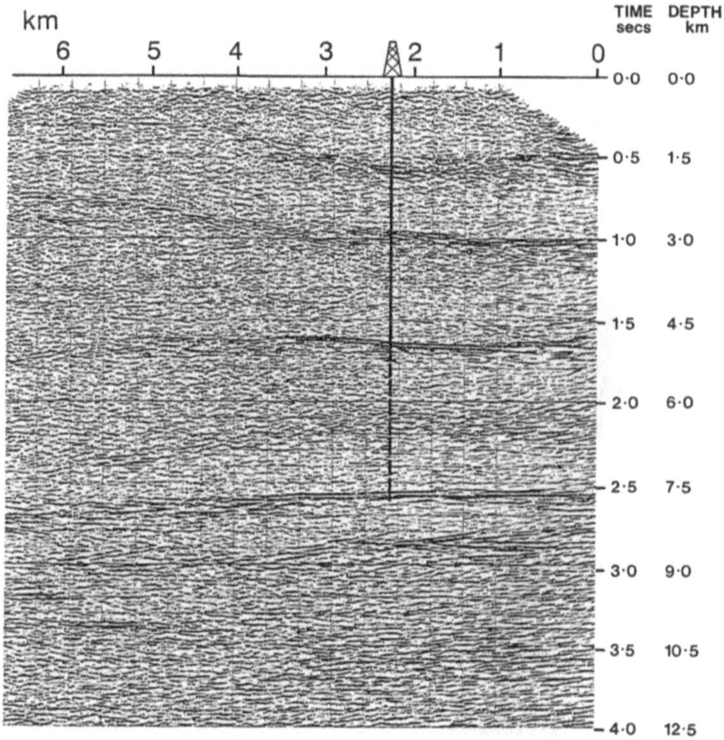

Fig. 3. Reflection seismic profile through Gravberg site

ing lithostatic pressures, perhaps even higher up in the well, however remote. Depth v estimated pressure and stress profiles for the Gravberg-1 well assumed at the planning stage are shown on Fig. 4.

The fracture gradient was also a matter of some uncertainty and again, in this geological context, it must be related to the in situ stress environment. It was considered that the fracture gradient was probably controlled by the minimum horizontal stress. This infers that at well pressures above the fracture gradient, a lost circulation situation will exist and this will control the mud density that can be used over any given interval of hole.

Temperatures. Surface temperature in the area ranged from -40°C in winter to +30°C in summer. The geothermal gradient was thought to be low at about 15°C/km and so not likely to give rise to any special problems with a total depth (TD) rock temperature, even at 7500 m, of only about 125°C.

Depth and TD Diameter Target. The depth of the Gravberg-1 well was to be at least 5 km and possibly up to 7.5 km. The target diameter at TD was to be 8-1/2 in (216 mm) which provided for a contingency hole diameter of 6 in (150 mm) to 6-1/2 in (165 mm).

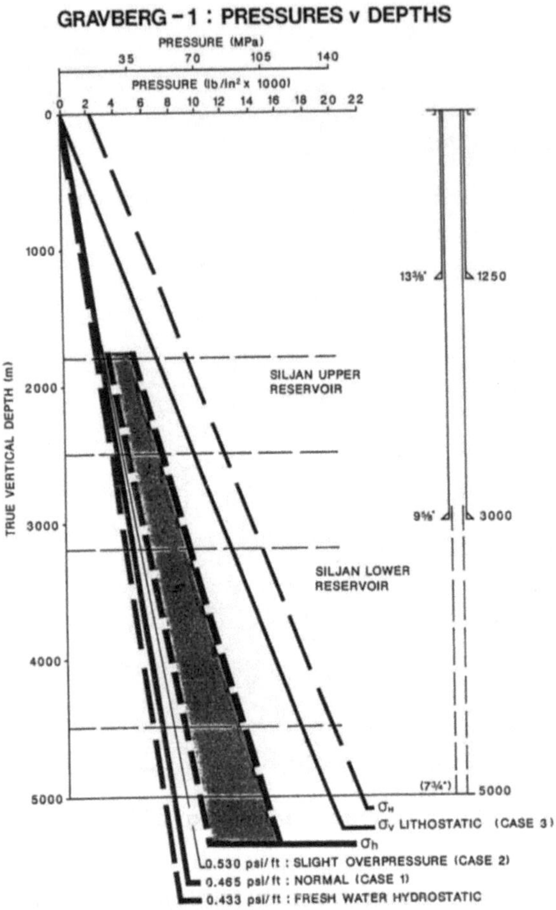

Fig. 4. Assumed pressures vs. depths

Scientific Investigations

A number of scientific activities had to be accommodated within the drilling programme, together with some special requirements with respect to gas sampling and monitoring. The following list summarises the main activities.

- Mud logging with additional gas monitoring and sampling
- Geochemical monitoring and sampling
- Drilling data monitoring
- Coring
- Sidewall coring
- Wireline logging
- In situ stress measurements
- Drill stem tests
- Vertical seismic profiling
- Well testing (if appropriate)

Main Aspects of the Well Plan

The main aspects of the original well plan are described briefly below. During the progress of the drilling, modifications were made to casing depth and the drilling fluids programme.

Bits. It was decided to adopt standard bit and casing sizes if at all possible to ease lead time problems and give the maximum flexibility in any amendment to the plan. Another important decision was that a standard bit type was adopted, bearing in mind that granite was anticipated from the surface to TD and over 20 km of granite drilling experience was available on which to base the selection. This had two advantages. Firstly it permitted better identification of any subtle changes in the geology and, secondly, where experimentation with drilling parameters such as rotary speed, weight-on-bit, etc was desirable, a standard bit would allow a good correlation without the added complication of interpreting the effects of different bit types.

Coring. For coring, the plan was to attempt to core with six cone roller type bits that had been relatively successful in other granite wells. This approach recognised the unsatisfactory situation with respect to coring with diamond bits in deep granite holes using oilfield type equipment.

Casing. The casing design was based on a pessimistic pressure scenario, but avoiding non-standard drift diameters and the very highest grade steel available in order to give some small tolerance to minor H_2S or free hydrogen, although it was recognised that this tolerance was token. Availability was also a consideration.

Cementing. The cementing philosophy was to be simple and strong, but such as to ensure adequate isolation for pressure control.

Drilling Fluids. Drilling fluids were the subject of much debate and gave rise to a dilemma. For high resolution monitoring of small quantities of gases to permit trends to be identified. the preference was for clean water. Hole cleaning and hole stability considerations make this scientific desire untenable. Hence the original plan was to use clean water as far as possible and then increase the density and viscosity as necessary using a water based mud. Mud weights at 5000 m were not expected to be more than 10.5 lb/gal (1.27 SG). However. ever mindful of the remote possibility of abnormally high pressures. provision was made for weighting materials to counteract pressures up to lithostatic.

Pressure Control Equipment. As well as blow-out preventers. a rotating head was included in the rig inventory to allow underbalanced drilling if pressurised formations were encountered to permit gas to flow to surface while drilling was in progress.

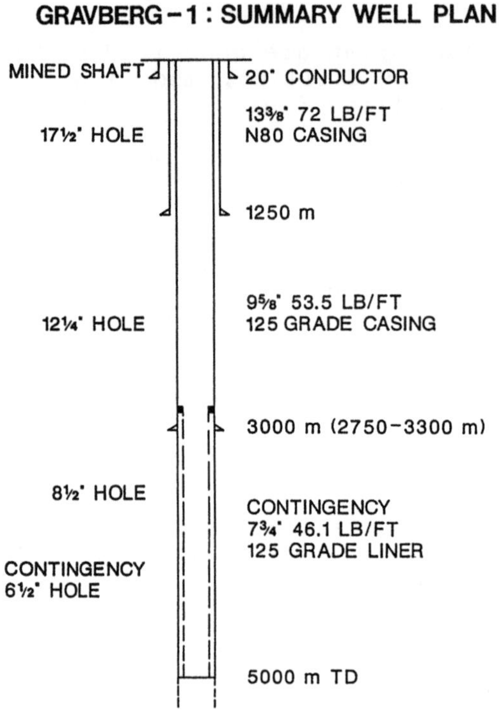

Fig. 5. Summary well plan

Rig Specification. The possibility of extending the well to 7.5 km prompted consideration of rigs capable of drilling to that depth, if the price differential was realistic in the context of the budget.

Summary Well Plan

The summary well plan shown in Fig. 5 illustrates the main features of the well design. Considerable thought was given to the initial entry arrangements and the near surface conditions. Reference was made to previous experience of spudding wells directly in fresh granite. The debate was whether or not to drill a short length of 26 in diameter hole necessitating special and expensive bits or whether to mine a shaft and pre-set a conductor. Following inspection of the core and records from a 600 m deep slim hole near the site including a review of the TV log, it was decided to mine a shaft 12-13 m deep and preset a 20 in (508 mm) conductor concreted into the shaft to form a rigid slot through which to commence drilling.

Time vs. Depth Programme

A time v depth schedule was developed recognising the optional scenarios with respect to pressures and mud weights. The original plan to 5 km is shown in Fig. 6. Case 2 is perhaps the most appropriate to the actual events that followed.

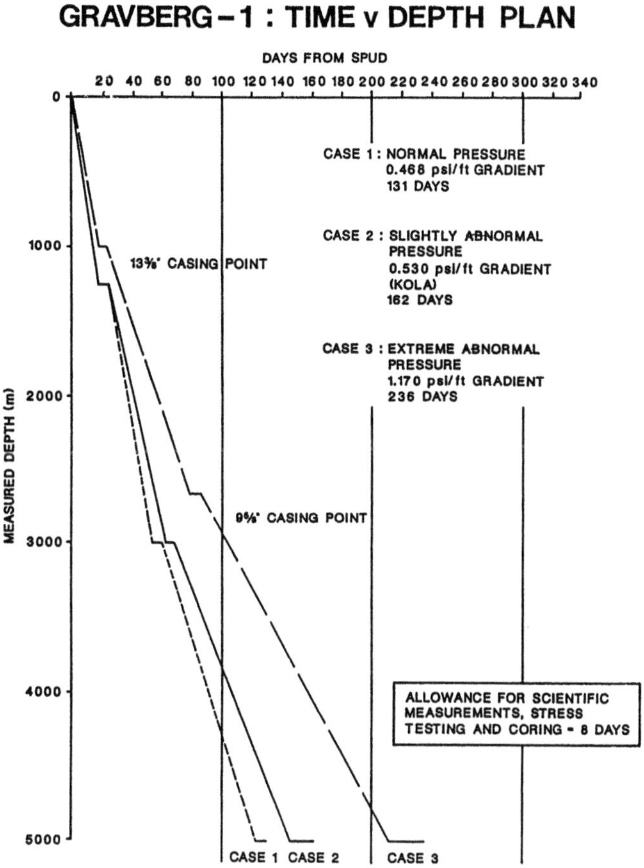

Fig. 6. Time vs. depth plan

Organisational Structure

A management, engineering and supervision organisation was developed to cover the site development, drilling and scientific activities as well as the normal administrative functions required for a major project of this kind. Additional scientific specialists were coopted on to the project team as required during the progress of the project.

Drilling Site

As part of the preparation activities, a drilling site was designed with all the facilities considered appropriate for the implementation of the drilling programme, including provision to cater for large fluid volumes and storage of drilling mud. The detailed layout could not be finalised until the drilling rig was selected. A reinforced concrete cellar and substructure foundation construction was incorporated in the design in view of the potential loads involved and the variable nature of the underlying glacial soils at the actual site selected. The mud pits were lined to minimise environmental pollution from see-

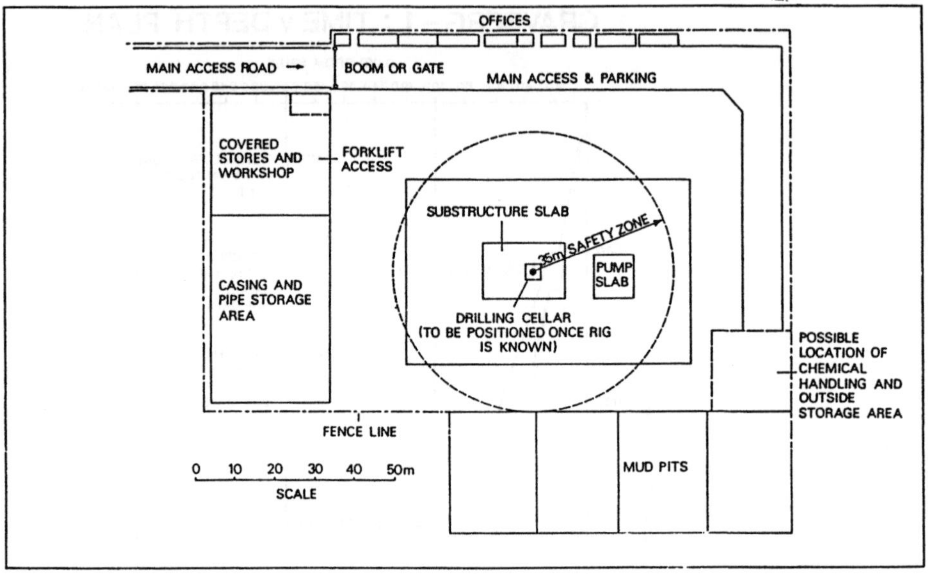

Fig. 7. Typical drilling site layout plan

page into the surrounding soils and watercourses. Water wells were drilled to provide an adequate supply of drilling water and potable water. A typical layout plan for a deep well location is shown on Fig. 7. At the actual site selected, the access was from the opposite side to suit the local conditions.

The access road from the nearest highway was reinforced and made suitable for heavy vehicles in all seasons.

Drilling Rig

During the procurement period and the process of evaluating tenders for the drilling rig, an opportunity arose whereby the Operator was able to purchase a suitable drilling rig capable of drilling to at least 7500 m, at an attractive price, bearing in mind the depressed state of the drilling industry at that time. This opportunity gave flexibility in contingency planning and the assessment was that it would give significant overall cost benefits or allow deeper drilling within the budget if progress was good. Such a decision meant that additional arrangements had to be made to manage the rig, contract crews and provide all the necessary consumables and spares, which involved additional organisation, but the cost benefits were attractive. It was decided, therefore, to purchase a relatively new Mid-Continent 1220UB diesel electric drilling rig capable of drilling to about 9 km. The rig was located in Elk City, Oklahoma, USA. Hence, within the pre-drilling period, the process of purchasing the rig had to be completed, additional equipment sourced, the rig and equipment shipped by land and sea to site and prepared for drilling in time for the scheduled spud date of 1st July 1986, all in the process of about four months. A photograph of the rig ready to commence drilling in shown on Fig. 8.

Arrangements were made to power the rig from the 10,000 V mains which resulted in further cost savings as electrical power is relatively inexpensive in Sweden. This necessitated the erection of a transmission line to the site and the installation of appropriate transformers and switchgear. The three diesel generating sets which form part of the drilling rig were retained for standby power.

Commencement of Drilling

After 18 months of planning and organisation, everything was ready for the official commencement of drilling of the Gravberg-1 deep well on 1st July 1986 on schedule.

Fig. 8. Drilling rig ready to commence operations, June 1986

Scientific Programme of Well Investigations – an Overview

H. SANDSTEDT[1]

Introduction

Vattenfall (the Swedish State Power Board) is carrying out an exploratory drilling program in the Siljan ring area in the central part of Sweden to investigate the presence of hydrocarbons in crystalline bedrock at depth. As a part of the research program, drilling of the well Gravberg-1 is in progress down to a planned TD of 7500 m. The well is drilled entirely in granite of pre-Cambrian age. The likelihood of finding commercial quantities of gas in crystalline bedrock is based on the theory of deep earth gas, developed by Professor Thomas Gold. It is also based on comprehensive field work consisting of seismic, gravity, magnetotelluric, geochemical and geological investigations.

The overall objective of the borehole is to discover gas in the structure and test the commercial viability of any finds. In order to achieve this goal, a number of investigations are performed during drilling. These investigations consist primarily of mud logging, core sampling, gas analyses, geophysical borehole logging, vertical seismic profiling (VSP) and geochemical sampling and analysis.

In addition to the objective of discovering gas in the structure, the planned investigations will ensure many geoscientific spin-off effects. The mere fact of a 7.5-km-deep hole being drilled in the Swedish bedrock will contribute vastly to the geological knowledge, and this will be reinforced by the extensive geological investigations that have preceded the drilling. There will now be an opportunity to link geophysical results with reality.

The drilling operation commenced 1st July 1986. At the time for this report, 1st September 1987, the drilled depth was 6600 m and the vertical depth 6337 m. The drilling operation has just been closed down for a longer furlough and drilling is expected to commence again in spring 1988.

This report gives a short summary of all different geological investigations carried out during the drilling operation and at different offsite laboratories.

---

1 The Swedish State Power Board, 162 87 Stockholm, Sweden

Gas Analysis

Gas analysis and gas sampling have been one of the most important parts of the investigations.

Gases have been sampled in three different ways on site, flowline, "bombs" and dunks. The flowline gases are liberated by a gas trap in the possum belly and drawn into the mud logging unit through a gasline. Total gas (all burning gases) is measured off the flowline as well as C_1-C_5 on a continuous basis. Mud "bombs" consisting of 150 cc steel cylinders which contain samples of the drilling fluid; these are collected before the drilling fluid goes into the well and before it enters the possum belly. Gases measured from the mud "bombs" on site are C_1 - C_5, as well as H_2, He, N_2, O_2. The advantage with the bombs is the possibility to take samples under 2 m of hydrostatic head with no contamination of air. The dunk system consists of a 10-l plastic container which is initially filled with drilling fluid and submerged with the bottleneck facing down. While drilling, gases present in the drilling fluid displace the fluid in the container until it is completely gas-filled. This process takes about 1 h, after which a sample of the gas which displaced the fluid is taken. The gas is then analyzed for C_1-C_5, H_2, He, N_2, O_2 and CO_2. The main purpose for the dunk sample is to take accumulated gas samples for analyses at different off-site laboratories. Radon has also been analyzed on-site.

Besides the work on-site, many different analyses have been performed with the main objective to evaluate the origin of different gases. As an example, detailed analyses of hydrocarbons, including unsaturates, carbon delta C-13 and isotopic composition of helium, could be mentioned.

Cutting Studies

Cuttings have been sampled every 10 feet in most parts of the borehole. Detailed cuttings studies have been crucial, as relatively few cores have been recovered. Studies of the cuttings at the well site include a lithologic description with emphasis on quantifying mineralogical, textural and pore space parameters which affect porosity, and permeability, that can indicate the potential presence of a fractured reservoir. Important mineral parameters are epidote, reddish feldspar, chlorite and other alteration products. Textural parameters include the presence of fractures, cataclastic and brecciated grains, vugs and drusy crystals. The study is also geared to identify minerals which affect wireline logs.

As a backup to the work on-site, more detailed analyses have been carried out in an off-site laboratory with more sophisticated equipment.

Additionally, the following analyses have also been carried out on site on a regular bases; fluid salinity, pH value, calcite content, and susceptibility.

Geochemical Analyses Off-Site

A broad geochemical program has been performed on different samples as gases, cuttings, mud, and cores. Gas analyses performed have already been described. The element concentrations have been analyzed on cuttings and mud. Of special interest have been the isotope ratios of the carbonate (calcite) in cuttings, delta C-13 and delta O-18 on the $CO_3^{-2}C$.

Drilling Parameters

In the mud logging unit, all drilling parameters such as depth, ROP, WOB, RPM, etc. have been monitored continuously. A good control on depth is essential for all sampling and testing. The drilling parameters are also of importance for the understanding of variations in the rock mass. For example, an increase in ROP may indicate more fractures and higher porosity.

Geophysical Borehole Logging

Because of a very limit amount of cores recovered from the borehole, the logging data has been extremely important for the evaluation of the well. The geophysical borehole logging program is designed to aid in providing the following information:

- Ensure a good geological description of the rock that has been drilled through. This is highly important when core sampling takes place intermittently.
- Provide information of structures in the rock.
- Localize fractures and fracture zones and also evaluate their orientation in the bedrock.
- Differentiate between open and sealed fractures.
- Measure primary and secondary porosity of the rock.
- Localize gas in the formations.
- Localize various anomalies in the vicinity of the borehole.
- Measure borehole geometry, inclination and orientation.
- Evaluate the quality of cementation behind the casing pipe.

Below 1500 m the following logs have been run in the borehole; spectrum gamma ray, litho density, compensated neutron, dual laterlog, electro magnetic propagation tool, spontaneous potential, sonic, micro electric scanner tool (selectivity over interesting parts of the borehole), dipmeter, borehole geometry tool and temperature log. One test has also been carried out with the aluminum activation tool.

To verify the gravity surface investigation in the area, a borehole gravity tool has been run between surface and 5000 m.

Vertical Seismic Profiling (VSP)

VSP employs the technique of using a surface source and recording the resulting energy at different levels in the borehole. The VSP investigation consists of both zero offset and fixed offset VSP.

The purpose of carrying out VSP in the Siljan borehole is threefold. First, it will give more precise measurements of subsurface velocities than were obtainable using surface measurements. This means that the VSP will give direct measurements of interval, average and RMS velocities as opposed to the indirect ones obtained using reflection seismology. Secondly, it is hoped that through analyzing the upgoing waves, the locations of the reflections on the stacked seismic sections can be predicted before they are drilled through. The latter point is extremely important in regards to the coring objectives of the project. If the depths to the reflecting horizons are well known, then the chances of taking cores in these horizons improve considerably. Thirdly, the data obtained by VSP will help greatly in correlating the stacked seismic sections with depths in the borehole.

VSP has been run at the following depths:

1. At 1250 m Zero offset VSP
2. At 2372 m Zero and fixed offset VSP
3. At 3932 m Zero and fixed offset VSP
4. At 5673 m Zero offset VSP

Coring

Coring in granite with a full-size drilling rig is normally difficult and this has also been the case in the Gravberg borehole. Normally a six-cone insert corebit has been used especially designed for granite. So far, 14 coring runs have been performed in interesting parts of the formation, 8 with no recovery and the others with a recovery between 1-40%. Twelve of the coring runs have been drilled with roller type bits. Except for normal rotary assembly, two coring runs were also carried out with a low speed downhole motor and both with no recovery. In an attempt to improve core recovery, two cores have also been drilled with diamond bits. One was a normal thick-walled oil-drilling diamond coring bit, and one was a thin-walled mining-type bit. After 0.6 m of drilling by the thick-walled bit, the recovery was nil, while the bit was totally worn out. The same bad result was also encountered with the thin-walled bit.

The cores are of great importance in measuring physical properties of the rock, for example the porosity, and in the interpretation of cuttings and geophysical borehole logs. In addition, the cores are also an important source for geochemical analyses. On the cores the following physical parameters will be analyzed as a standard; density, porosity, magnetic susceptibility, magnetic remanence, induced polarization and resistivity. As well as shedding light on the main issue of the project, i.e., finding gas, the cores will ensure an increased understanding of the actual meteorite impact and geological conditions at great depths.

Hydraulic Tests

To receive hydraulic information both drill stem- and leak off tests have been performed at various depths. An extensive formation test program including several zones of interest is planned at TD.

Four DST have been performed between 1300 and 3000 m. The purpose of the tests are the following; formation water characteristics, to liberate possible gases, evaluate formation pore pressure and permeability and flow capacity.

Twelve LOT have been performed between 1250 m and 5000 m. The purpose of the tests is mainly to test hole stability if drilling into high pressure zones and to verify the existence of and evaluate permeable zones.

Drilling of the Gravberg-1 – Review and Operational Problems

P. L. Moore, T. L. Brittenham[1]

Introduction and Historical Update

The objectives of the Deep Gas Test Well were twofold: first, to discover commercial quantities of natural gas, and second, to gather scientific data on deep crystalline rock and the Earth's crust. Both objectives require that a large proportion of time be devoted to scientific investigation of the wellbore. Many site-specific drilling problems have been encountered due to the nature of the project, namely drilling in a structure created by a meteorite impact that caused massive fracturing in the rock.

Drilling operations officially commenced on the Gravberg-1 Well July 1st, 1986. Surface casing of 13-3/8" was set at 1250 m on July 27th, 1986 in 17-1/2" hole. Drilling proceeded in a 12-1/4" hole with a freshwater drilling fluid system to 3932 m. Drilling fluid was viscosified with polymer at this point in preparation for a logging suite.

The hole was subsequently drilled to 4167 m, where 9-5/8" intermediate casing was run and cemented on November 22th, 1986. This section of the hole required several cement plugs and increased mud weight (9.5 ppg, 1.14 S.G. maximum) to facilitate drilling.

Drilling continued in an 8-1/2" hole. A cement plug was set at 5683 m to improve hole stability. An incompatibility between the drilling fluid and a lubricant additive were encountered and eventually alleviated.

The hole was deepened to 6081 m on February 25th, 1987, where the pipe became differentially stuck and was subsequently twisted off at 4691.97 m. Fishing operations were unsuccesful and the hole was sidetracked at 4636 m on March 24th, 1987.

A 7-3/4" drilling liner was run at 5799 m on May 10th, 1987 in the 8-1/2" hole. The liner was stuck at 5453 m during running operations. After several hours of attempting to free the liner, it was cemented in place and operations resumed.

---

1 Southern International, Incorporated, 4401 NW 4th Street Suite 121, Oklahoma City, OK 73107, USA

Drilling then proceeded with 6-1/2" hole. The hole was deepened to 6636 m MD, 6310 m TVD on July 1st, 1987 when the decision was made to spot a cement plug to alleviate hole problems of breakout, high torque, and pipe sticking, causing excessively long trip times. The drillstring became stuck and twisted off at 4883 m. The parted string was fished successfully, but could not be freed. The drill string was eventually cut at 6138 m with a Pengo Cutter. The hole was successfully sidetracked at 5991 m on July 29th, 1987. At this time, the mud system was changed to an invert emulsion oil-base system. The oil-base mud system reduced or alleviated most of the drilling problems.

The 6-1/2" hole was deepened to 6600 m MD, 6337 m TVD on September 1st, 1987 at which point the project was furloughed with the wellbore ready to resume operations.

Operational Review

Time Breakdown and Statistical Analysis. A time Breakdown for the Gravberg-1 is provided in Fig. 1. The data reveals an average progress rate of 15.27 m per day, inclusive of all activities that took place in the wellbore. This breakdown also shows the effects of unstable hole and wellbore breakout and the steps required to cope with these problems. Two complete months were devoted to scientific investigation of the wellbore.

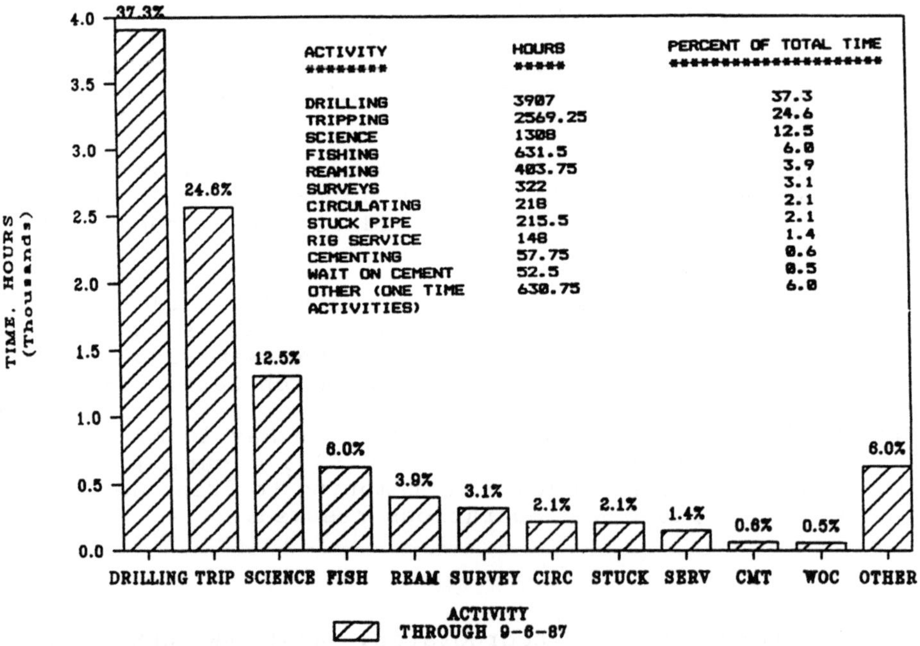

Fig. 1. Gravberg-1. Time breakdown

The statistics tend to mask the difficulties that were encountered while fishing. Eleven separate fishing jobs consumed a total of 631.5 h, an average of 57.41 h per job. Considering that six of the jobs were successful on the first fishing attempt, the problems encountered on the remaining five are emphasized.

The statistical analysis (Fig. 2) shows the effects of a remote location, a crystalline rock environment, and the necessary scientific bias on the drilling efforts.

The downtime statistic shows the excellent support the drilling effort received from the Vattenfall organization and local Swedish service industry. It also underlines the efforts of the drilling crews under sometimes adverse conditions, such as -40°C ambient temperature.

A final wellbore configuration schematic is provided in Fig. 3.

Casing Point Selections. Table 1 presents a summary of casing strings run in the Gravberg-1 including size, grade, weight, connections, length, and reason for setting casing.

Table 1. Casing program. 0-1250 m Surface casing

Size	Grade	Weight	Connection	Length	Hole Size
13-3/8"	N-80	72#/FT	Buttress	1250 m	17-1/2"
Reason: Well control					
0 - 4167.27 m Intermediate casing					
9-5/8"	LS-125	53.5#/ft	Buttress	3117.74 m	12-1/4"
9-5/8"	P-110	53.5#/ft	Buttress	1051.36 m	12-1/4"
Reason: Well control. Isolate breakout zone at 3925 m-3940 m with diameter up to 80 inches (203 cm). Stabilize hole so that drilling may continue. Protect upper portion of hole.					
3724 - 5453 m Drilling liner					
7-3/4"	LS-125	46.1#/ft	HYDRILL SFJP	1729.0 m	8-1/2"
Reason: Isolate unstable hole and protect upper portion of hole, leave thorium zones and lost circulation zones open for testing at later date. Originally planned to set liner at 5799 m, but the liner could not be worked to bottom.					

Depth and days:	6,600 m in 432 days
No. of bits:	138
No. of coring attempts:	14 and 1 sidewall coring
Logging suites:	10
Drill stem test:	4
No. of trips:	209
No. of trips for holes in drillstring:	14
No. of successful fishing jobs:	9 out of 11
No. of intentional sidetracks:	2
No. of cement jobs for hole stability:	8, aborted 1
Downtime due to mechanical breakdown:	90-1/4 hours & 8-1/2 electrical power down

Fig. 2. Statistical analysis

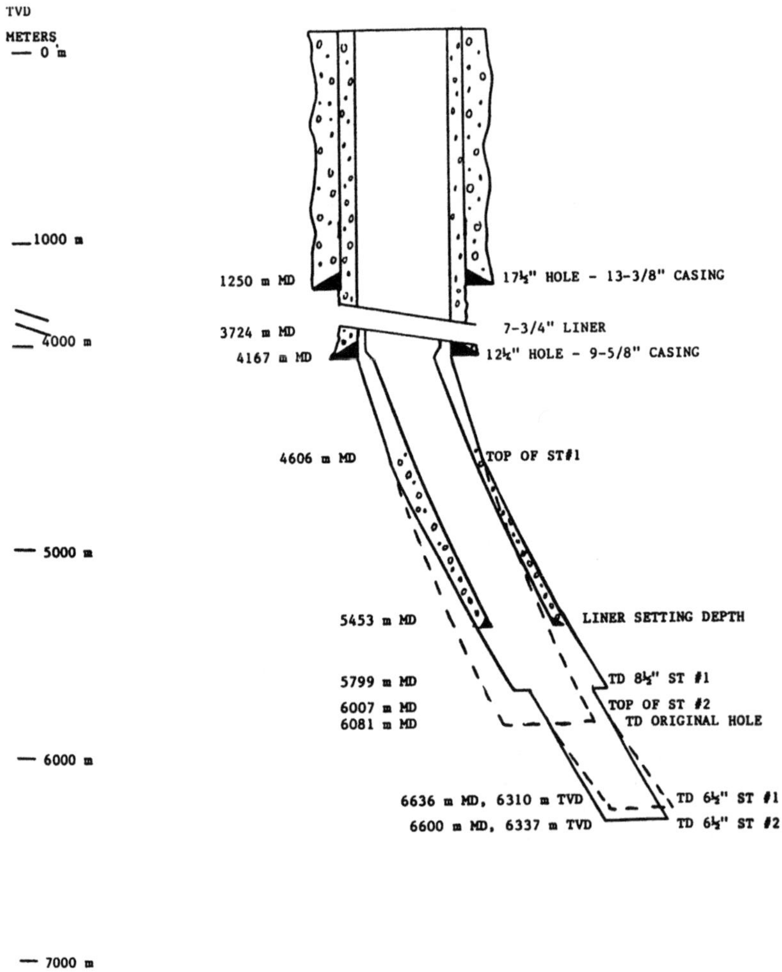

Fig. 3. Gravberg - 1. Wellbore schematic

Drilling Fluid Program

The Gravberg-1 was drilled with four different mud systems, three freshwater-base, and one oil-base. A brief recap is presented in Table 2.

Several problems with the drilling fluid systems were encountered while drilling the Gravberg-1. The number of cement plugs required to regain hole stability repeatedly led to calcium ion contamination of the mud systems. Pre-treating the system before cementing was not always effective, and therefore required massive post-treatments with sodium bicarbonate and water. This fluid system required constant maintenance.

Additions of drilling fluid lubricant to reduce downhole torque also presented problems. The first mud lubricant used contained contaminants that caused the barite in the mud to become oil-wet, flocculate, aggregate, and drop out of the mud system. This problem was solved by the addition of surfactants to reverse the wetting characteristics of the barite and changing mud lubricants.

Mud foaming was often a problem, but could be dealt with most of the time mechanically, thereby reducing the excess chemical content of the drilling fluid.

Circulation was lost only once in this well. Mud density at the time of the loss was 13.5 ppg (1.62 S.G.), the highest ever achieved in the well. Circulation was regained at 11.5 ppg (1.38 S.G.). This event occurred at a depth of 6047 m.

The drillstring became differentially stuck three different times. The first time at 5905 m the pipe was freed by spotting diesel pills across the stuck portion of the pipe. The second time at 6081 m diesel pills proved ineffective and hydrostatic reductions to 8.6 ppg (1.03 S.G.) were required to free the

Table 2. Drilling fluid summary

Depth	System	Portion of Hole
0 - 3932 m	Freshwater	Original hole
3932 - 4167 m	Freshwater and polymer	Original hole
4167 - 6081 m	Freshwater and bentonite	Original hole
4636 - 5799 m	Freshwater and bentonite	Sidetrack 1
5799 - 5938 m	Freshwater	Sidetrack 1
5938 - 6636 m	Freshwater and polymer	Sidetrack 1
5991 - 6600 m	Invert emulsion oil-base	Sidetrack 2

drillstring. The third time, also at 6081 m, and the ensuing mechanical difficulties led to Sidetrack No. 1.

After the drilling liner was set at 5453 m, freshwater was utilized to facilitate the measurement of gas contained in the drilling fluid. No special problems were encountered until drilling below 6400 m when hole stability and cleaning became a major problem. Additions of polymer and lubricants were ineffective and the drillstring was stuck again.

Permission was finally obtained to use an invert emulsion oil-base mud system after the second sidetrack was successfully begun. The oil-base mud provided excellent lubricity and hole stability was substantially improved. Hole angle began to drop and bit life increased from about 12 h to over 40 h. Pipe sticking and tight hole were eliminated. Penetration rates dropped dramatically (60%) at first, but were improved as the emulsion was relaxed with water additions. When the well was furloughed, the Gravberg-1 was drilling ahead with no problems.

Bit Selection and Performance. Overall bit performance for the Gravberg-1 was very good with an average of 70.23 m per bit and 2.40 m/h. All drill bits were tungsten carbide insert type with a zero offset. Table 3 provides average bit performance data for the six major sections of hole.

Gauge problems were not excessive for a crystalline rock well and gauge problems disappeared when the oil-mud system was utilized.

Cones were left in the hole three times in 138 bit runs.

Table 3. Drill bit performance

Hole Section	Average Meterage	Average Penetration Rate (m/h)
17-1/2"	102.24	2.46
12-1/4"	64.37	2.73
8-1/2" OH	63.80	2.02
8-1/2" ST-1	91.72	1.88
6-1/2" ST-1	62.77	3.28
6-1/2" ST-2	36.49	2.05
6-1/2" ST-2[a]	47.65	1.10

[a] Excludes two unsuccessful sidetrack attempts

Routine Operational Problems

Downhole Conditions. Downhole conditions for the Gravberg-1 were generally very poor. Drilling through highly fractured, unstable granite was difficult at best. Severe breakout of the formation led to problems for cementing, tripping, running casing, and drilling. The unstable granite was also responsible in part for the deviation problems and high torque encountered.

Low formation pressure (9.2 ± o.5 ppg, 1.10 ± 0.06 S.G. EMW) led to differential sticking situations.

The low geothermal gradient (15.78°C/km) encountered had both advantages and disadvantages. Low temperatures at depth (100°C maximum) allowed the use of conventional surveying tools, mud and cement systems, and logging tools. Low temperatures created some difficulties in downhole cement curing.

Severe hole problems were encountered below 6400 m in Sidetrack #1. The freshwater-polymer mud system in use at the time could not stabilize the hole at such depths, provided inadequate hole cleaning and insufficient lubricity. Round trips for bit change reached 24 h. In contrast, trip times averaged 15 h in the same section of hole drilled with oil-base mud in Sidetrack #2. The BGT log also shows less breakout and decreasing hole angle in Sidetrack #2.

Remote logistics and arctic weather conditions provided the bulk of routine operational problems. Careful planning was required to obtain supplies on a timely basis. The low temperature gradient downhole combined with -30°C weather conditions made surface operations somewhat difficult at times with the water-base mud system. In summary, only 90 h downtime was experienced in 15 months of routine operations.

Coring. This section provides a summary of the coring activities at the Gravberg-1, specifically the poor core recoveries and the work that was done towards improving recovery.

Core recovery was poor. Attempts were made to core 103.9 m in 14 runs with a total recovery of 8.59 m. This is an 8.27% recovery rate based on recovery length. Overall average recovery was 0.614 m per run. Seven runs (50%) had no recovery. Of these, two did not reach the bottom of the hole to commence coring.

In an attempt to improve recovery, cores were attempted using three different size catchers, three different length barrels, three different rotary speeds, three different bit throats, and with both stabilized and unstabilized bit bodies. Running parameters, catcher size, modifications or the lack thereof did not appear to be the basic problem. One conventional diamond coring run cut 2 ft, destroyed the bit, and recovered no core. A large diameter core attempted with a single wall barrel and a diamond bit was unsuccessful and recovered no core. The only variable left is the intrinsic nature of the rock.

The following general observations were noted from the core that was recovered.

1. All core recovered indicated that the rock was either fractured or had visible fractures contained.

2. Most of the dolerites have a flat inclination. One core preserved a contact between dolerite and granite with an apparent dip angle of about 10°.

Severe Deviation Problems. Deviation was a difficult problem in the Gravberg-1. Mechanical solutions to drop angle were tried with no success. Slick, pendulum, and packed hole assemblies were all tried with different bit weights and rotary speeds with no effect.

Increased mud density was tried to stabilize the stresses in the rock and appeared to work in Sidetrack #1. While drilling 8-1/2" hole from 5450 m to 5750 m with 12 ppg (1.44 S.G.) mud, angle dropped from 34° to 26°. Hole angle then began to increase just prior to switching over to freshwater. In the same section of hole in the original wellbore, hole angle increased steadily while drilling with 11 ppg (1.32 S.G.) to 13.5 ppg (1.62 S.G.) mud.

An intriguing correlation is noted between the silica content of the rock and hole angle. As can be seen on the graph shown in Fig. 4, as silica content decreases, hole angle increases. Hole angle appears stable at a silica content above 72% in the original hole.

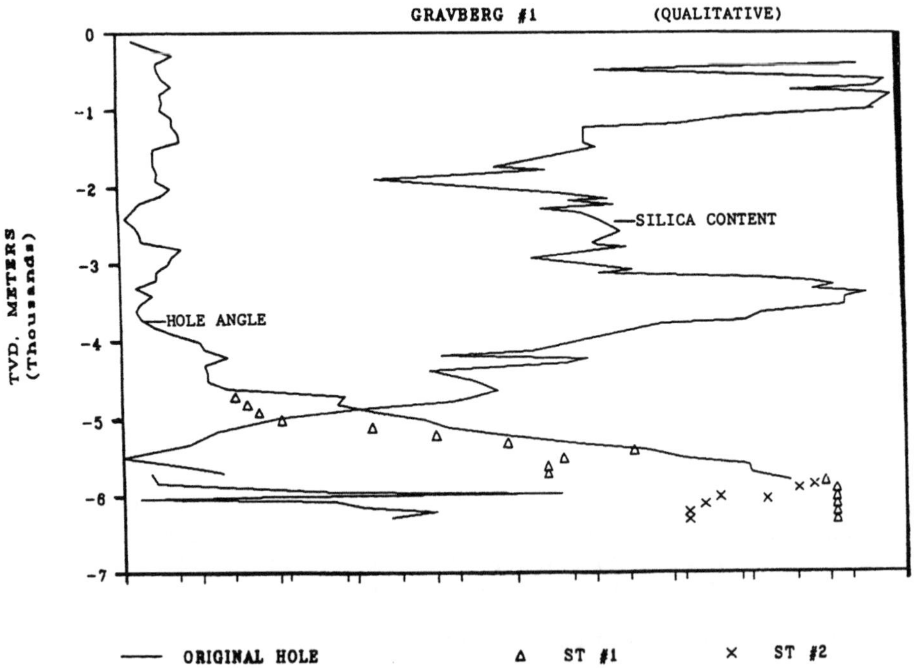

Fig. 4. Silica content and hole angle comparison

Sidetrack #2 was drilled with oil-base mud. Generally, angle decreased throughout this section of hole, but it is difficult to determine if the mud was a primary, secondary, or irrelevant factor. Silica content varied, but was consistently below 72%. Interestingly, at 6050 m TVD, silica content had a low "spike" of 63.3% from 65.9% and hole angle "spiked" upward to 41° from 38°.

It appears that extreme hole angle has no definitive technological solutions. The natural forces involved are too great to be overcome by bottom hole assembly sideforces.

Hole Instability. Hole instability was a major problem encountered while drilling the Gravberg-1. Many times downhole torque would be so high that the drill pipe could not be rotated. Downhole motors proved ineffective at these times also. Cement plugs placed on the bottom were required to stabilize the hole so that drilling could continue.

Breakouts also complicated cement jobs. Pump volumes were often indeterminate because hole volume could not be measured or calculated accurately.

Mechanical sticking was often a problem due to the unstable borehole walls. Polymer sweeps were required prior to connections on Sidetrack #1 after the drilling had passed 6400 m. Making a drill pipe connection could take up to 3 h. Tripping out of the hole past 6400 m became time-consuming, as single joints had to be "pumped" out of the hole to prevent sticking. Drilling fluid at the time was a freshwater-polymer mud having a density of 8.6 ppg (1.03 S.G.).

Increased mud density was tried to stabilize the stresses in the rock and appeared to work in Sidetrack #1. While drilling 8-1/2" hole from 5450 m to 5750 m with 12 ppg (1.44 S.G.) mud angle dropped from 34° to 26°. Hole angle then began to increase just prior to switching over to freshwater. In the same section of hole in the original wellbore, hole angle increased steadily while drilling with 11 ppg (1.32 S.G.) to 13.5 ppg (1.62 S.G.) mud.

Sidetrack #2 was drilled with oil-base mud. The BGT log for this section shows much less breakout than that observed in the previous holes. Angle also dropped steadily in this section of hole. The log section provided (Fig. 5) shows near gauge hole for parts of Sidetrack #2.

Obviously, oil-base mud at least partially stabilized the hole and will allow deepening of the Gravberg-1 to 7500 m TVD, which might not have been possible with water-base mud.

Cementing Problems. A statistical analysis of the cementing operations undertaken on the Gravberg-1 is presented in Table 4 followed by a discussion of the problems encountered while attempting to perform cement operations.

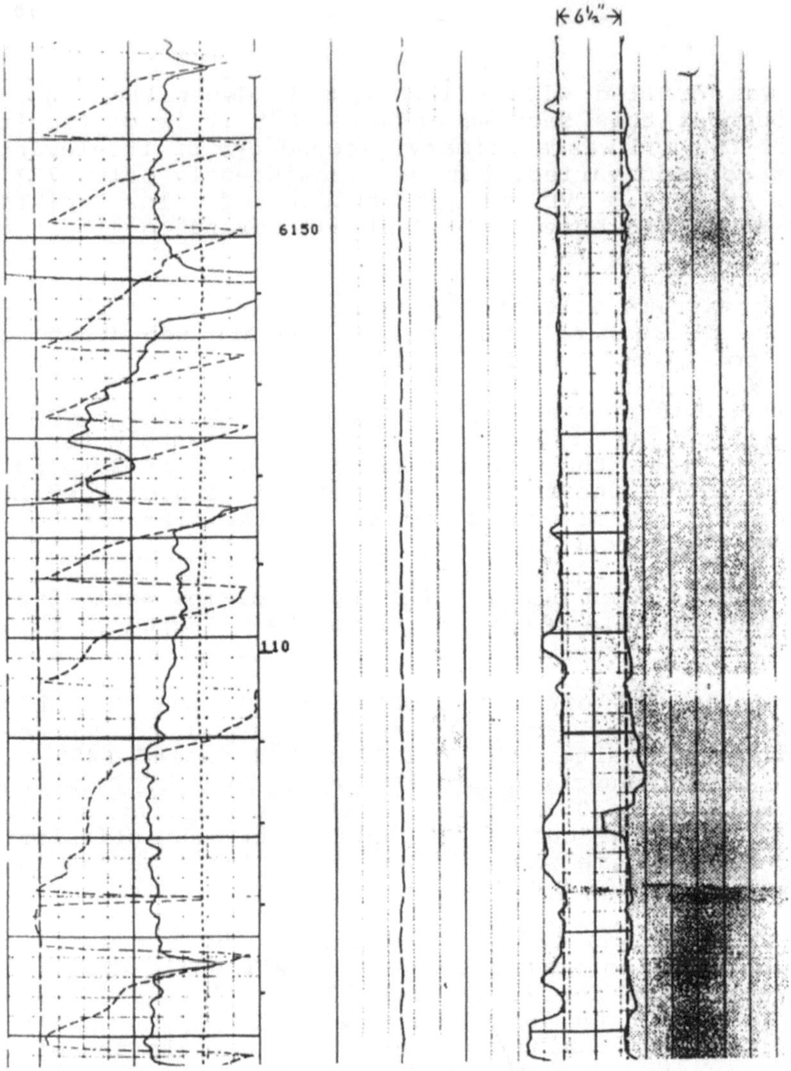

Fig. 5. Sidetrack #2, drilled with oil-base mud

The reason that two jobs failed and four succeeded was never precisely determined. Cement, mixing water, and additive samples were taken and analyzed. No specific cause was found. Surface samples of mixed cement slurry were taken during all

Table 4. Statistical Analysis of Cementing Operations

1. No problems were experienced with casing cement jobs.
2. Cement did not set up properly on two of six jobs (33%) pumped for hole stability.
3. Four out of six jobs (67%) pumped for hole stability were successful.
4. One job for hole stability was aborted (not included in above total).
5. Multiple (3) plugs needed to achieve Sidetrack #2 were not a cement related problem; simply could not kick-off in an irregular-shaped crystalline wellbore.

jobs and all samples set up properly on surface. Wellbore temperature and thickening time tests were run prior to each job to obtain a recommended slurry recipe. Recipes were followed each time with the aforementioned mixed results.

Too little information is known about cementing in deep crystalline wellbores to determine exact solutions to the problems encountered in the Gravberg-1. More experience and trial-and-error solutions will eventually prove invaluable.

Conclusions

Resolution in these problem areas will lessen the mechanical constraints of such ultra deep projects.

These problems can be investigated most effectively by a continuation of the Siljan borehole and should be enthusiastically supported by the Crustal Drilling Community.

Acknowledgments. Our special thanks go to Charles P. Willis and Dorphus A. Van Buskirk.

Scientific Papers

The Gravity Field of the Siljan Ring Structure

D. Dyrelius[1]

Introduction

The Siljan Ring structure in the central part of the Swedish province Dalarna is an annular depression with an average diameter of about 40 km, containing preserved Paleozoic sediments and surrounding a central dome of Proterozoic granites. The geology of Dalarna was described by Hjelmqvist (1966) and Fig. 1 is a simplified extract from his geological map. An up-to-date account of the geological setting of the Siljan Ring structure is given by Collini and Karlsson (this Vol.).

A meteorite impact origin of the Siljan Ring structure was first suggested about 25 years ago (Fredriksson and Wickman 1963, Wickman et al. 1963) and is today generally accepted. Bottomley et al. (1978) found an $^{40}Ar-^{39}Ar$ age of 362 Ma for the impact, which is in accordance with the strong erosion of the structure indicated by the absence of the Paleozoic cover outside the annular ring.

In many cases negative gravity anomalies, caused by disrupted rocks, have been found in association with inferred impact structures on the Earth (e.g., Innes 1961, Dence 1964, Pohl et al. 1977, Sweeney 1978). A gravity study of the Siljan Ring structure was initiated several years ago and showed a strong negative anomaly close to the annular ring in the NNE part of the structure. The anomaly was interpreted by a mass deficit extending to a depth of about 5 km. This and subsequent results have been presented in project reports (Dyrelius 1984, 1985, Agustsson and Dyrelius 1985) to the Deep Earth Gas Project conducted by the Swedish State Power Board.

Two critical problems of interpretation were identified: the separation of regional and residual gravity anomalies and the assignment of rock densities for the gravity modeling. This paper presents a revised regional-residual separation, and an interpretation constrained by new density data from logging of the deep well in Gravberg.

---

1 Uppsala University, Department of Geophysics, Section of Solid Earth Physics, Box 556, 751 22 Uppsala, Sweden

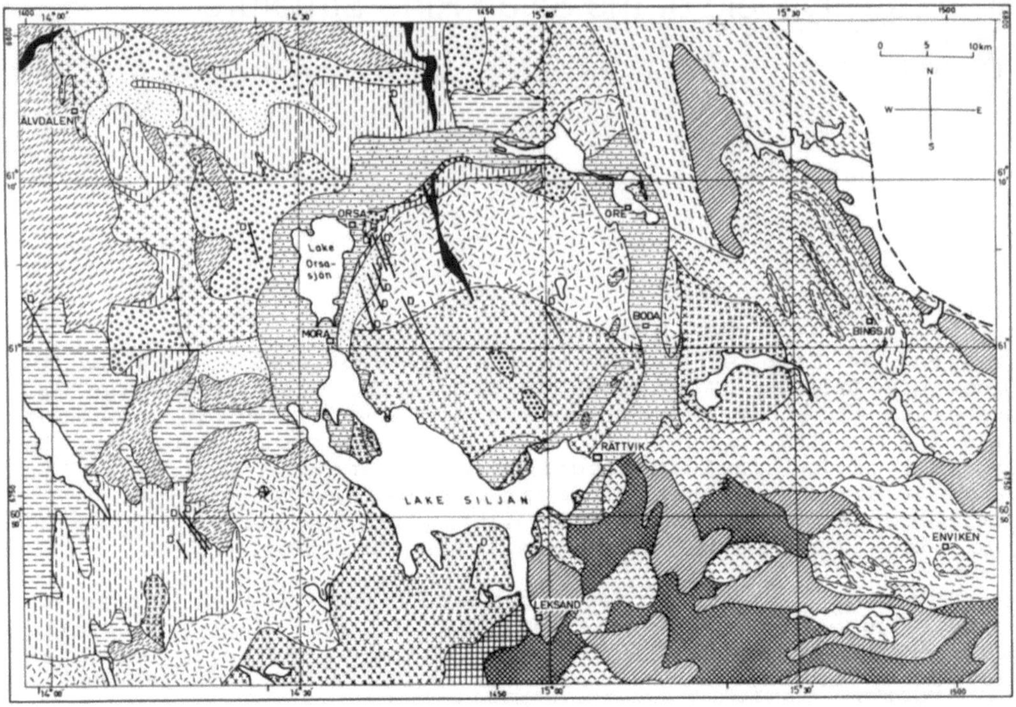

- Dala granite in general
- Dala granite (red Siljan type)
- Järna granite
- Ordovician-silurian (sandstone, shale, limestone)
- Dykes of Diabase
- Migmatite
- Digerberg rocks: sandstone and conglomerate
- Porphyry
- Porphyrite
- Urgranite
- Late-Svecofennian granites
- Leptite
- Quartztite
- Sandstone
- Gabbro and diorite

Fig. 1. Geology of the Siljan area. (Simplified after Hjemqvist 1966)

The Regional Gravity Field

The regional gravity field (terrain corrected Bouguer anomalies) in a wide area around the Siljan Ring structure is shown in Fig. 2. Apart from the obvious local anomalies related to the structure, also strong regional features can be observed, for example the steep gradient along the eastern (right) edge of the ring. This gradient marks the border between the leptite and granite region of southeastern Dalarna, with its usually deformed Svecokarelian rocks, and, to the west of this (including the Proterozoic part of the Siljan structure), the porphyry and granite region of central Dalarna (the Dala Series). The latter granites are related to the Värmland granites, to which the gravity minimum in the lower left part of Fig. 2 is associated. This low, in combination with the antipodal high of unknown origin in the upper right part of the map, complicates the separation of regional and residual gravity fields at the structure.

Clearly there is a prominent linear trend of about 0.3 mgal per km from west to east (left to right) over the area. However, it is not suitable to represent the regional trend with a first-order surface polynomial due to the biases caused by the mentioned complexities of the gravity field. Instead, a fifth-order polynomial was fitted by weighted least squares to the gravity stations between 30 and 120 km distance from the center of the Ring structure. Weighting with respect to the local density of gravity stations diminished the bias due to varying station density in different parts of the area, and exclusion of the gravity data close to the structure eliminated the effect of the interesting anomaly itself on the estimated regional trend. The fifth order was chosen because it accounts quite well for the major features of the field without including more local anomalies.

The Residual Field

By subtraction of the fifth order regional trend the residual gravity field of the Siljan Ring structure (Fig. 3) was obtained. The most prominent feature here is the gravity low with a peak value of about 15 mgal located at the northern periphery of the ring. The amplitude is about 5 mgal smaller than for previous residual fields, which is due to a difference in the level of the estimated regional field. The change primarily eliminates the previous indications of a general mass deficit extending over the whole area, but does not significantly alter the interpretation of the northern, main anomaly. The present version of the residual field has also been less smoothed than previous fields.

From the main negative anomaly in the north there are two offsets: one towards the SE, then continuing to the SW, and a weaker one running SW. These are separated by a group of local highs in the S and SSE parts of the Ring structure. Along the margin of the map area there are several local maxima. Both the positive and negative local anomalies reach amplitudes between 5 and 10 mgal.

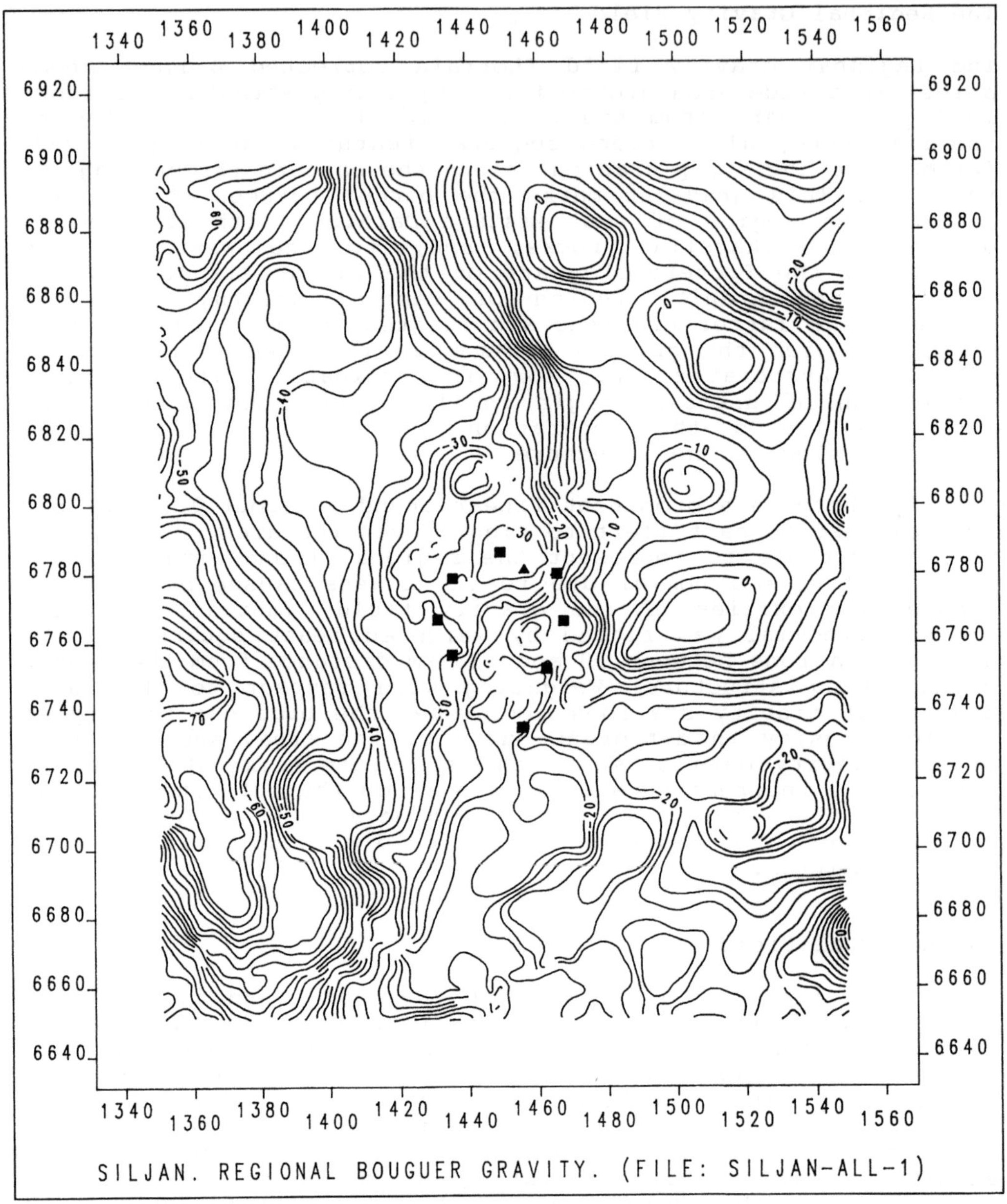

Fig. 2. The regional Bouguer gravity field in the surrounding of the Siljan Ring structure. 2 mgal equidistance between isolines (numbers in mgal). Black squares mark places on the Paleozoic belt (cf. Fig. 1), black triangle shows the location of the deep well at Gravberg. Numbers along the margins are map coordinates in km (Swedish system)

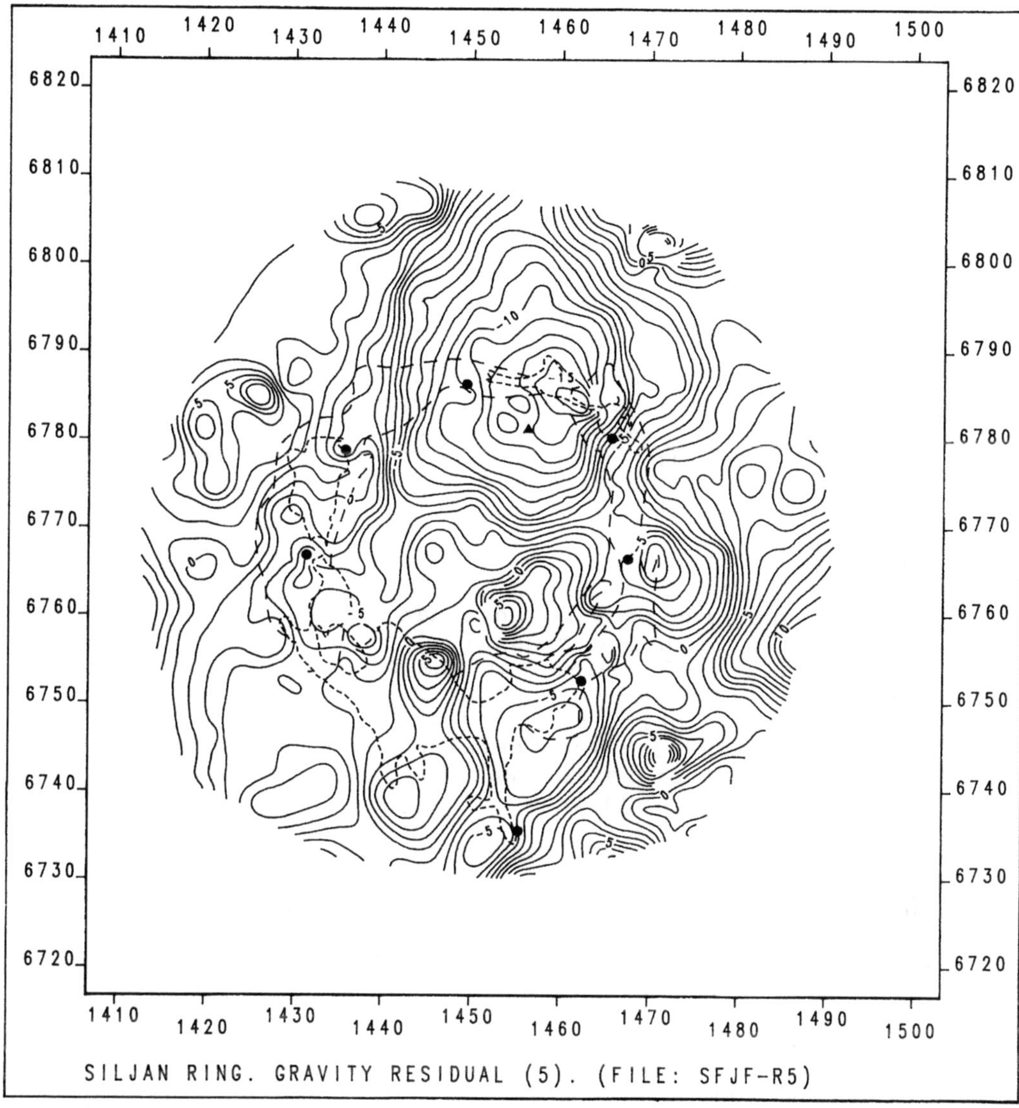

Fig. 3. Residual Bouguer gravity anomalies in the Siljan Ring structure. 1 mgal equidistance of isolines (numbers in mgal). **Black dots** show places, **black triangle** the deep well at Gravberg. Lakes are shown by **short-dashed contours**, the Paleozoic belt by **long-dashed contours**. Coordinates in km (cf. Fig. 2)

Several features of the residual Bouguer field can be correlated to geology as was shown by studies of detailed gravity profiles in the area (Agustsson and Dyrelius 1985). In the central part of the structure two types of Dala granites have traditionally been recognized (Hjelmqvist 1966), the felsic Siljan type in the northern half and the more basic Järna type in the south. However, the border between the two types is probably not as distinct as it appears in Fig. 1. The main negative anomaly and its offsets may be related to some extent to the lighter Siljan granite, while the positive anomalies in the southern part of the structure can be correlated to the Järna granite in combination with the gabbro intrusions shown on the geological map (Fig. 1).

Usually only a weak decrease of gravity was found in relation to the Paleozoic ring, showing that no large volumes of low density sediments exist within this belt, but in some places deeply eroded valleys filled with glacial deposits (and water in the lakes) give rise to very local negative gravity peaks with amplitudes up to about 3 mgal.

Low density porphyries and high density porphyrites belonging to the Dala series were also found to correlate quite well to gravity lows and highs respectively. In the Svecokarelian part, east of the Siljan Ring structure, leptites and old granites appear to give rise to gravity maxima.

Rock Densities

Data on rock densities provide invaluable constraints on gravity interpretation of upper crustal structures. In the present area only densities of sparse surface samples and samples from the eight test drill-holes within the Siljan Ring structure were previously available. Recently well-log densities from the deep drilling at Gravberg were obtained, and the main objective of the present interpretation was to introduce these data to constrain the gravity modeling.

Surface samples, collected inside and outside the structure during the gravity field work (Dyrelius 1984, Agustsson and Dyrelius 1985), show a low mean density of the Siljan granite (2604 kg/m^3) and a normal (2668 kg/m^3) of the Järna granite. Most other rocks in the area have rather normal, or even somewhat high densities. However, the surface sampling, with a total number of samples of only 130, is insufficient for mapping the detailed density distribution in the area and for adopting any other background density than the normal Bouguer density of 2670 kg/m^3 used in the reduction of the gravity data. One may argue that the Svecokarelian rocks east of the Siljan structure probably have a higher mean density than this, as indicated by the regional gravity field, for example the positive anomaly to the east and northeast of the structure. In fact, that has to a great extent already been accounted for in the nonlinear regional trend, subtracted from the observed Bouguer gravity field.

Additional density data were obtained by core samples from test drilling in the structure. Most boreholes reached depths between 300 and 700 m in the granite, and the obtained mean densities range from 2591 to 2672 kg/m^3, including mean porosities between 1.3 and 2.1%. These rather high porosities lower the rock densities by 20 to 30 kg/m^3 considering water as pore filling.

Density logging of the deep well at Gravberg has provided new data down to about 5600 m, but due to extensive break-outs and irregular walls of the hole, the data are poor in many places. The weighted 200-m averages shown in Fig. 4 were calculated by giving significant weight only to reliable data. The weights presented in the top diagram of Fig. 4 give the relative proportion of reliable data in each 200-m interval. This shows that below 3500 m only 20 to 30% of the data are useful. Still, comparison with core samples has shown good agreement with the log data. Also the previously mentioned surface samples and test drilling samples (mean densities shown at the left axis of Fig. 4) support the log densities. The density distribution with depth shows interesting variations with densities below 2600 kg/m^3 down to 1 km depth, then rising to about 2630 kg/m^3 at 1.5 km, remaining at that level down to 3 km where the density drops rapidly below 2600 kg/m^3 again. From about 4 km depth the density then rises steadily to a level of about 2640 kg/m^3 between 4.5 and 5 km, then appears to increase again towards the end of the data, up to a normal density of 2670 kg/m^3 or more.

Strong subhorizontal seismic reflectors have been found in the area (Juhlin this Vol.) and related to dolerite intrusions. These are shown in Fig. 4 and appear to be correlated to the density variation. A qualitative agreement was also found with

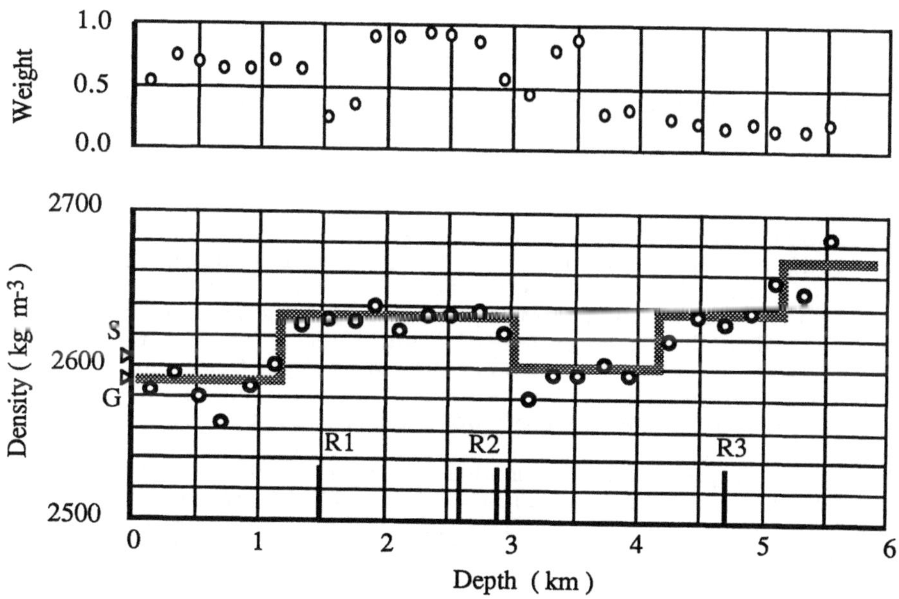

Fig. 4. Density data from the deep well at Gravberg. **Lower diagram** weighted 200-m averages of the density log (circles), density from test drilling (triangle **G**), density of surface samples (triangle **S**), density model (grey line), location of seismic reflectors (R1 to R3). **Upper diagram** relative amount of reliable data, used to calculate the mean values in each 200-m interval of the density log

seismic velocities obtained from a preliminary velocity analysis (Christoffer Juhlin, Uppsala, personal communication 1987), at least down to 3 km depth. Considering all these data, the model of density variation with depth shown in Fig. 4 was constructed and used in the gravity modeling.

The Gravity Model

The gravity model used for the interpretation is shown in Fig. 5. The element (Fig. 5a) is a rectangular prism, defined by a horizontal, rectangular polygon with vertical sides, and possibly divided into layers of different density. In the present case it is the depth of the bottom surface that is adjusted by an iterative least squares procedure (Dyrelius 1972, Dyrelius and Vogel 1972). Figure 5b shows the arrangement of prisms used to interpret the residual gravity of the Siljan Ring structure. The densities (contrasts) assigned to each prism can be seen in the cross-section in Fig. 6. The northern half of the model is used to represent the main negative anomaly, in the southern half the larger body is modeling a southern area of low gravity, while the smaller, high-density body represents basic or ultra-basic intrusions giving rise to the strong local, positive anomalies.

The fixed interfaces in the main model are obtained from the density model in Fig. 4. However, a very low density body (2270 kg/m^3) of 200 m thickness was added to the top of the central prism to account for a local anomaly peak (not clearly seen in this profile), due to an about 120-m-deep valley with glacial deposits in combination with 500-m-thick Paleozoic sediments.

Results and Conclusions

The result of the model interpretation is shown in Fig. 6 as a vertical cross-section from south to north along coordinate y = 1455 km (cf. Fig. 5b and Fig. 3). The bottom depths of the model have been obtained by the adjustment procedure, and the calculated model gravity field shows a good fit to the data, in particular to the main negative anomaly. Some further improvements may be achieved by readjusting the horizontal contours of some prisms but this is of no importance for the main result.

The deepest part of the model reaches 6.8 km which exceeds the bottom of the density model (Fig. 4) by more than 1 km. However, at this depth the resolution of the model is rather low and the difference from the density model is therefore not very significant. The important and reasonable conclusion is that the main anomaly is well explained by a body of light Dala granite (Siljan type) extending to a maximum depth of 6-7 km. The density of the body varies with depth as found by density logging of the deep drill-hole at Gravberg, i.e., as 1 to 2 km thick, alternating layers of low and intermediate density. The layer interfaces are to some extent correlated with sills of dolerite which also show up as strong, subhorizontal seismic reflectors. An increased porosity of the granites in the area contributes to a smaller part of the total mass deficit.

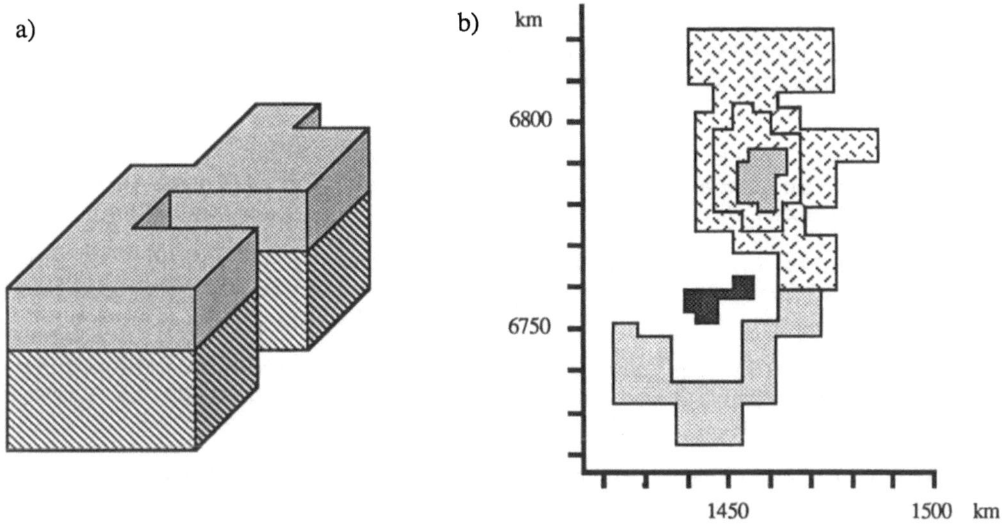

Fig. 5 a, b. The gravity model. a Model element: a rectangular prism with adjustable bottom depth and possible density layering (here only two layers). b Model used to interpret the gravity anomaly in the Siljan Ring area, composed of elements like in a. Coordinates are as in Fig. 3

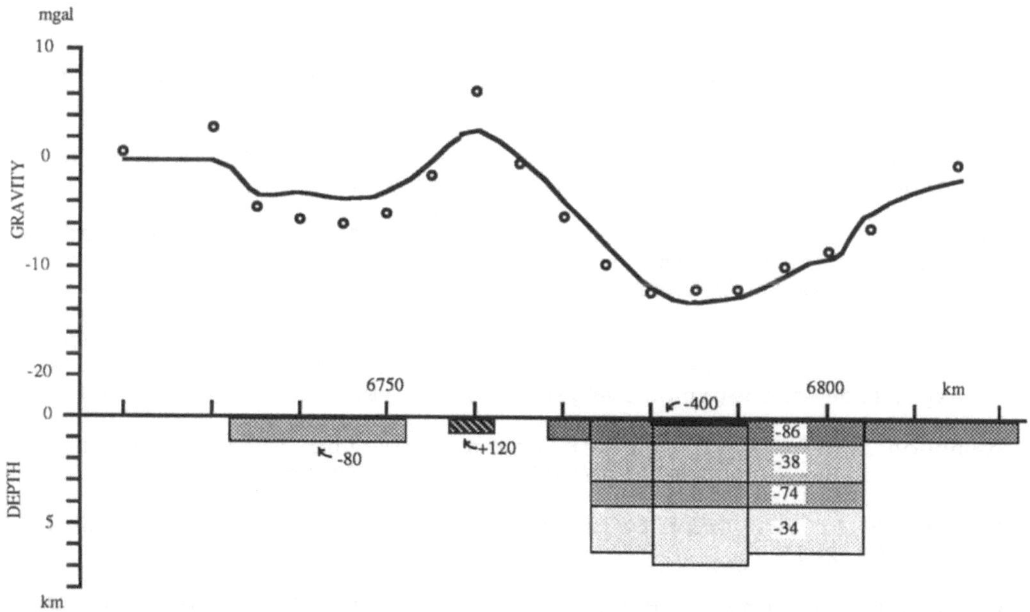

Fig. 6. Model interpretation of the residual Bouguer anomaly in the Siljan Ring structure, shown as a cross-section along a south-north profile (y = 1455km). Density contrasts of the model are given in kg/m^3. Observed gravity is shown by circles, calculated gravity as a solid curve

References

Agustsson K, Dyrelius D (1985) A gravity investigation of the radial density distribution in the Siljan Ring Structure. Rep Vattenfall, Gas Project

Bottomley RJ, York D, Grieve RAF (1978) ^{40}Ar-^{39}Ar Ages of Scandinavian impact structures: I Mien and Siljan. Contrib Mineral Petrol 68:79-84

Dence RA (1964) A comparative structural and petrographic study of probable Canadian meteorite craters. Meteoritics 2:249-270

Dyrelius D (1972) An iterative method for three-dimensional gravity interpretation. Univ Uppsala, Dep Solid Earth Phys, Rep No 9, 11 pp

Dyrelius D (1984) The gravity anomaly of the Siljan Ring structure: Review of measurements, calculation and interpretation. Rep Vattenfall, Gas Project

Dyrelius D (1985) The gravity anomaly of the Siljan Ring structure: Part 2. Additional interpretation. Rep Vattenfall, Gas Project

Dyrelius D, Vogel A (1972) Improvement of convergency in iterative gravity interpretation. Geophys J R Astr Soc 27:195-205

Fredriksson K, Wickman FE (1963) Meteoriter. In: Lundholm B (ed) (NFR) Svensk Naturvet 16:121-157

Hjelmqvist S (1966) Beskrivning till berggrundskarta över Kopparbergs län. Sver Geol Unders Ca 40:217

Innes MJS (1961) The use of gravity methods to study the underground structure and impact energy of meteorite craters. J Geophys Res 66:2225-2239

Pohl J, Stöffler D, Gall H, Ernstson K (1977) The Ries impact crater. In: Roddy DJ, Pepin RO, Mevill RB (eds) Impact and explosion cratering. Pergamon Press, New York, pp 303-404

Svensson N-B (1973) Shatter cones from the Siljan structure, central Sweden. Geol Foeren Stockholm Foerh 95:139-143

Sweeny JF (1978) Gravity study of great impact. J Geophys Res 83:2809-2815

Wickman FE, Blomqvist N G, Geijer P, Parwel A, v Ubisch H, Welin E (1963). Isotopic constitution of ore lead in Sweden. Ark Mineral Geol 3:193-257

Electrical Conductivity Structure Around the Gravberg Well

L. B. Pedersen, P. Zhang and T. Rasmussen[1]

Introduction

The electrical conductivity (or inverse resistivity) of crystalline rocks shows a large variation related both to mineralogy and to fluid concentration and salinity.

Even though dry granitic rocks in the laboratory may exhibit extremely high resistivities (Kariya and Shankland 1983) field measurements normally gives values in the range 10^{-2} S/m to 10^{-5} S/m. The reason is that the volume of rock effectively investigated in surface measurements is of the order depth of investigation cubed, which should be compared with the small volume investigated in laboratory measurements. With such a large volume, the effect of fracture zones and other fluid-filled planar features in the rock will dominate the electrical conductivity at upper crustal conditions of pressure and temperature (Olhoeft 1986).

The VLF (Very Low Frequency) map of the Siljan area shown in Fig. 1 gives an indication of the lateral conductivity changes in the upper few hundred meters of the crust. The normal conductivity away from fracture zones is around 10^{-4} S/m, whereas fracture zones may show up as regions of 10^{-2}-10^{-3} S/m. The blue linear features in Fig. 1 are believed to represent such fracture zones embedded in more resistive less fractured rock mass. The zones are separated by a few kilometers. Their preferred direction is apparently N-NE, but this is probably influenced by the fact that measurements were done with the Rugby transmitter in England which is in that same direction. It is not clear how the picture would look if another transmitter lying at a perpendicular azimuth were used.

Magnetotelluric Measurements

A total number of 85 MT stations have been measured in the area. Sixty five of them have been analyzed up to now (see Zhang et al. 1987b). Of special interest in this context are the results along the seismic N-S profile passing by the drill-site at Gravberg. The profile is denoted by A-a in Fig. 2.

---

[1] University of Uppsala, Department of Solid Earth Physics, Box 556, 751 11 Uppsala, Sweden

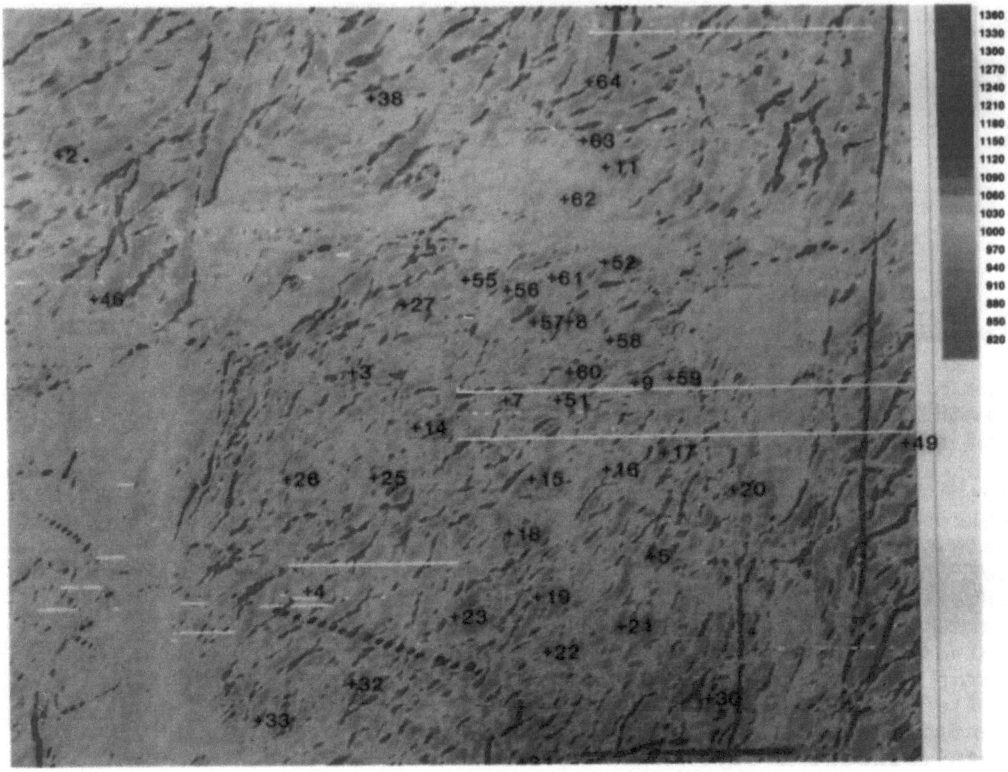

Fig. 1. Airborne VLF map centered on the Siljan impact structure. Transmitter at Rugby, England, oriented southwest. Flight altitude and line spacing were 30 m and 200 m, respectively. The color scale corresponds to total field values in steps of 30 promille. The normal field is defined as 1000 promille. Highly conductive fracture zones and power lines appear as dark areas. Magnetotelluric stations in the area are denoted by numbers

The stations close to profile A-a were projected onto the line. A two-dimensional regional model striking perpendicularly to the line A-a was constructed. The influence of the near surface lateral inhomogeneities (in the upper kilometer) were represented by a real distortion matrix, thereby enabling a fit to all four complex elements of the impedance tensor in the period interval 0.1-10 s (Zhang et al. 1987a). For longer periods, such fit was no longer possible, due to strong anisotropies related either to conductive structures away from the Siljan structure or to anisotropy of electrical conductivity in and around the Siljan structure proper. Therefore the lower crustal and upper mantle conductivities were found by fitting to the rotationally invariant determinant of the impedance tensor. Thus this part of the model represents an effective or average one.

The upper crustal model along profile A-a is shown in Fig. 3 superimposed on the seismic line drawing along approximately

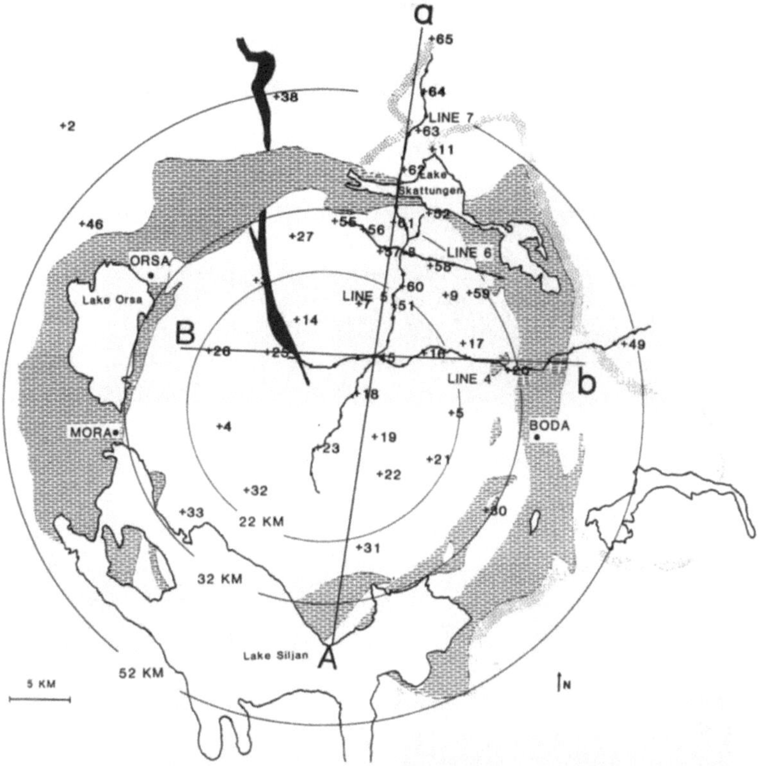

Fig. 2. Simplified geologic map of the Siljan impact area, showing the peripheral lakes, the sedimentary ring (dashed lines), the border zone separating the older Svionian rocks in the east from the Dala granites (small circles), diabase dikes (black areas), the four seismic reflection profiles, and the magnetotelluric stations together with the two profiles A-a and B-b. The circles with diameters corresponding to 22 km, 32 km, and 52 km refer to actual transient crater diameter, expected transient crater and the accepted rim diameter of the impact structure. (Map after Hjelmqvist 1966)

the same profile. Even though the electricical conductivity model is subject to some uncertainty, it contains some dominating features which are well resolved. The first is the coincidence of the uplifted part at station 250 on line 5 and the region of enhanced conductivity of 10^{-3} S/m. This coincidence is more clearly noted on the W-E profile. Secondly, the upper crustal reflectors which are visible on the whole S-N profile north of station 400 on line 5 are neutral from an electrical point of view. Notice also that, at the position of the Gravberg well, in the junction between line 5 and line 7 on Fig. 3, the predicted depth to the shaded area of conductivity 10^{-3} S/m is about 10 km. The unshaded part of the model has a conductivity of 10^{-4} S/m.

The shaded areas have been interpreted as regions of increased fluid concentration (Zhang et al. 1987b). They seem to be re-

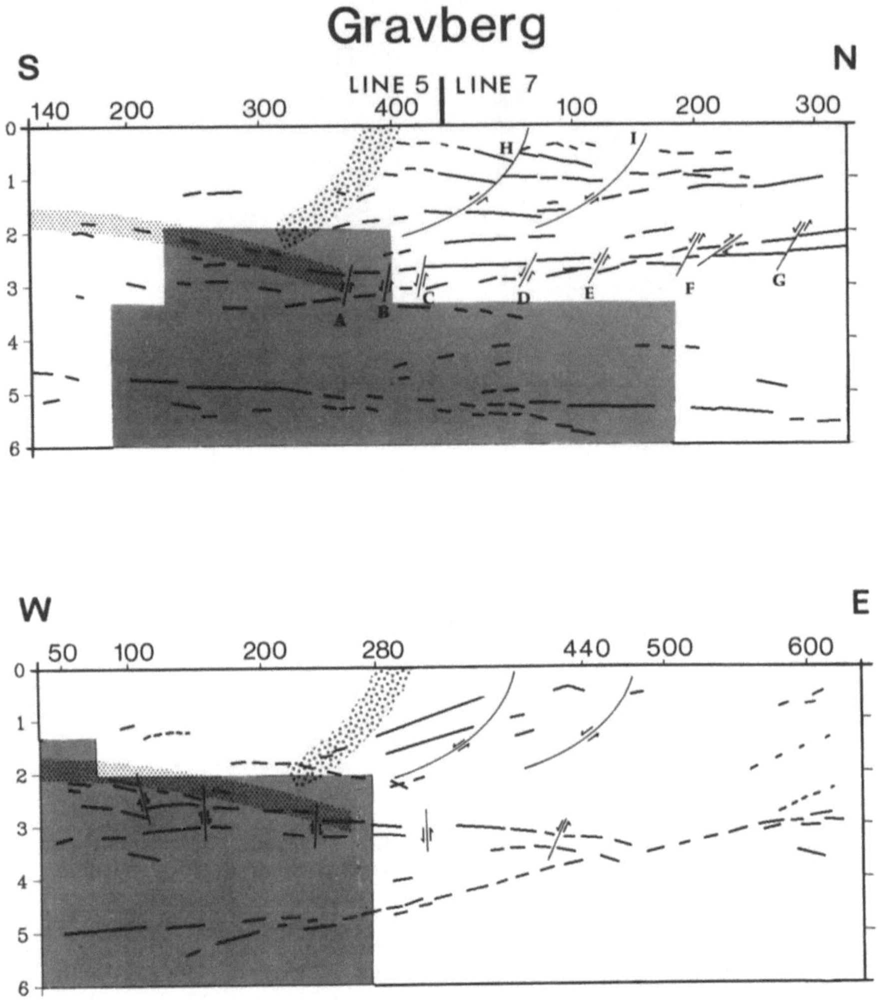

Fig. 3. Reflection seismic interpretation approximately along the profiles A-a and B-b, overridden by the impact models of electrical resistivity. Station 15 has been chosen as a fixed point for which the reflection line and the MT-profiles coincide exactly. The **shaded areas** have resistivities of 1000 Ωm. The **stippled area** is the seismically observed transient crater limit and the dotted area refers to the central uplift. The vertical scale is two-way travel time. The MT-model has been converted into this by a standard granite slowness of 1/6 s/km. The horizontal scale is roughly 5 km per 100 units of geophone group spacings (50 m)

lated to the impact structure and a tentative explanation for the increased fluid concentration is that it is a result of fluid migration after the impact event from deeper levels in the crust where the electrical conductivity is still higher (around $3x10^{-3}$ S/m) and thus the inferred fluid concentration is higher.

Results from Drilling

The position of the borehole is indicated in Fig. 3 where the predicted depth to a region of enhanced conductivity is estimated at 10 km. The model represents the whole section down to 10 km as uniform with a conductivity of 10^{-4} S/m. Before going into a comparison with the results from the borehole, it is necessary to explain the difference between current flow in horizontal and vertical directions. If it is assumed that the electrical resistivity variations measured in the borehole represent variations in a horizontally stratified medium, then horizontal currents will predominantly flow in the well conducting channels of the section and the effective average resistivity $\bar{\rho}_{hor}$ will be expressed by

$$1/\bar{\rho}_{hor}(z) = \frac{1}{2\,\Delta z} \int_{z-\Delta z}^{z+\Delta z_0} \frac{1}{\rho(\zeta)}\, d\zeta \tag{1}$$

where z is the vertical depth and $2\Delta z$ is the averaging length. The well-conducting "layers" of the section will then contribute strongly to $\bar{\rho}_{hor}(z)$.

On the other hand, if the currents are vertical, then the resistive layers of the section will determine the current strength for a given electrical field strength. Then the effective average resistivity $\bar{\rho}_{ver}(z)$ will be

$$\bar{\rho}_{ver}(z) = \frac{1}{2\Delta z} \int_{z-\Delta z}^{z+\Delta z} \rho(\zeta)\, d\zeta \tag{2}$$

In the magnetotelluric case, currents are predominantly horizontal, especially around the Gravberg well, which is away from major lateral conductivity changes. Therefore, if the vertical variations observed in the borehole have a lateral extension of the order 5-10 km, then it will be correct to use Eq. (1) for a comparison between measured average resistivity and that which is modeled on the basis of the magnetotelluric data.

The borehole resistivity data are described elsewhere in this volume (Rissler this Vol.); here we shall only show average resistivities $\bar{\rho}_{hor}$ and $\bar{\rho}_{ver}$ as calculated from Eq. (1) and (2) on the basis of the original resistivity log. The two curves are shown in Fig. 4. From the variation of $\bar{\rho}_{hor}$, we note that the fracture zones connected with the dolerites (Juhlin this Vol.) show up as low values around 1500 m, 2500 m and most notably around 4900 m. It is remarkable that $\bar{\rho}_{hor}$ averaged over the total section is equal to 10,000 Ωm within a few percent. In this context there is no space for going into details, but it should be stated that this coincidence is fortituous in view of the severe near-surface distortion effects that have been observed at several of the MT stations, and even considering that we have good reason to believe that the whole crustal section is anisotropic, as we shall discuss later. This aniso-

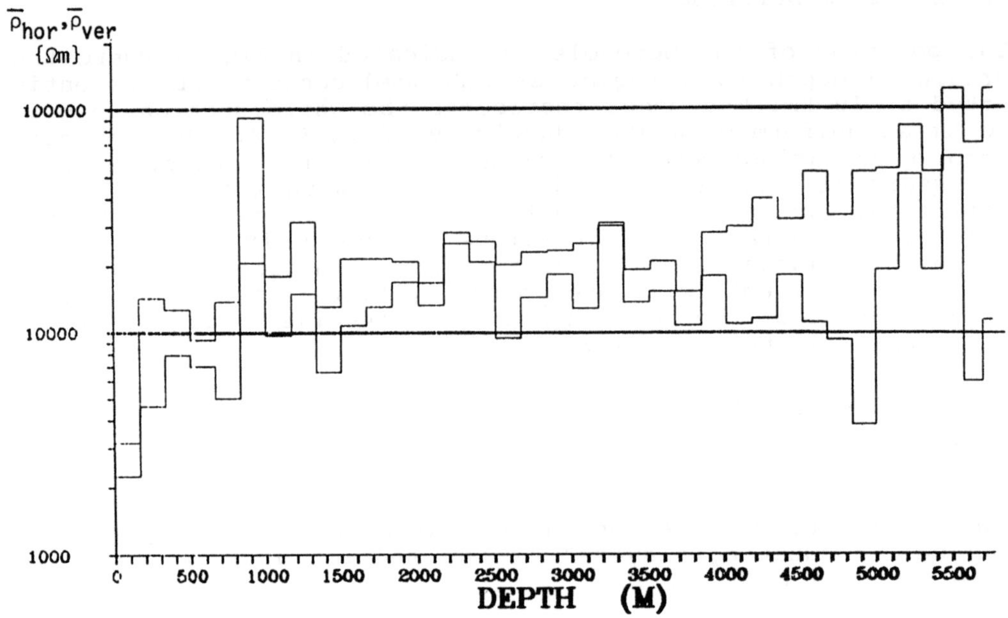

Fig. 4. Average horizontal and vertical electrical resistivities obtained from the Gravberg well. The lower curve is the horizontal average.

tropic effect has been compensated by our use of a distortion tensor as described in Zhang et al. (1987a), thereby enabling an isotropic representation of the upper crust.

The curve representing $\bar{\rho}_{ver}$ in Fig. 4 shows some remarkable features. The general trend is an increase in average vertical resistivity with depth. Especially this effect is clearly noted below 4000 m depth. Note also that a resistive section exists around 900 m. A comparison between $\bar{\rho}_{ver}$ and $\bar{\rho}_{hor}$ below 4000 m reveals that even though the intact highly resistive rock becomes more and more resistive, there exist highly fractured zones with free water, giving rise to the same resistivity as found in the zone at 1400 m. In this moment it is not clear whether the low resistivities are a result of increased pore fluid conductivity due to an increase in pore fluid salinity, or whether the frequency of open fractures is greater at depth. Otherwise it is difficult to explain why the resistivity of open fracture zones does not increase with depth in response to the increase in lithostatic pressure.

Anisotropy in Electrical Conductivity

As a result of the extensive magnetotelluric studies in the Siljan area (totally 85 stations spread over 2500 km^2), we have noticed some very clear signatures of our two measuring directions N-S and E-W and even more so when the impedance tensor is

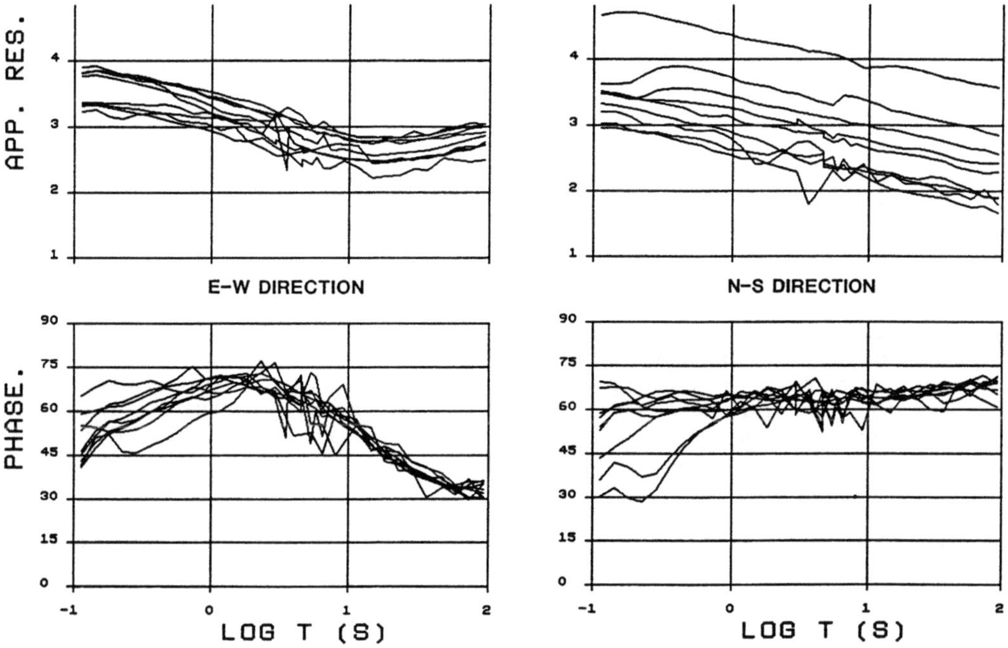

Fig. 5. Apparent resistivity and phase curves in the N-S and E-W directions for stations north of station 15 on profile A-a. Data points have been connected with straight lines

rotated through 15^0 to the East from North as plotted in Fig. 5. Here we may see some common features of stations along A-a north of station 15. Notice that the phase curves show some distinct differences between the two directions. Even though the apparent resistivity curves spread considerably due to static shift effects and to a lesser degree due to the upper crustal anomaly in the Siljan structure, it is noted that the N-S direction generally shows a lower resistivity than the E-W direction. Even more characteristically, it is noted that the E-W apparent resistivity decays faster than the N-S apparent resistivity around a period of 1 s. Modeling has revealed that this decay is associated with an anisotropy in the lower crust with a better conducting direction N 105^{0}E, which is about 90^0 off the better conducting direction in the upper crust. Similar anisotropies in the lower crust were found by Rasmussen (1987) in a study of the Transscandinavian Granitic Belt north of lake Vänern. The Siljan structure lies at the eastern border of the belt, and it is reassuring that the same effect can be traced over distances of several hundred kilometers, thus reducing the probability that a structural explanation must be sought.

The borehole breakouts (Rissler this Vol.) have given a consistent direction of approximately N 15^{0}E. This is interpreted as the minimum stress direction according to the method of Bell and Gough (1979). Our direction of best electrical conductivity for the upper crust is coincident with that direction. This means that the present state of stress probably has had no influence on the distribution of open fractures in the upper crust. The upper crustal fractures must thus be a result of tectonic movements prior to the impact event 360 M ago, since the direction is conserved in the interior of the impact structure itself. The VLF-map in Fig. 1 even to some degree shows this direction as the dominant direction of fractures. But, as explained before, this may be a result of the transmitter used in the study. The alignment of the well-conducting direction in the lower crust with the inferred direction of the principal horizontal stress component on the other hand can be explained as a response of the ductile lower crust to the present stress field. An interpretation in terms of free fluids in these vertical fractures seems to be the most simple explanation. The questions as to how the fluids have been generated - as a result of prograde metamorphism or as a result of outgassing of the mantle - and how such fluids can exist at lower crustal conditions over a billion years or more are now being debated extensively in the literature. The contribution of the present work is to lend further support to the hypothesis that free fluids actually exist at lower crustal pressure and temperature conditions, at least in the Transscandinavian Granite Belt.

Conclusions

Magnetotelluric investigations at 85 stations spread in the area have allowed an identification of major directions of upper crustal and lower crustal electrical conductivities. These directions coincided with the present horizontal principal stress directions. Old fracture systems oriented approximately N-S are still significant carriers of electrical currents in the upper crust, independent of the present stress field. Fracture systems in the lower crust seem to be aligned with the present principal horizontal stress direction, under the assumption that enhanced conductivity in the lower crust is a result of the existence of free fluids.

The borehole resistivity data indicate that 10,000 Ωm is a reasonable value for the upper crustal average horizontal resistivity. The resistivity of intact rock gradually increases to about 10^5 Ωm at a depth of about 5 km. Open fluid-filled fractures are still found at these depths with nearly the same average horizontal resistivity as at shallower depths.

It is suggested that future investigation in the Gravberg well should include measurements of pore fluids, their salinity and electrical conductivity. It would also be extremely valuable to investigate whether the inferred direction of vertical fluid-filled fractures can be found, for example by electrical logging or by other methods of fracture mapping.

Acknowledgments. The Swedish State Power Board has funded substatial parts of the magnetotelluric research in the Siljan area. Their support is gratefully acknowledged. The Swedish Geological Survey is thanked for permitting us to publish the VLF map of the Siljan area.

References

Bell JS, Gough DI (1979) Northeast-southwest compressive stress in Alberta: Evidence from oil wells. Earth Planet Sci 45:475-482

Hjelmqvist S (1966) Beskrivning till berggrundskarta över Kopparbergs län. Sver Geol Unders CA40:217 pp

Juhlin C, Pedersen LB (1987) Reflection seismic investigation of the Siljan Impact Structure, Sweden. In J Geophys Res 92NoB13

Kariya KA, Shankland, TJ (1983) Electrical conductivity of dry lower crustal rocks. Geophysics 48:52-61

Olhoeft GR (1986) Electrical conductivity. Review paper presented at 8th workshop on electromagnetic induction in the Earth and Moon. Univ Neuchatel, Switzerland

Rasmussen TM (1987) Magnetotellurics in south-western Sweden: Evidence for electrical anisotropy in the lower crust. J Geophys Res, in press

Zhang P, Roberts RG, Pedersen LB (1987a) Magnetotelluric strike rules. Geophysics 52:267-278

Zhang P, Rasmussen TM, Pedersen LB (1987b) Electrical resistivity structure of the Siljan Impact Structure. J Geophys Res, in press

Early Results of the Electric Wireline Logging in the Gravberg-1 Well

G. Rissler-Åkesson[1]

Introduction

The evaluation of electric wireline logs in a sedimentary environment is very well established and has for many years been used for hydrocarbon detection and reservoir description with very good results.

In the crystalline environment, however, the wireline logs have been used primarily for uranium ore prospecting and very few wells drilled in a crystalline environment have been logged with more than a few different logging instruments. The logging suites in most crystalline wells have not been designed to give a comprehensive evaluation of the rock environment, but rather to function as a limited complement to the data given by continous coring.

The Gravberg well offered a new situation with several difficult factors to be considered:

- The objective of the well was to find ambiogenic methane
- A potential reservoir was expected to be of very low porosity
- The reservoir porosity was expected to be in the form of fractures
- The well was to be drilled to depths requiring special logging equipment
- Only a limited number of cores were planned to be drilled.

The Basic Well Evaluation Program

The Gravberg well was drilled with the objective to explore commercial quantities of natural gas of ambiogenic nature in crystalline rock. The evaluation program was designed to use the combination of various direct and indirect sources of information on the downhole conditions of the rock drilled through.

The basic well evaluation program consisted of the following major parts:

---

1 MLC MicroLine Consulting AB, Snäckerskärsgaten 9, 421 Västra Frölunda, Sweden

- Mudlogging on site with the major components
  Cutting analysis
  Gas analysis
  Drilling parameters recording

- Electric wireline logging

- Selective coring (10 drill cores planned)
  (Replaced by sidewall drill cores later in the program)

- Well seismic profiling

The Role of Electric Wireline Logging

The major role of the wireline logging was to provide continuous measurements of various physical properties related to the down hole conditions of the rock. The logging programs have been designed to give information in the following areas:

- Lithology
- Mineralogy
- Fracture zone identification
- Permeability indicators
- Hydrocarbon identification and saturation
- Fracture and feature orientation
- Acoustic data for surface seismic correlation and correction
- Density for surface gravity study
- Depth correlation for cuttings
- Borehole geometry inclination and orientation
- Casing cement bond quality
- Production control

The Logging Programs

The programs on the different logging occasions have been adapted to the requirement at the time of logging. The latest logging technology has been used in order to give the best possible information on the rock drilled through.

Examples of this are:

- The formation scanner tool recording an image of the borehole wall has been used over very long intervals.

- The Gravberg-1 well is the first crystalline rock well that has been logged with the so-called Geochemical Logging Tool, or Gamma Spectroscopy-Aluminum Activation tool string. This tool records various important elemental yields from which the mineralogy can better be reconstructed.

- Full digital sonic recordings have been made and advanced Tube Wave processing has been applied in an attempt to differentiate between permeable and non permeable fractures.

- Sidewall drill cores have been taken in the deeper parts of the well.

The various logging tools that have been used in the well are listed in Table 1. The recording principle and the main registered data are also indicated in the table.

Table 1. Logging instruments used in the Deep Gas Project

NATURAL GAMMA RAY SPECTROSCOPY
Spectral analysis of natural gamma ray activity
Uranium potassium thorium total gamma ray activity

LITHIO DENSITY LOG
Backscattered gamma ray from formation radiated by cesium source
Formation bulk density
Photoelectric effect

COMPENSATED NEUTRON POROSITY
Measurement of energy loss in the formation of fast neutrons radiated by AmBe Source
Hydrogen index (porosity)

GAMMA SPECTROSCOPY-ALUMINUM ACTIVATION
Measurement of resulting gamma ray spectrum from neutron activation
Element y. of Si Ca Fe S Cl H Ti Gd Al Mn (O C)

DIGITAL SONIC LOG
Waveform recording of ultra sound generated downhole and transversed through the formation
Characters of pressure, shear and tube waves

DUAL LATEROLOG RESISTIVITY
Measurement of the formation resistivity using two different depth of investigation resistivity of formation detection of fractures

FORMATION MICROELECTRIC SCANNER
Microresistivity scanning of the borehole wall with image orientation rel to north and vertical
Detection of fractures
Orientation of fractures
Shape and orientation of bore hole

HIGH RESOLUTION TEMPERATURE
Continuous mud temperature using thermistor
Formation temperature gradient
Indication of permeable zones

ELECTROMAGNETIC PROPAGATION TOOL
Measurement of attenuation and propagation time of 1.1 GHz electromagnetic waves
Dielectric constant for detection of hydrocarbon in freshwater

Table 1. (cont.)

BOREHOLE TELEVIEWER
Borehole wall scanner using high frequency acoustic signals transmitted from the tool and analyzing reflected sound
Fracture identification and orientation acoustic caliper casing wear inspection

WELL SEISMIC PROFILE
Downhole geophone record sound events at surface
Depth to time transform for surface seismic look ahead of the bit

SIDEWALL DRILL CORE
Mechanical core drill tool for up to 12 sidewall cores per trip
Cores from the rock drilled at depths selected from logs

BOREHOLE GRAVITY TOOL
Measurement of the earth gravity field in selected stations
Gravity anomaly study large-scale bulk density

Ten different logging suites have been run so far in the well. Several of these logging suites have been undertaken for technical drilling reasons and these logging programs have been of limited nature. The depths for the different logging suites and their scope are indicated in Fig. 1. Well depth is in meters.

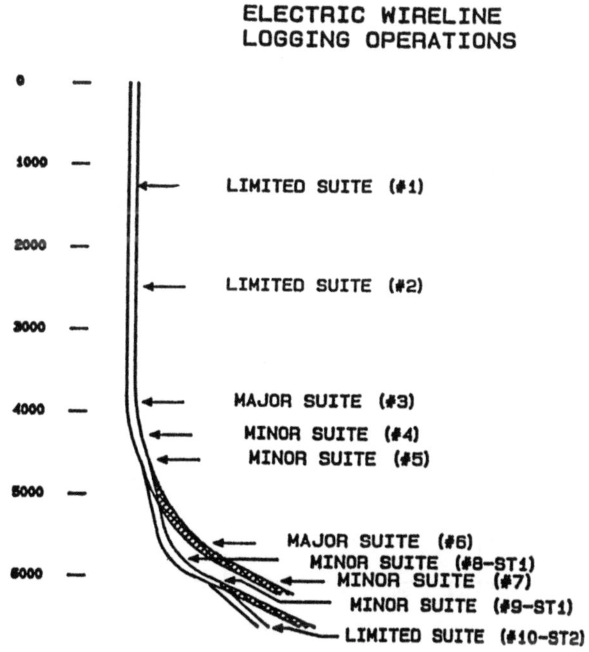

Fig. 1. Logging suites in the Gravberg-1 well

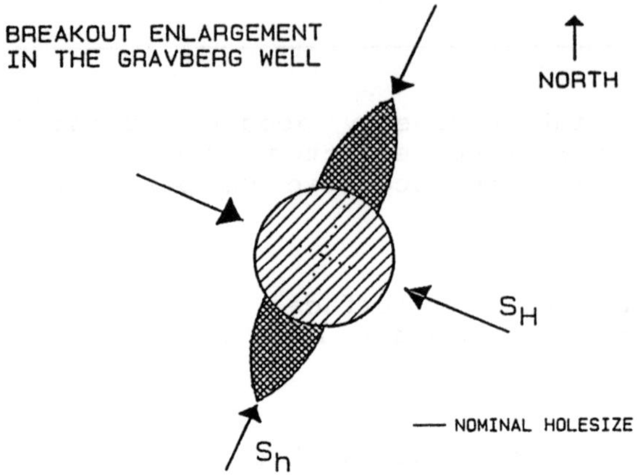

Fig. 2. Shape of borehole due to breakout enlargements with indication of maximum horizontal stress and minimum horizontal stress directions (S_H resp S_h)

Some Early Results from the Wireline Logging

Breakouts. The Gravberg-1 well has been severely enlarged by breakouts interpreted to be due to large horizontal stress differences. The well has an ovalized shape from 1500 m depth down to TD with only short intervals of borehole drilled in gauge.

The larger diameter of the hole, considered to be in the direction of the minimum stress field, is oriented North 10-20°East

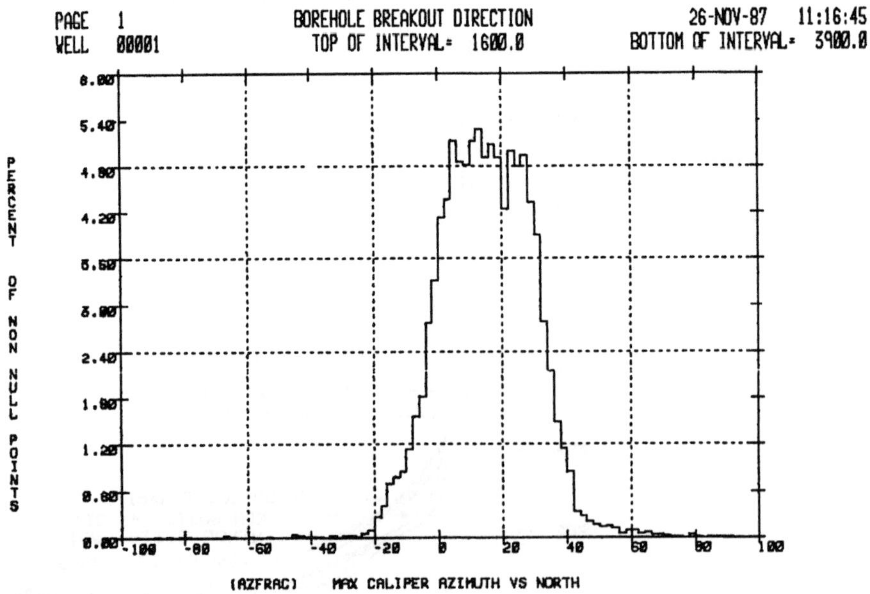

Fig. 3. Histrogram of maximum caliper direction relative to North. Scale in deg. relative to North

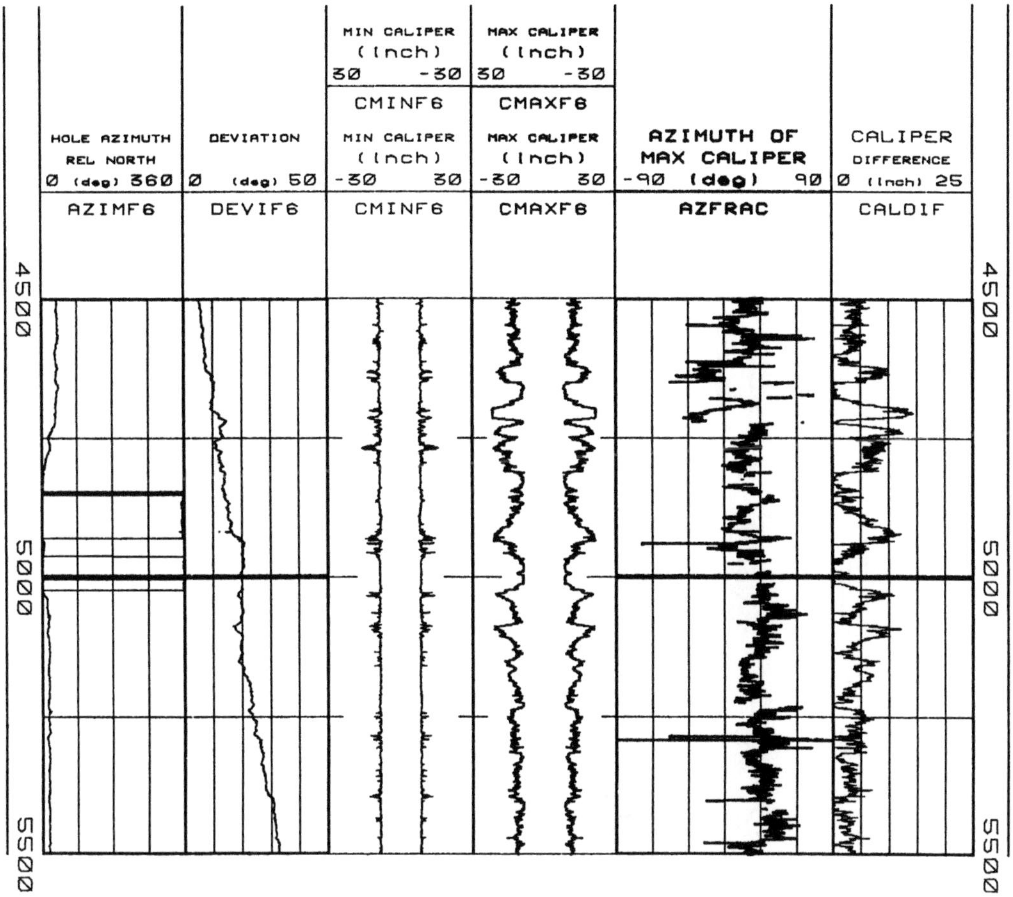

Fig. 4. Log section display of data related to the shape of borehole over an interval of Gravberg well

(Fig. 2). This direction is more or less constant over the whole interval drilled (Figs. 3 and 4). The sections that are drilled in gauge are generally the intervals that have the best indications of natural fractures. This could be explained by the fact that these zones have already been released from the anistropic stress condition by the fractures. The magnitude of breakouts has been very large, and this has caused severe problems for logging and log interpretation. Many of the drilling problems and hole deviation problems can also be attributed to the breakouts.

Fracture Zones. Several intervals with high fracture intensities have been registered in the well. These intervals are often associated with dolerite intrusions. The strong seismic reflectors correspond to dolerite intrusives in combination with heavily fractured zones. Figure 5 is an example of FMS image on a fractured interval. The boundaries of the fracture

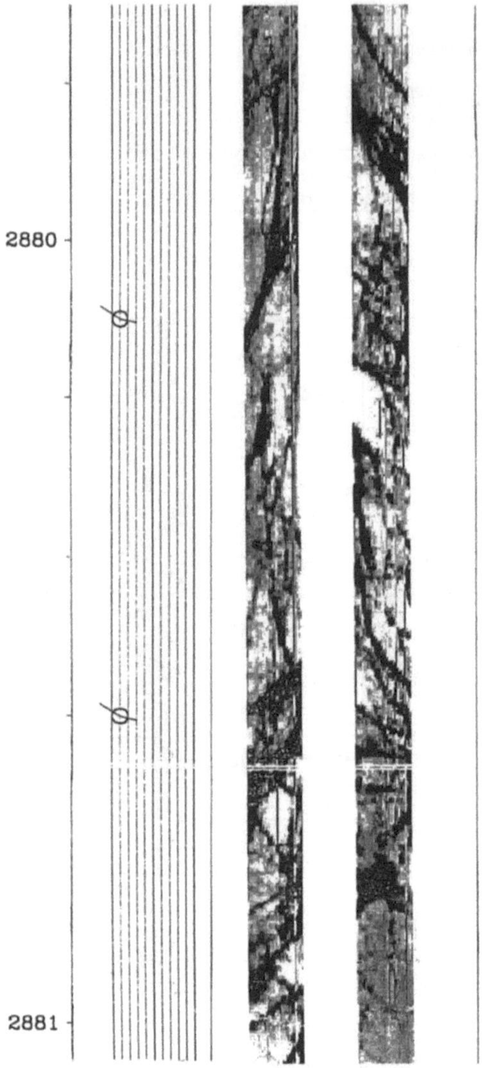

Fig. 5. Example of Image on fractured interval recorded by the Formation Scanner Tool

zones have in some cases been oriented for the shallower zones to have a more or less horizontal distribution or a 10-15° dip towards North.

Porosity Distribution. The porosity found in the Gravberg well is primarily in association with the fractured zones. The highest porosity measured on core (side wall) is approx 4%. This corresponds well with the porosity estimated from logs at the same depth. Other sections where the core is not available have log porosities of more than 10% over short intervals.

General Lithology. The well is drilled through a granitic sequence of various types of granitic types (for details, see Karlsson et al this Vol.) Intrusions of dolerite are registered at various levels of the well and the major intrusions are found at the levels indicated by the surface seismics as strong

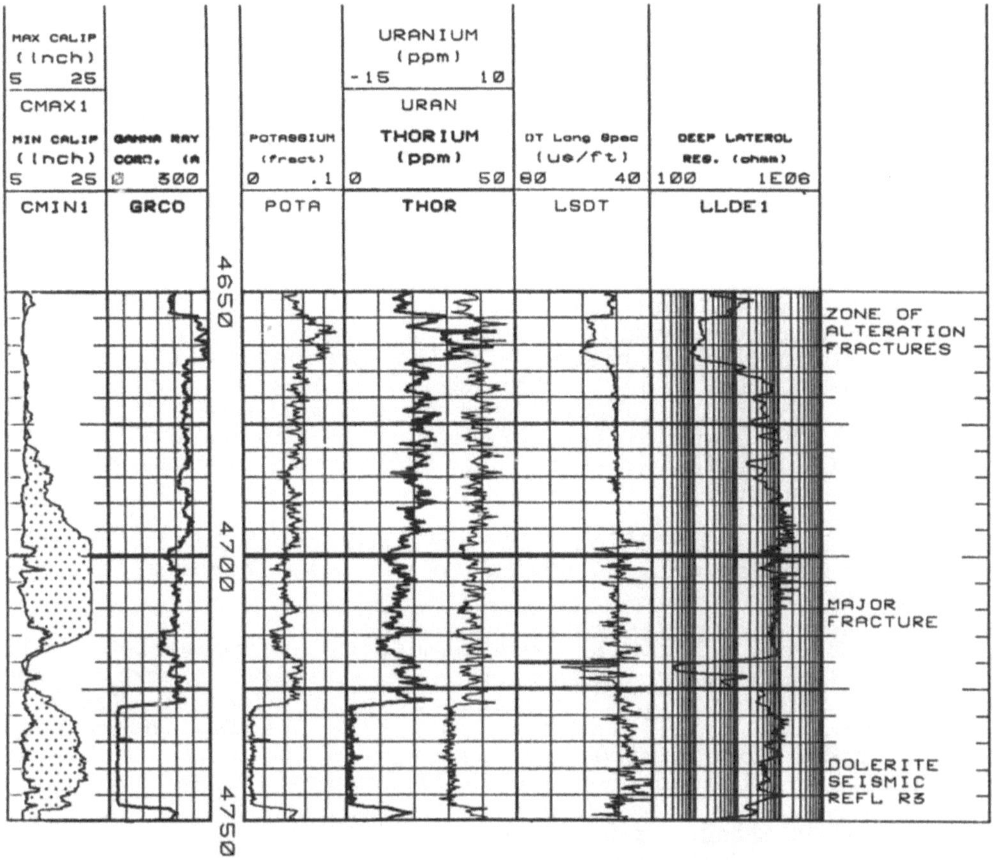

Fig. 6. Log section display of related data showing indications of fluid migration pathway between two fracture intervals. The major dolerite intrusion identified as the 3rd seismic reflector can be seen at the lower part of the section

reflectors. Below 5400 m depth intrusions of a granitic type with very high Thorium content (80 ppm) starts to emerge. Some sections of the well show indications of fluid migration pathways. Figure 6 is an example of this. Here the gradual change in the natural radioactivity between two intervals of intense fracturing suggest such a fluid migration path way. These observations are supported by observations made on Volatile Trace Elements (Reed et al. this Vol.)

Experiences from Applying Wireline Logs to the Crystalline Environment of the Gravberg Well

Important Recent Developments. The geochemical logging tool string is a very important development for helping to solve the sometimes very complicated lithology encountered in crystalline rock. The tool and evaluation software has been used with great success and there is no reason why this tool should not be an important part of crystalline logging programs in the future (Ellis et al. this Vol.). However, further work is needed in order optimize the interpretation in the crystalline environment. Formation scanner tool imageing is a very powerful tool for fracture and feature identification and orientation. The images are often superior to the borehole televiewer in quality and the tool is not limited by hole size and well fluid in the same way as the BHTV. The present drawback of the limited coverage will be reduced with a 4-pad version of the tool. (Cull this Vol.).

Major Areas where Improvements are Needed. Improved methods of collecting quality data in holes with breakouts and generally enlarged borehole are needed. In the Gravberg well, several attempts have been made to use kits for so-called short axis logging in order to rotate the tool in preferred directions relative to the shape of the borehole. This has been only partly successful, mainly due to the additional problem of high borehole deviation.

Borehole effect corrections in ovalized borehole is another area where further work is needed to establish reasonable principles.

The evaluation of logs in the sedimentary environment is well established and many computerized evaluation programs are available to help in the interpretation process. However, further work is needed in the crystalline logging interpretation in order to optimize the use of wireline logs in solving the different problems of the crystalline environment.

Further improvements as regards the databank on tool responses in different minerals are needed.

Interpretation of the Seismic Reflectors in the Gravberg-1 Well

C. JUHLIN[1]

Introduction

As part of the exploration program for deep earth gas in the Siljan ring impact structure by Vattenfall (The Swedish State Power Board), four seismic reflection profiles were shot during 1985. Figure 1 shows the location of these profiles and a simplified picture of the geology of the area. For a more detailed description of the surface geology the reader is referenced to Collini this Vol. The results from the seismic profiles and their interpretation in relation to impact mechanics has been discussed elsewhere (Juhlin and Pedersen 1987) and will not be dealt with in this paper.

The exploratory borehole, Gravberg-1, is located in the northern Siljan ring area where lines 5 and 6 cross (Fig. 1) at about station 480 on line 5. Line 5 (Fig. 2) reveals a relatively transparent crust in the upper 2 s in the central portion of the structure and a considerably more reflective crust in the northern section where the borehole is located. To date, 6337 m of rock measured vertically have been penetrated consisting primarily of granitic rock, but also of basic and acidic intrusives (Karlsson this Vol.). Rock cuttings inspection, wireline logging, vertical seismic profiling (VSP), mud logging and monitoring of drilling parameters all aid in the interpretation of the surface seismic data. The rest of the chapter will deal with the results from these borehole investigations and how they relate to what is observed on the surface seismic line 5 in the vicinity of the rock column which has been penetrated by the bit.

Surface Seismic Line 5

The presence of five sub-horizontal high amplitude reflectors in the northern Siljan ring area is apparent on line 5. These are all within the upper 3 s of two-way travel time (TWTT) or approximately 8.5 km. There is a trend to more lateral continuity and greater amplitude the deeper in the crust that the reflector is found.

---

[1] Vattenfall and the University of Uppsala, Department of Solid Earth Physics, Uppsala, Sweden

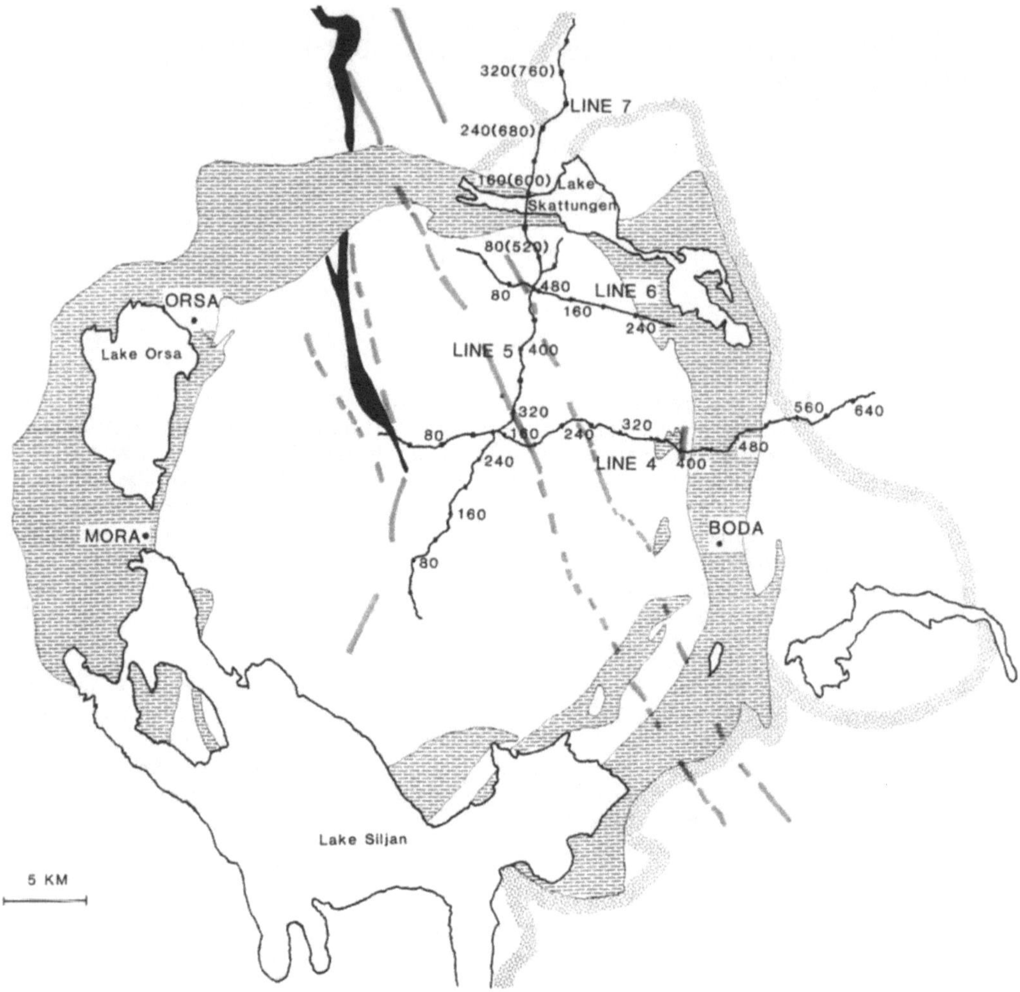

Fig. 1. The Siljan ring area and the location of surface seismic profiles. The Gravberg-1 well is located where lines 5 and 6 cross one another. The **(black)** NNW-SSE-trending features are dolerites. Shown also are the sedimentary ring **(dashed lines)** and border zone **(stipple)** separating the Dala granites from the older Svecokarelian rock to the east

In addition to the high amplitude reflectors there are numerous low amplitude dipping and sub-horizontal reflectors in the upper 3 s. These reflectors generally do not show any great lateral continuity and some can only be correlated over 10 or 20 traces (corresponding to 250 to 500 m). Special mention should be made of two dipping reflectors at 1.85 s below stations 5488 to 5496. Another zone of interest is the series of diffuse dipping events at 2.0 to 2.2 s between the third and fourth reflectors. The higher amplitude here is attributed to the processing in part (a 700-ms automatic scaling window was used); however, there appears to be lateral continuity of the

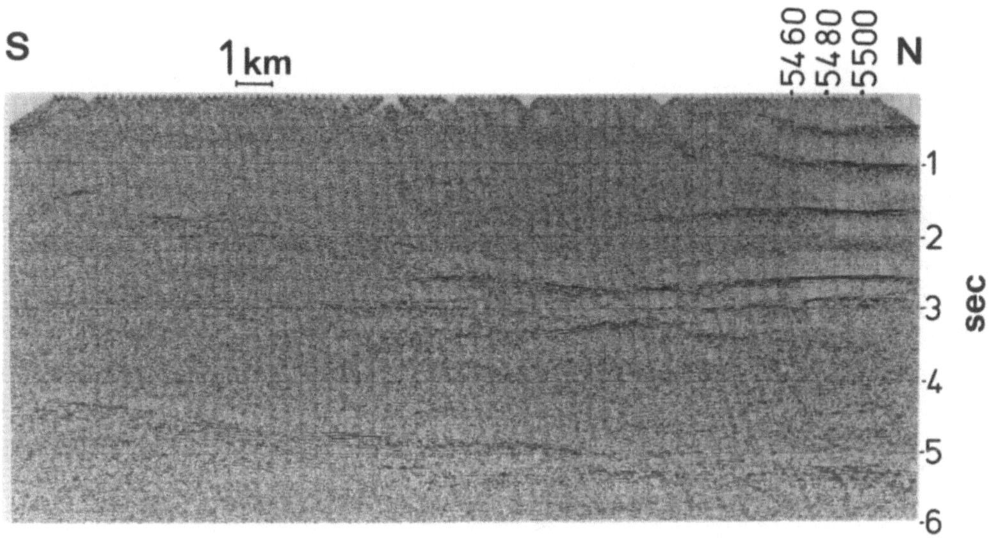

Fig. 2. Stacked section from surface seismic section line 5. The borehole is located in the north.

signal. The structure inferred from the seismic is a zone of weak reflectivity with an apparent dip of about 20° to the south.

Vertical Seismic Profiling (VSP)

In order to interpret the results from the borehole in relation to the surface seismics, VSP is essential. VSP serves two purposes in this aspect. First, it allows a direct time-to-depth conversion, allowing events occuring at a specific depth in the borehole to be converted to time in the seismic section. This is considered to be a more representative conversion than simply integrating the sonic log, since the wavelengths involved in VSP surveys are of frequencies similar to those employed in surface seismics, while the sonic tool wavelengths are several orders of magnitude shorter. Secondly, the existence of reflectors observed on the surface seismics are verified. This eliminates the possibility that the observed reflectors are processing artifacts, sideswipe, multiples or some other event not related to the true subsurface geology. In addition, VSP also makes it possible to predict events ahead of the bit. This attribute was used on several occasions to pick coring points and to some extent help in planning the drilling program.

Four VSP surveys were run in the course of the drilling at the following depths; 1250, 2372, 3932 and 5683 m. The four runs have been spliced together to form one data set, which covers the rock column measured vertically down to about 5500 m. The following processing steps have been applied to the data to produce an upgoing wavefield which can be compared directly to the surface seismic and can be used to correlate events:

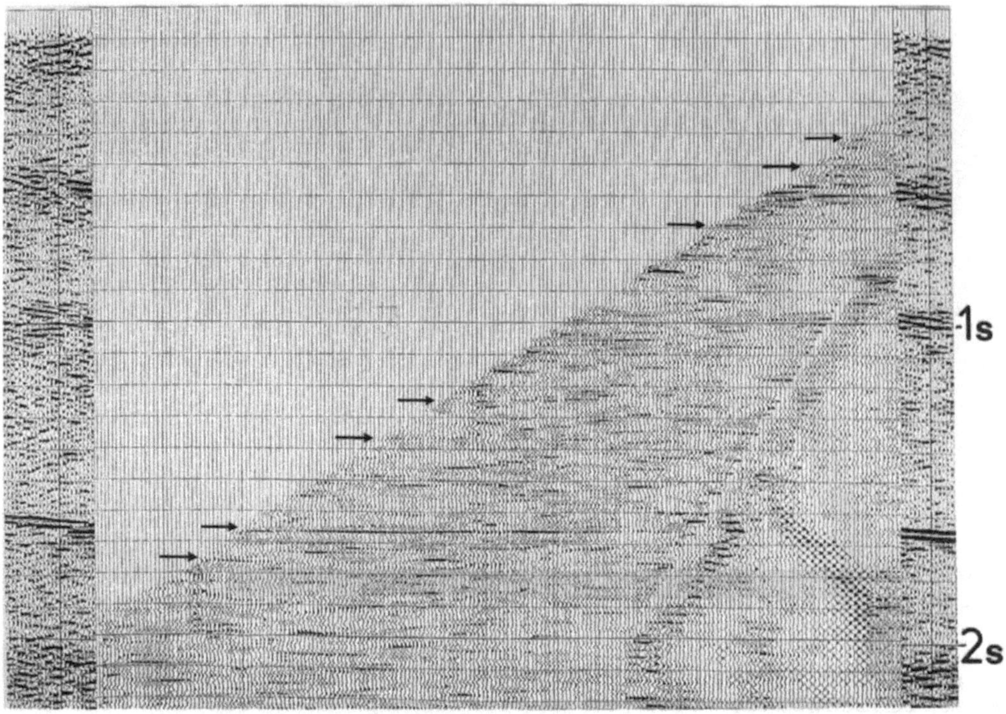

Fig. 3. The upgoing wavefield from the VSP is located in the center with the surface seismics on the edges split approximately where the borehole is located. The **arrows** point to events on the VSP which can be correlated to the surface seismic that may be fracture zones

1. Demultiplex
2. Bandpass filter 8-80 Hz
3. Amplitude decay compensation at X = 1.0
4. Median stack
5. Pick first arrivals
6. Median seperation, 1st pass 13 levels, 2nd pass 9 levels
7. Hydrophone/Gun statics plus 9 ms
8. Filter zero-phase 8-80 Hz
9. Display normal polarity

The results of the processing are shown in Fig. 3. The first reflector at 0.55 s shows up distinctly on the VSP and it is quite clear that it is coming from the subsurface. The same is true of the third reflector at 1.65 s. The second reflector does not show up as clearly on the VSP, but it can be correlated with some weaker events and is also an expression of the subsurface geology.

The arrows (Fig. 3) mark events on the VSP which can be correlated with weak reflections on the surface seismics. Although on the surface seismics these events appear insignificant and may be interpreted as noise initially, it is felt that they are

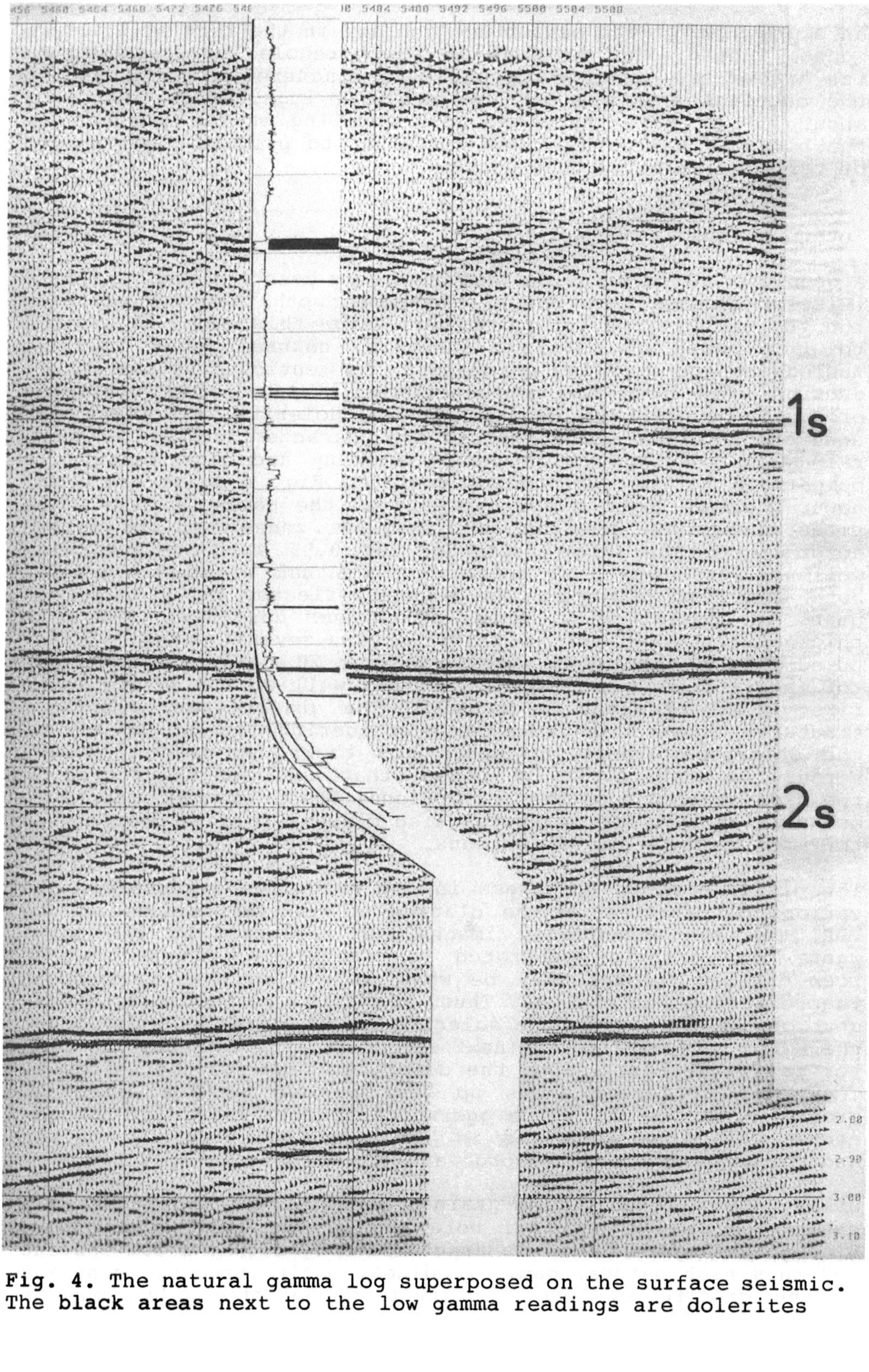

Fig. 4. The natural gamma log superposed on the surface seismic. The black areas next to the low gamma readings are dolerites

an expression of the subsurface geology in the form of fracture zones. Indeed, the sections in the borehole corresponding to the arrows are zones which are more fractured as exhibited on the cuttings analysis and the wireline logs. Below 5300 m or about 1.8 s TWTT, there was poor coupling of the geophone to the borehole wall and it was difficult to produce clear images of reflections below this depth.

Interpretation

A complete set of wireline logs have only been run down to 5600 m (Rissler-Akesson this Vol.), the same depth that VSP has been run to, so that the interpretation below this depth is limited to data based on cuttings analysis, natural gamma logging, mudlogging and drilling parameters. As mentioned in the introduction, the rock column penetrated to date has consisted primarily of granite; however, several dolerite intrusions have been encountered. The dolerites are characterized by low natural gamma readings, high sonic velocity and high density in comparison to the surrounding granite. Fig. 4 shows the total natural gamma log and the geometry of the borehole. The dolerites themselves show up as low gamma zones and are marked accordingly. The first reflector (at 0.55 s) correlates very well with a 60-m-thick dolerite which had fractured granite both above and below it. The second reflector (at 0.95 s) appears to consist of a series of thinner dolerites (4-9 m in thickness). In between the dolerites are several zones of fractured granite varying in thickness up to 20 m. The third reflector (at 1.65 s) again correlates very well with a single dolerite, which is about 20 m thick. The granite was especially fractured above this dolerite. Considering the occurrence of the dolerites, their thickness and their impedance contrast versus the granite, it is evident that they are the source of the high amplitude of the reflections. The fractured granite in association with the dolerites also plays a role in determining the amplitude of the reflections.

The differences in thickness in the dolerites may explain the various differences in the clarity of the reflections on the VSP. The VSP essentially images the reflector at the point where the borehole penetrated it; therefore, the section of rock drilled through will be what determines the strength of the reflection on the VSP. Thus, the first reflector is strongest on the VSP, since the dolerite was thickest at this point, the third reflector is the next strongert and the second reflector is the weakest, since the dolerites here are the thinnest. The third reflector shows up best on the surface seismics, probably because it is the one which is most laterally homogeneous, while the dolerites in the upper two probably thicken and thin, diverge and converge and are more faulted.

Below 5400 m several fine-grained granite rock sequences were penetrated in the original hole and in the sidetracks. These fine-grained granites are characterized by high thorium content and thus high natural gamma readings. Their impedance contrast to the granite is less than that of dolerites, but may be

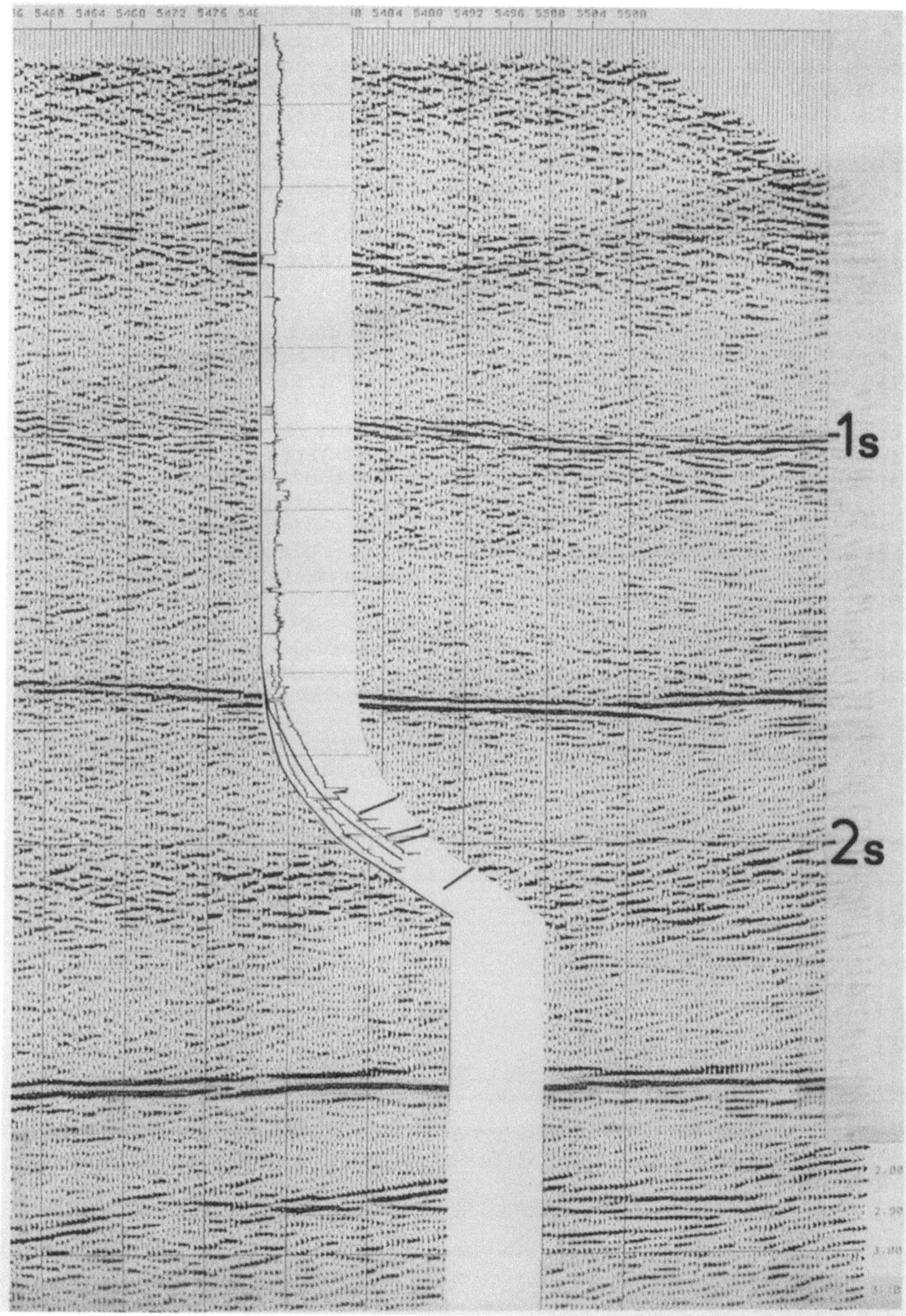

Fig. 5. The natural gamma log again superposed on the surface seismic. The black areas next to the high gamma readings are high thorium zones representative of the fine-grained granites encountered at these depths

enough to produce reflections if they have lateral continuity. The fact that they were penetrated in both the original hole and in the sidetracks indicates that they do have some lateral continuity. An apparent dip can be calculated and the fine-grained granite intrusions appear to correlate with two weakly dipping reflectors at about 1.85 s TWTT (Fig.5). Since these reflectors dip, their location on the surface seismic section is critically dependent upon the migration velocity used and exact correlation is not to be expected. It is quite significant if the surface seismic reflection method is capable of imaging features of this impedance contrast and thickness for future interpretations of the upper crystalline crust.

Neither acoustic logs or VSP has been run below 5600 m; thus, it has not been possible to verify if the zone of diffuse correlating events between 2.0 and 2.2 s TWTT is representative of the subsurface. Several other events during the drilling indicate the rock corresponding to this interval being more fractured and even containing some open porosity. These events are the presence of relatively larger amounts of hydrogen and helium gas between 5800 and 6200 m, the loss of circulation of 150m^3 of drilling fluid at about 5900 m, differential sticking of the drillstring (indicating increased permeability) at about the same depth, and loss of the water fraction when oil-based mud was used. Temperature logs have also indicated flow into some zones at these depths. It should be the goal of future seismic processing of the data to establish if these signs of open porosity express themselves on the seismic data, and if so to what extent they can be traced away from the borehole.

Table 1 lists the various types of rocks encountered in the well so far and estimates of their densities and velocities

Table 1. Rocks encountered in the Gravberg-1 well

Rock	Density	Velocity
Granite (G)	2.65 g/cc	6.0 km/s
Dolerite (D)	3.00 g/cc	6.6 km/s
Fine Grained Granite (FGG)	2.65 g/cc	5.5 km/s
Fractured Granite (FG)	2.60 g/cc	5.2 km/s

From the impedance contrasts in Table 1 reflection coefficients can be calculated for the different rock contacts encountered in the well

G/D 0.11
FG/D 0.19
FG/G 0.08
FGG/G 0.04

It is apparent that all of the impedance contrasts are sufficiently large to produce reflections if the contact has lateral continuity. It may well be that it is the lateral continuity of a contact which is the determining factor if we observe a reflection in crystalline crust or not. Since all the reflections observed in the area appear to be the result of thin features in relation to the wavelength of the seismic energy, the thickness of the feature also plays an important role in determining the amplitude of the reflection.

Conclusions

The drilling and subsequent borehole investigations in the Gravberg-1 well have contributed important new information concerning the nature of reflections in crystalline crust. Dolerites, fine grained granites and fracture zones all have impedance contrasts sufficiently high in comparison to granite to produce reflections. Determining factors as to whether a reflection is observed in the Siljan area are the lateral continuity of the contrast and the thickness of the contrasting feature. Dolerites are the main cause of the high amplitude subhorizontal reflectors penetrated so far, however, fractured granite in association with the dolerites may contribute to their strength. Since the high amplitude reflectors show great lateral extent, the dolerites must have intruded in the form of sills in either pre-existing fracture zones or fractured the surrounding granite as they intruded.

Fine grained granite intrusions may also be responsible for dipping reflectors below 1.8 s TWTT. The occurrence of these intrusives correlates with events observed on the surface seismic. The zone between 2.0 and 2.2 s TWTT may be representative of more fractured rock and the data should be looked at more carefully to determine if this is indeed the case. Finally, fracture zones may be responsible for some of the low amplitude and subhorizontal weak events observed on the surface seismic.

Reference

Juhlin C, Pederson LB (1987) Reflection seismic investigations of the Siljan impact structure. J Geophys Res Vol 92, No. B13, pp 14113-14122

Geochemical Studies, Gravberg-1 Well, Sweden; Integration of Well Site and External Laboratory Analyses

J. R. CASTAÑO[1]

Introduction

The program for geochemical analysis in the deep borehole at Gravberg incorporated a comprehensive and highly sophisticated series of geochemical measurements made at the well site. In addition, the on-site measurements are closely coordinated with the work carried out at eight off-site laboratories performing various analyses. Due to the nature of the project, many devices were redesigned in order to provide the precise data required. From the work done on the shallow boreholes and from the surface geochemical studies, we anticipated gas readings in the ppm levels, and, because of changing condition, largely due to changes in the mud properties, the sampling protocol was modified a number of times in an attempt to get better data. The changing character of the drilling fluid created major problems in the interpretation of the gas data, because of contamination and because of the different retentive characteristics specific to each of the different types of drilling fluid used: freshwater, bentonite-barite, polymer and oil-based.

Methods for Sampling and Analysis

On-Site

At the well site, hydrocarbon gases are monitored by three different systems, each of which offers specific advantages (Table 1). These are the flowline "autosampler", "bombs" and "dunks". The flowline gases are liberated by a gastrap in the possum belly and drawn into the mud logging unit through a flowline. The "autosampler" can be adjusted to take a sample every 5 min (which is the usual sampling interval), or a longer period such as ten minutes was occasionally used. Mud "bombs" consist of 150 cc steel cylinders which contain samples of the drilling fluid, they are collected before the mud goes into the well ("mud in"), and before the fluid enters the possum belly ("mud out")

Initially, the "dunk" system consisted of a 10-l plastic container which was filled with drilling fluid and submerged with the bottleneck facing down. Gases present in the drilling fluid

---

[1] 722 Oder Lane, Houston, Texas 77090, USA

Table 1. Gas sampling protocol

Method	Type-frequency	On-site	Off-site	$HC(C_1-C_5)$	Inorg. gases N_2, O_2, CO_2, H_2, He
Autosampler	Spot 5 min		*	*	
Steel cylinder	Spot, 10-20 ft Mud in Mud out	*	*	*	*
Dunk tank	Accumulative 20 ft +4-12 h	*	*	*	*
Headspace Gas + cuttings	Spot 40 ft +100 ft		*	*	*
H_2S	Continuous	*			
Radon	Spot 20 ft Mud in Mud out	*			

On-site: Baroid

Off-site: MKS, Sweden
DGU, Denmark
GGC, USA
GCA, West Germany

On-site lab measures saturate HC C_1 - C_5. Off-site labs also measure unsaturates.

displace the fluid in the container, and the gas is withdrawn through a septum. To begin with, the dunk tank was sampled every several hours. Later, the samples were collected every 20 ft, in order to coincide with other samples. When the change was made to a mudbased drilling fluid, two major modifications were made: (1) the dunk container started out air-filled, which allowed better breakout of the gases, and (2) a second dunk operated in parallel to the first one; the second container was sampled on a time basis, ranging from 4 to 12 h. Although the second system cannot be considered as quantitative, it does often provide sufficient gas volume for off-site laboratory analysis. Obtaining sufficient quantities of gas for isotopic analysis was a serious problem, which was at time overcome by using the long-term dunk samples.

A second chromatogram was installed and was operating for depth below 4300 m. This unit uses nitrogen as a carrier gas and it analyzes the dunk tank gas for hydrogen and helium. The other chromatograph uses helium as a carrier gas, and has about 1/10 the resolution for H_2 as the second unit (250 ppm vs 20 ppm).

Radon gas is measured every 20 ft in the mud in and mud out. Because drilling mud outgasses at a slower, and more variable rate than water, radon values are probably less accurate when weighted mud was used.

In sum, the autosampler takes frequent spot analyses, but no samples are retained. The steel cylinders give the best quantitative data, as the sample size is always the same. However, because of the small levels of gas encountered, the "bombs" often did not record values above background levels. The dunk samples are inherently less accurate in terms of the collection system, yet they give the largest gas volumes. The chromatograph has a sensitivity of 10 ppm for hydrocarbon gases, which is better than most field units.

Off-Site Laboratories

The off-site laboratories make analyses that parallel the ones done at the well. In addition, they analyze canned cuttings (for headspace and desorbed gas analysis) and dried cuttings and cores. These laboratories have made numerous isotopic analyses of methane, ethane, propane, deuterium, and helium. They have also performed specialized studies of an experimental nature which are detailed in separate chapters in this volume.

Some Factors that Can Affect Gas Readings

As shown in Table 2, a number ot factors affect gas readings. Bit size greatly influence the amount of rock excavated, and the pumping rate determines how much fluid passes through the several detection and collection devices. Because we use a vacuum degassing system for the mud bombs, these samples are less affected by fluid type than the autosampler or dunk readings.

Because of the aerated nature of the drilling fluid, N_2 and O_2 data have proven to be of little or no value. Furthermore, as O_2 is scavenged by the drilling fluid and the cuttings, we cannot use N_2/O_2 ratios to determine nitrogen content. CO_2 data has proven to be difficult to interpret because of contamination.

Oil-based mud was used as the drilling fluid in Sidetrack 2. Radon measurements and the use of the steel cylinders was discontinued for ST-2 because it would have been impossible to keep the instruments uncontaminated.

Trip gas is the gas that accumulates in the well when no drilling fluid circulation takes place. Gas can come into the borehole by diffusion or through fractures. Clearly, time is a very important factor in trip gas readings, and although trip gas

Table 2. Factors affecting gas measurements

Bit size	%	Pump rate (GMP)	
17.5"	100	600	
12.25"	49	500	
8.5"	24	320	
6.5"	14	180±	
Collection method	**Circulation rate**	**Hole size**	**Fluid system**
Dunk tank		* *	* *
Autosampler	*	* *	* *
Mud bombs	*	*	* *
Headspace gas	*	* *	* *
Gas solubility		$CO_2 > C_1 > H_2$ $C_1 > C_2+$	

measurements have proven to be of very great value, they cannot be compared directly with readings taken during drilling operations.

Contaminants were a considerable source of concern, because of the need to measure hydrocarbons in the low ppm levels. They also became the concern of secondary effects, that is, the possibility that some hydrocarbons and hydrogen could be created artificially in the borehole. The character and the amount of additives, lubricants, and the incidence of abnormal amounts of metal, rust, rubber and other material introduced into the hole have been monitored and catalogued.

Results and Discussions

As pointed out previously, the various factors that affect gas measurements make it difficult to directly relate gas readings throughout the entire well even when using the same measurement system. For example, methane in the headspace in canned cuttings samples dropped dramatically from 100-200 ppm to an average of 10-30 ppm when a changeover from freshwater to a weighted mud was made (Fig. 1). The methane/ethane ratio is also affected very strongly across the same interval (Fig. 2). The ratio is 3:4 in the freshwater system, and increases to 20-50 after a weighted mud was introduced. Differences in solubility and relative retentive capacity of the two fluid systems for methane and ethane cause these large changes. On the other hand, the methane/ethane ratio of desorbed cuttings is independent of the mud system used (Fig. 3) and therefore can be used

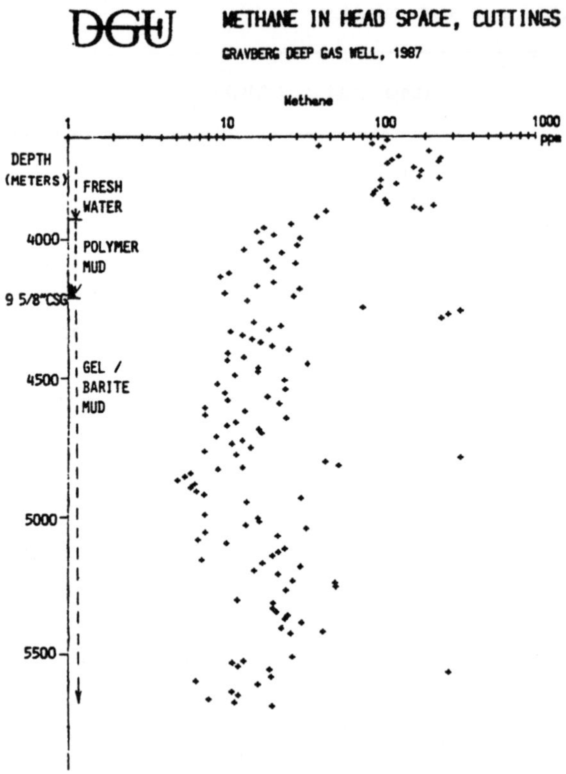

Fig. 1. Methane in Head-space gas

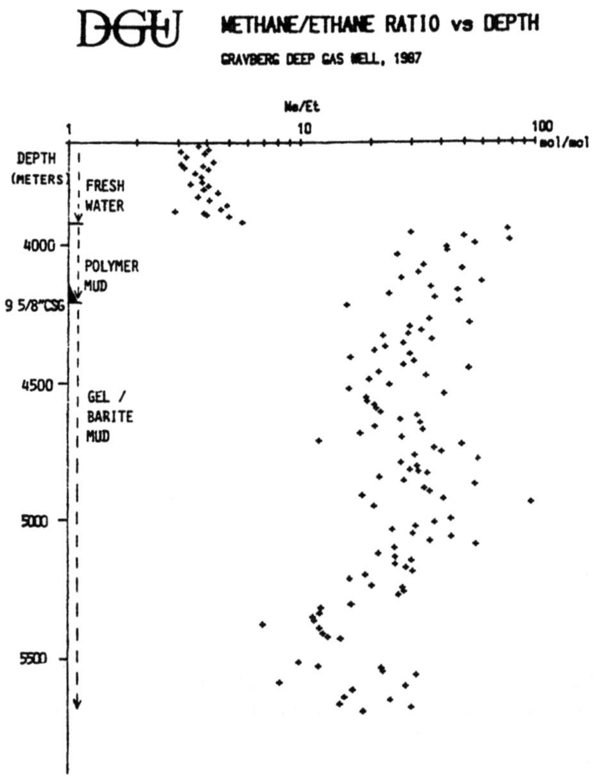

Fig. 2. Methane/ethane ratio in sandspace gas

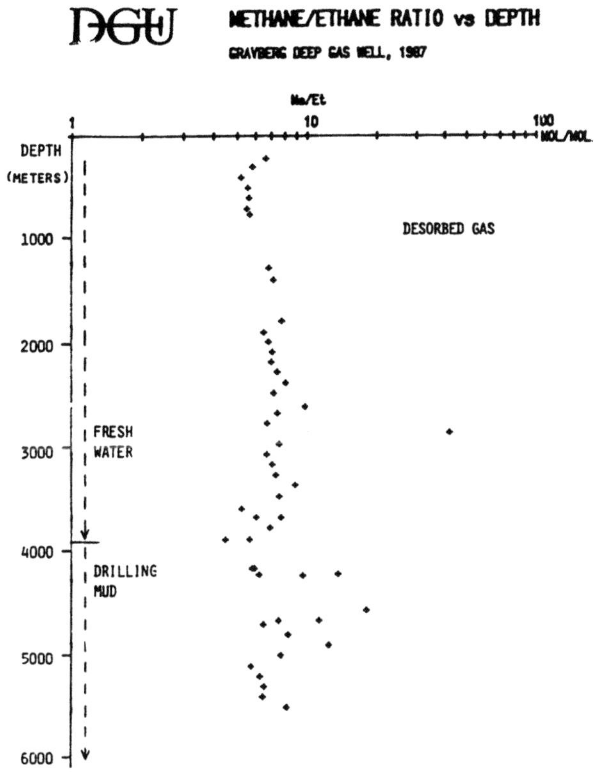

Fig. 3. Methane/ethane ratio in desorbed cuttings

for the entire well for comparative purposes (refer to Laier this Vol.).

The highest concentration of radon gas was associated with fracture zones above and below a dolerite intrusion (Fig. 4). There does not appear to be a correlation of radon with zones of high radioactivity or with zones rich in hydrocarbon gases. It is concluded that the radon comes primarily from the fractured intervals. The fracture intensity was established from a description of the cuttings in which a number of different fracture indicators were measured. These include the presence of open and healed fractures, cataclastic fragments, and altered minerals such as epidote and sericite. However, there are many other fracture zones in the well that do display radon anomalies.

All of the dolerite (diabase) intrusions exhibit an increase in hydrocarbon gas compared with the adjoining granite (Fig. 5). In Sidetrack-1, we were able to analyze the R-3 dolerite at closely spaced intervals because the sidetrack had started a short distance above it, and we knew where we would encounter the dolerite. As shown on Fig. 5, the dolerite contains far more hydrocarbon gas than the adjoining granite, although the amount of hydrogen is about the same or perhaps a bit lower in the dolerite than in the granite. Although the show profiles differ somewhat for the other dolerite units, the pattern is

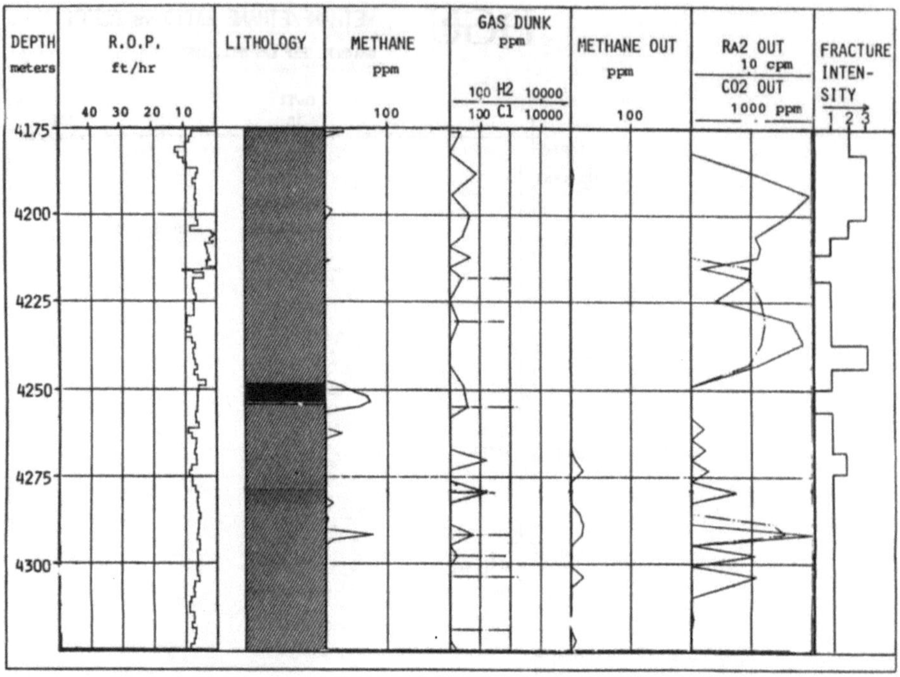

Fig. 4. Occurrence of radon gas associated with a fracture zone surrounding a dolerite intrusion. In the lithology column dolerite is colored black. Everything else is granite

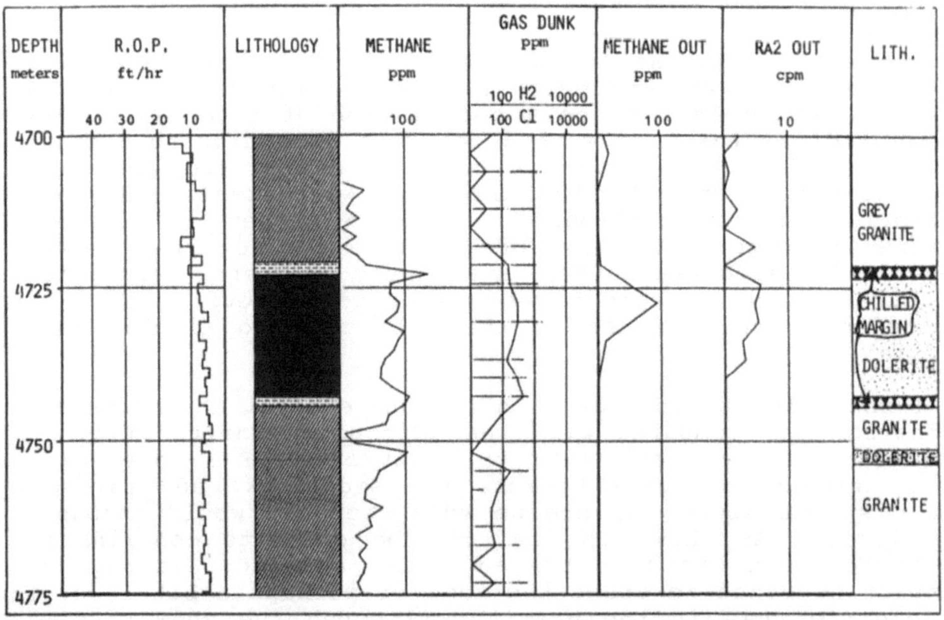

Fig. 5. Gas analysis log in the vicinity of the R-3 dolerite

very much the same: the dolerites are the zones that contain the major gas shows. The dolerite gases are very rich in methane (98% is typical), with only a small amount of ethane and propane. Unsaturated hydrocarbons are rare.

In contrast with the dolerite dry gases, the hydrocarbon gases recovered from the granite are wet gases, they usually are present in much smaller concentrations. In turn, the gases in the granite have about 50-70% methane, and they contain fairly high concentrations of ethane (methane/ethane ratio is 3:8) and propane. Further, these gases are rich in unsaturates; ethane/ethene and propane/propene ratios are both about 1. The high proportion of ethene and propane is highly unusual compared with natural gases in other parts of the world. Very small amounts of C_2 and C_3 unsaturates may be found in gases of bacterial origin, but otherwize they are rare indeed in natural gases and petroleum deposits.

Isotopic analysis is extremely useful to help differentiate the gases (Fig. 6). On Fig. 6 it is shown that the gas in the dolerites is isotopically heavier than the gas in the granites: the

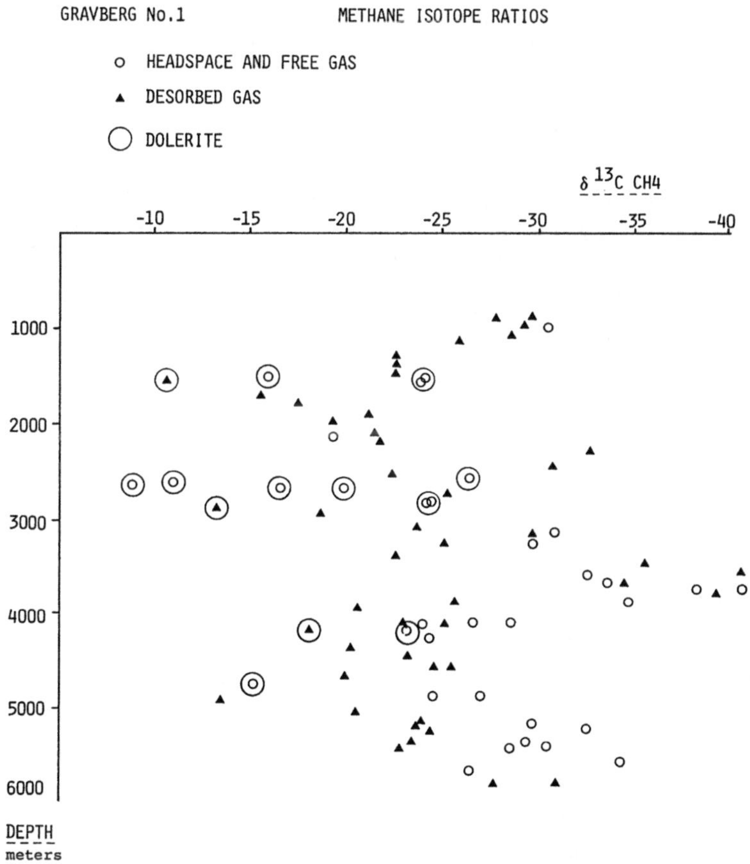

Fig. 6. Methane isotope ratios

dolerite gases are typically -10 to -20‰, while the gases on the granites are isotopically lighter, more commonly running from -21 to -32‰. There is probably some mixing, as some gases in the granite are isotopically heavy.

The gases present in the dolerite are very likely of magmatic origin. They are isotopically very heavy, and they are similar in isotopic composition to gases found in the East Pacific Rise. The gases found in the granite are distinct from the dolerite gases: they are isotopically lighter and they contain large proportions of ethane, propane, and their unsatuated counterpasts. The few isotopic analyses of ethane and propane done until now indicate that the C^{13} for methane is isotopically heavier than for the ethane and propane. For example, this isotopic relationship is unlike that of thermogenic gases formed from the breakdown of kerogen. It is, however, similar to gases found in inclusions in igneous rocks, and to gases suspected of having a magmatic origin.

Hydrogen has been measured throughout the entire well, in quantities exceeding methane usually by at least one order of magnitude. In a number of intervals, hydrogen, as measured from dunk tank samples, often exceeds 1000 ppm. Helium has been noted sporadically throughout the well, although because of instrument limitations we could not measure helium at levels shallower than 4300 m. Work done at Studsvik and by Global Geochemistry verify that helium is present only occasionally and at low ppm levels in most of the well.

In Sidetrack-1, between 6050-6450 m (measured depth), hydrogen and helium are present in the greatest amounts, consistently, as seen anywhere in the well. Hydrogen typically was measured in the 5000 to 10000 ppm range, helium 100-600 ppm, and methane 60-160 ppm. During the drilling of this well segment, it was common practise not to fill the borehole completely while making trips to change the bit. At these times, the pressure in the hole was reduced by up to 10%. Gases were carefully monitored during trips, and we established that more gas came into

Table 3. Cuttings analysis, 5779-91 m. Desorbed fine cuttings of granite

Gas content (ppb by wt.)		Isotopic composition, ‰
C_1	154	-27.8
C_2	59	-29.9
C_2	25	
C_3	34	-30.9
C_3	20	
C_4	18	

the well when the pressure was relieved, and that the longer duration trips generally yielded more gas.

For the interval 6050-6450 m several cross plots were prepared to show the relationship between hydrogen, helium, and methane (Figs. 7, 8 and 9). On all three figures, 4 or 5 points are shown as black dots; their values were added later and were not used in making the regression calculation. The data were smoothed using a 3-point moving average before being plotted.

The data show little or no correlation between H_2 and C_1 (r = 0.268) (Fig. 7) and He and C_1 (Fig. 8) (r = 0.0026). The correlation between Hc and H_2 is 0.737, which would be improved if we included the five points presented as black dots (Fig. 9). Given the methods of data collection, I regard this as a good correlation between He and H_2.

These data are of particular interest, because although there is much speculation as to the possibility that methane and hydrogen could be artifacts of the drilling process, there is no doubt that any helium encountered must be resident in the formation. Regardless of the ultimate origin of the helium, it is certain that it is present in this particular granitic rock section, and, because the helium is in situ, then the hydrogen also must be (at least partially) in place also. In the Kola superdeep borehole in the USSR, there is a negative correlation between hydrocarbon gases and helium, while hydrogen typically exceeds the methane concentration by about an order of magnitude.

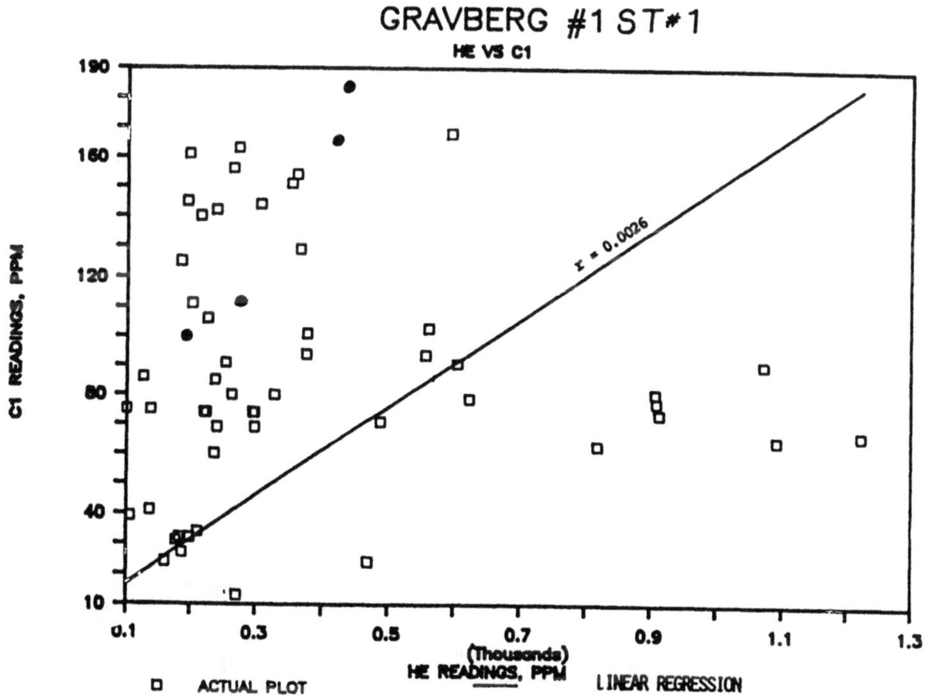

Fig. 7. H_2 vs C_1, ST-1, 6050-6450 m

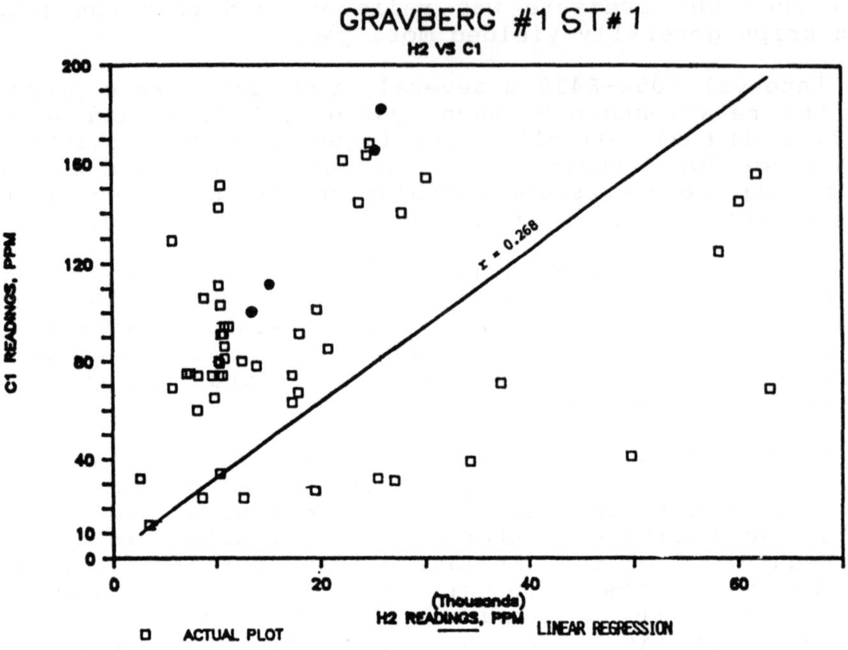

Fig. 8. He vs C_1, ST-1 6050-6450 m

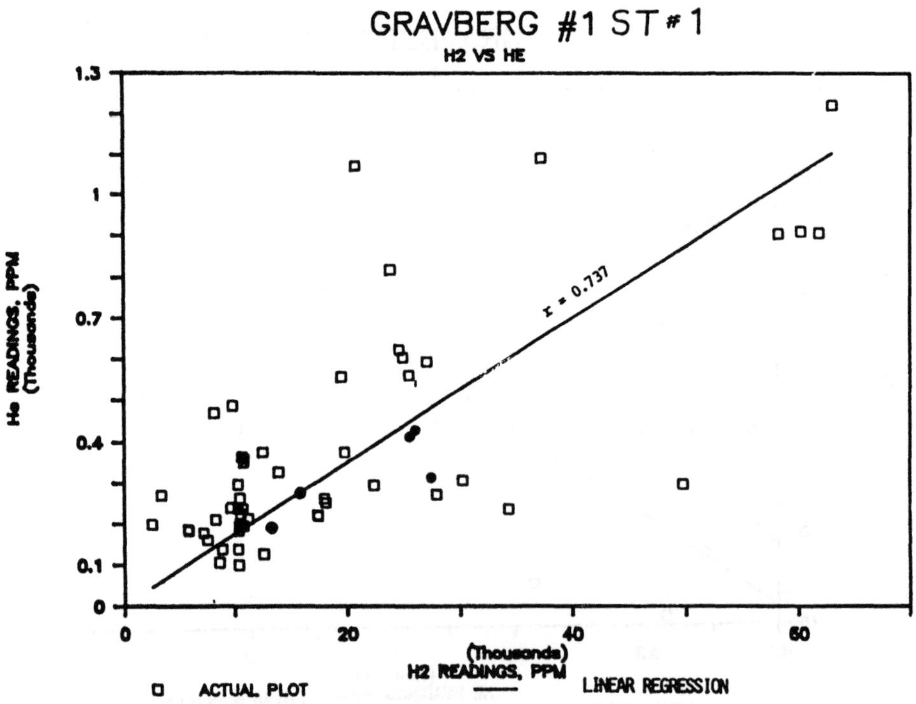

Fig. 9. H_2 vs Hc, ST-1, 6050-6450 m

Summary and Conclusions

Two different types of hydrocarbon gases have been encountered. The major gas shows have been found in the dolerites. These gases are dry, contain 98% methane, they are isotopically very heavy, and they resemble gases in chrondritic meteorites or gas recovered from the East Pacific Rise. It seems highly likely that they are of abiogenic origin, although we cannot say if they originate as part of crustal or mantle processes. The gases in the granites are wet gases, for they contain only 50-70% methane, and an abundance of light hydrocarbons including C_2=, C_3= and C_4=. They are isotopically quite diverse, but for the most part the isotopic composition is similar to that of high maturity coal gases. At this point, I would have to say that the origin of the HC gases in the granite is uncertain, and will require more investigation in order to obtain a resolution.

It is important to note that hydrogen and helium are correlated, which points to the fact that at least some of the hydrogen is coming from the formation and is not an artifact of the drilling process. As described by Jeffrey et al. (this Vol.), the helium is crustal in origin, probably created by radioactive decay of minerals in the granite.

In the area of the well when we observed a radon anomaly, the concentration of radon is associated with fracture zones adjacent to a dolerite intrusion.

These results must be considered as being preliminary, as much additional work is being carried out which may modify the interpretations presented.

Acknowledgments. The writer conducted these studies as part of his assignment as Chief Scientific Officer at the well site under the auspices of the Gas Research Institute. The cooperation of all those connected with the project is gratefully appreciated.

Analyses of Gases in the Gravberg-1 Well

A.W.A. JEFFREY[1], I.R. KAPLAN[1] and J.R. CASTAÑO[2]

Introduction

Gas samples analyzed at Global Geochemistry Corp. were collected at the Gravberg-1 drillsite in three ways:

1. Sealing fluid plus cuttings in metal cans leaving an air headspace
2. Filling a 150 ml steel cylinder with drilling fluid
3. Injecting gas from the wellsite dunk tank system into a 150 ml steel cylinder.

Analysis of the gas content of the headspace of cans collected by procedure 1 provided the most complete data set, since samples were collected at narrowly spaced intervals. Degassing fluid collected by procedure 2 and analysis of the isolated gas provided the most quantitative data in terms of gas released per unit volume of fluid. Samples collected by procedure 3 usually provided larger volumes of gas and were used primary for isotopic analysis of gas components.

The analytical procedure involved complete molecular composition, specifically, N_2, O_2, Ar, CO_2, C_1-C_5 hydrocarbons, H_2 and He. Isotopic analysis included $^{13}C/^{12}C$ ratio of methane, ethane and propane, H/D ratio of methane, H_2 and H_2O, and $^3He/^4He$ ratio of He.

Results and Discussion

Methane concentrations in the canned headspace samples from the original hole (OH) are shown in Fig. 1. Above 5 km, high methane spikes are associated with the dolerite intervals, whereas the granite sections are characterized by generally lower concentrations. Methane/ethane ratios in gases from the dolerite intervals are also higher and methane in these gases is enriched in ^{13}C (less negative $\delta^{13}C$ values) compared to gases from the granite sections (Fig. 2). Figure 2 also shows a suite of three gases, rich in C_2-C_5 hydrocarbons (wet), with ^{13}C enriched meth-

---

[1] Global Geochemistry Corp., 6919 Eton Avenue, Canoga Park, CA 91303, USA
[2] 722 Oder Lane, Houston, TX 77090, USA

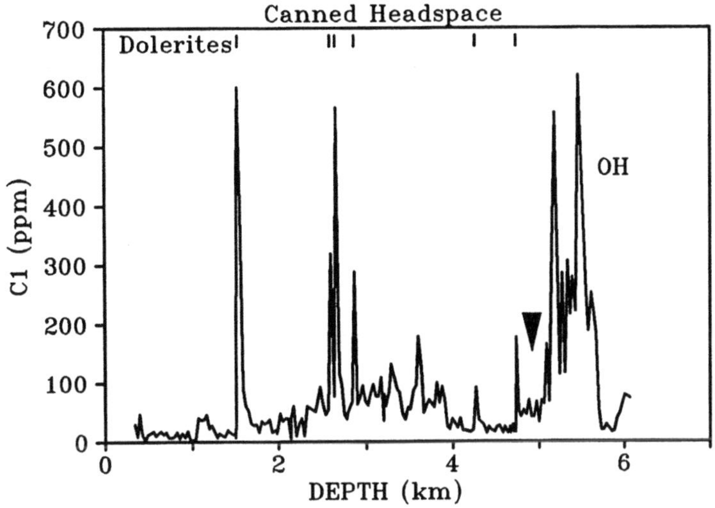

Fig. 1. Methane concentrations in canned headspace samples from the Gravberg-1 original hole (OH). **Arrowhead** identifies start of addition of IDLUBE

ane, which are apparently related to addition of carbide to the well. Carbide reacts with water to form acetylene, which is used to measure the time for drilling fluid to travel from the bottom to the top of the hole. Small, but significant amounts of saturated hydrocarbons appear to be a byproduct of the operation, but can be distinguished isotopically and compositionally. The dolerite hydrocarbons can be characterized as low

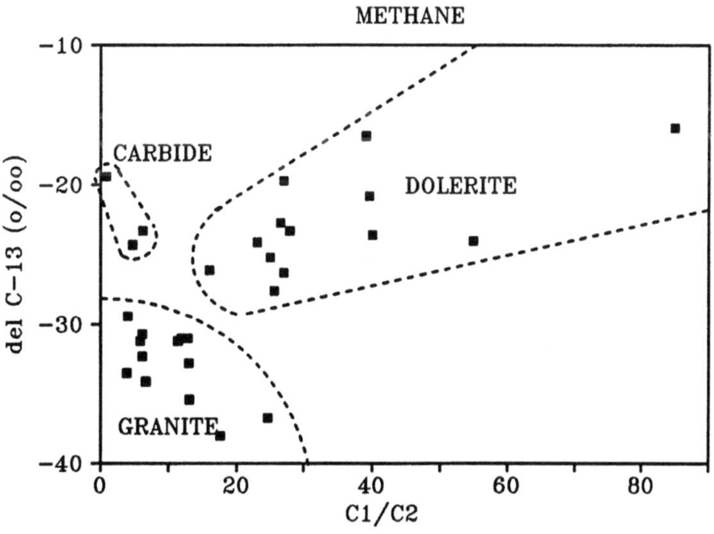

Fig. 2. Methane $\delta^{13}C$ versus methane/ethane ratios in gases from the Gravberg-1 well

in C_2-C_5 hydrocarbons (dry), almost lacking in unsaturated hydrocarbons, with methane enriched in ^{13}C. In contrast, gases from the granite intervals are wet, have significant concentrations of ethene and propene, and methane is enriched in ^{12}C.

These gases are plotted in a characterization scheme developed primarily by Schoell (1983) based on methane $\delta^{13}C$ and gas wetness (Fig. 3). The molecular and isotopic characteristics of abiogenic gas are taken from the analysis by Welhan and Craig (1983) of gases from mid-ocean ridge vents at the East Pacific

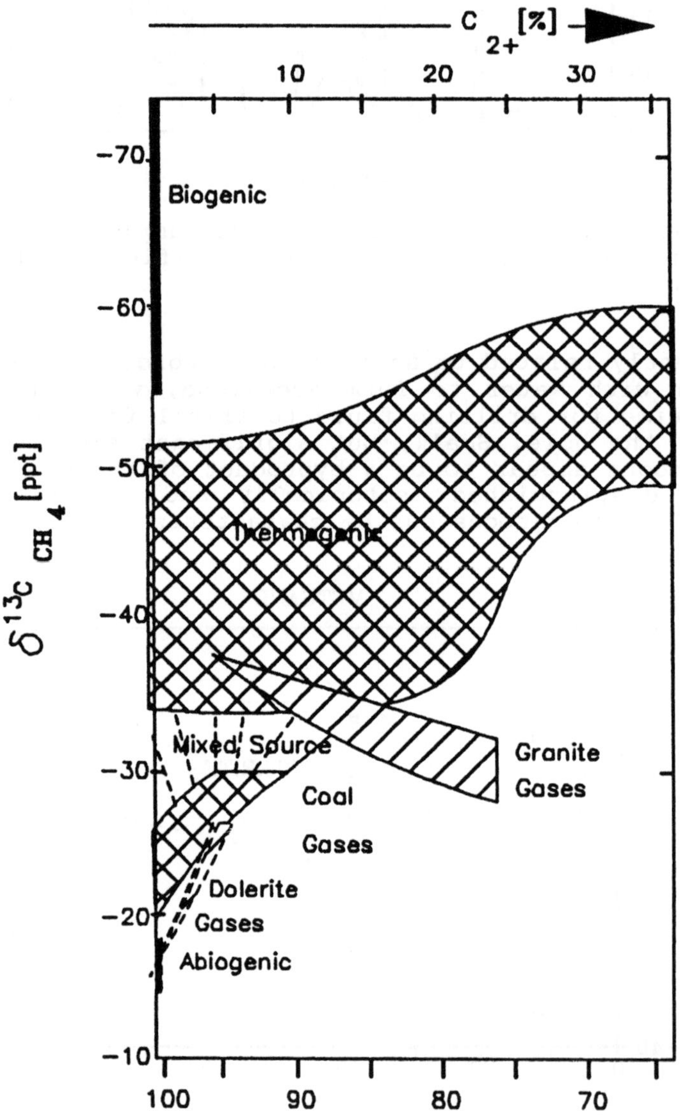

Fig. 3. Hydrocarbon genetic characterization scheme. (After Schoell 1983)

Rise. Dolerite gases fall close to the abiogenic fields, whereas the granite gases are closer to the thermogenic field (derived from the thermal breakdown of organic matter). However, it is still not possible to rule out a nonindigenous origin, from interaction of the drilling fluid or the drilling bit with the rock, for the small amounts of hydrocarbons in the granite.

Hydrocarbons of nonindigenous origin were identified below 5 km in the OH, represented by sharp increases in methane (Fig. 1) and butene concentrations. These increases correlate with the start of addition of IDLUBE, a drilling additive believed to contain fatty acids, and used as a lubricant. This was confirmed when the sidetrack hole (ST-1) was drilled through the same interval. Methane concentrations in ST-1 were low (Fig. 4), typical of other granite intervals, and butene concentrations were <10 ppm. However, n-pentane concentrations increased sharply below 5.5 km in ST-1, apparently correlated with the start of addition of Torque Trim, another lubricant additive. This additive did not appear to contribute methane, but did correlate with a marked decrease in methane/ethane. The compositional data indicate that Torque Trim promotes formation of n-pentane, n-butane, propane and ethane. These findings point to the necessity of closely monitoring drilling additives to effectively distinguish indigenous and nonindigenous hydrocarbons.

Hydrogen was observed in many intervals and normally in higher concentration than hydrocarbons (Fig. 5). The H_2 occurrences did not appear to the associated with the dolerites; however, there appears to be some indication of correlation of high H_2 levels and around 5.2 km in the OH with additives. Below 6 km, however, high H_2 concentrations, clearly associated with He, ap-

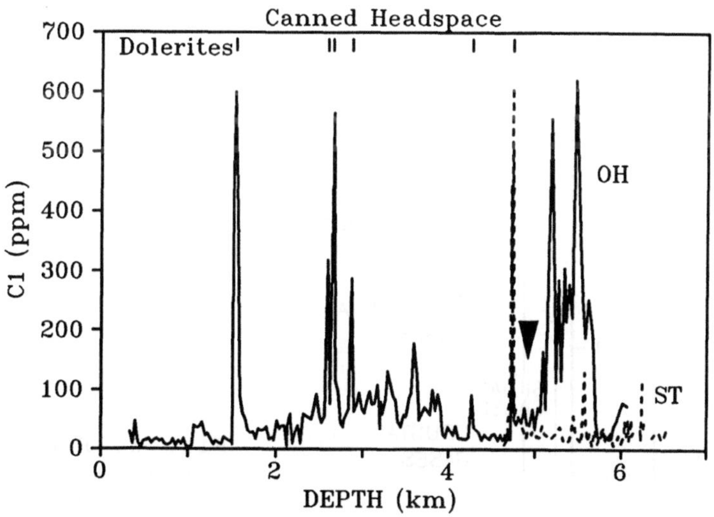

Fig. 4. Methane concentrations in canned headspace samples from Gravberg-1 original hole (OH) and side-track hole (ST-1)

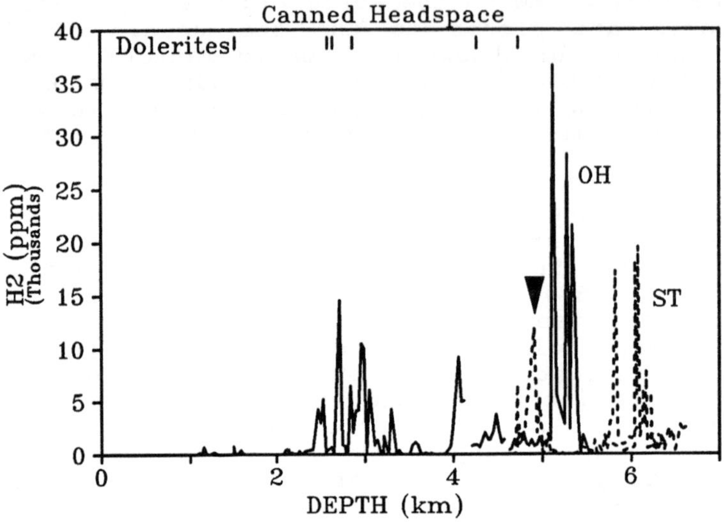

Fig. 5. Hydrogen concentrations in canned headspace samples from the Gravberg-1 original hole (OH) and side-track hole (ST-1). Arrowhead identifies start of addition of IDLUBE

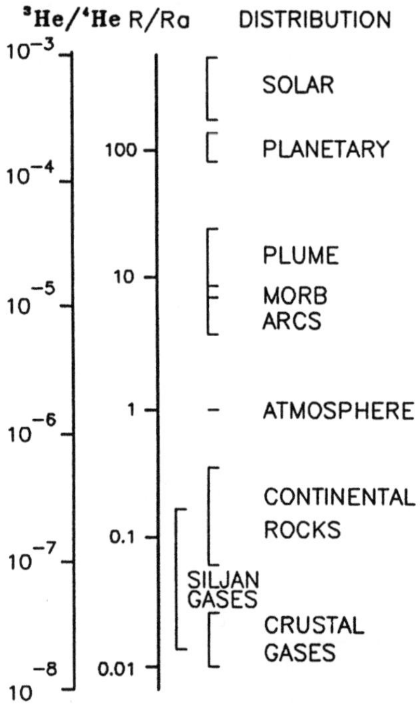

Fig. 6. Range of helium isotope ratios in nature. (After Craig unpublished data)

pear to be indigenous to the granite. Many of the highest concentrations were measured in gases accumulated during drilling lapses which would appear to preclude drilling operations as the source.

Helium in these intervals is enriched in ^4He and depleted in ^3He and is consistent with a crustal rather than a mantle origin (Fig. 6).

The source of the hydrogen is more difficult to identify, mainly because the hydrogen isotopic ratio is not as definitive as the helium isotope ratio. Isotopic exchange of H_2 with H_2O is relatively rapid and varies with temperature. The H/D ratio of H_2 in intervals below 6 km induates an equilibrium temperature of ~85°C, close to the formation temperature. However, this could result from equilibration of H_2 of either deep (high temperature) or shallow origin with H_2O at the formation temperature, or perhaps even with H_2O in the drilling mud. Consequently, a deep origin for the H_2 cannot be ruled out, but the close association with He of relatively shallow, crustal origin is significant.

References

Schoell M (1983) Genetic characterization of natural gases. AAPG Bull 67:2225-2238

Welhan JA, Craig H (1983) Methane, hydrogen and helium in hydrothermal fluids at 21°N, East Pacific Rise. In Hydrothermal processes at seafloor spreading centers. In: Rona PA, Bostrom K, Laubier L, Smith KL jr (eds) NATO-AIR vol. Plenum, New York, pp 391-409

Light Hydrocarbons in Drill Cuttings from the Gravberg-1 Borehole

T. LAIER[1]

Introduction

Light hydrocarbon analysis of drill cuttings is a well-known technique in conventional oil and gas exploration.

Analysis of the gases released by acid treatment of the cuttings is generally considered to give the most representative results. Investigation of head space gases in a sealed can containing cuttings is more simple and faster, but the results are less representative. The drilling fluid may alter the composition of the hydrocarbon gases, especially if Diesel oil is added, as is often the case. Another disadvantage of head space analysis is that it is impossible to estimate the amount of gas in the rock, since the amount of gas released before the cuttings are canned is unknown. An unknown amount of gas may also be lost upon storage if the cans are not completely sealed.

Head space analysis is, however, more efficient than mud gas analysis when drilling in a low permeable rock because it allows more time for the release of the gases to be measured. Therefore, in order to obtain a good overview of the hydrocarbons in the rocks of the Gravberg-1 well, it was decided to make many head space analyses, plus a few desorbed gas analyses by acid treatment of cuttings.

Cuttings for light hydrocarbon analyses were taken every 6 m and bactericide was added before the lid was placed on the tin can.

It was considered a good idea to make a few light hydrocarbon analyses on core samples in order to minimize the interference from the drilling operation. Unfortunately, only a few short cores were taken in the deep well. However, a preliminary study of the Siljan area included seven shallow wells (Vattenfall 1985) from which adequate core samples could be chosen.

The stable carbon isotope ratio of methane depends on its origin. The $^{12}C/^{13}C$ ratio is probably the best means to descriminate abiogenic methane ($\delta^{13}C \sim -7‰$) from biogenic ($\delta^{13}C$: $-60‰$ to $-80‰$) and thermogenic ($\delta 13C$: $-25‰$ to $-50‰$) methanes.

---

[1] Geological Survey of Denmark, Thoravej 8, 2400 Copenhagen, Denmark

Biodegradation will affect the $^{12}C/^{13}C$ ratio, but this is not considered to be a problem in the deep crystalline rock. Therefore, stable carbon isotope analyses were made on as many samples as possible. Stable isotope analysis requires much more material than gas chromatographic analysis.

Some of the analytical results were rather confusing. Consequently, a series of laboratory experiments were carried out to elucidate the origin of some of the hydrocarbon gases encountered in the rocks of the deep well.

Head Space Analyses

The cuttings were stored for 2 weeks to 2 months before analysis. The tin can was punctured and 0.2 to 1 ml of the head space gas was analyzed for C1 to C5 hydrocarbons on a gas chromatograph. The relative amounts of hydrocarbons in most head space gases were similar to those found in the desorbed gases.

Head space analyses were made on every sample, i.e., for every 6 m from 756 m to 1445 m. The results of these analyses showed that a head space analysis for every 12 m would be sufficient. Therefore, only every other sample was analyzed here after.

The results of all analyses of the Gravberg-1 well before the side track operations are summarized in Fig. 1. The depths of the dolerite intrusions are also shown in Fig. 1.

Head space analyses of samples from the dolerite intrusions shows higher hydrocarbon contents and generally higher methane/ethane and ethane/ethene ratios.

The atmosphere contains a background concentration of about 2 ppm of methane and for this reason, one has to be careful not to attach too much significance to the low results.

The shift at 3930 m corresponds to a change in the drilling fluid from just freshwater to a polymer freshwater mud system. The cuttings from below 3930 m were definitely larger than those from above that depth and this may be the reason for the lower concentration of hydrocarbons. At present, we cannot explain the tenfold increase in the methane/ethane ratio of the head space gas.

If the amount of methane in the can was sufficient, an evacuated 100-ml sample loop was connected to the can and the gas was flushed into the loop by injecting water into the tin can. Stable isotope analyses were then made on the methane. The results of the stable isotope analyses of head space methane are shown in Table 1.

Generally, the head space methane is a little lighter isotopically compared with the methane of the desorbed gas from the same depth. This difference may relate to the diffusion of gases out of the cuttings as the lighter isotope will diffuse more readily.

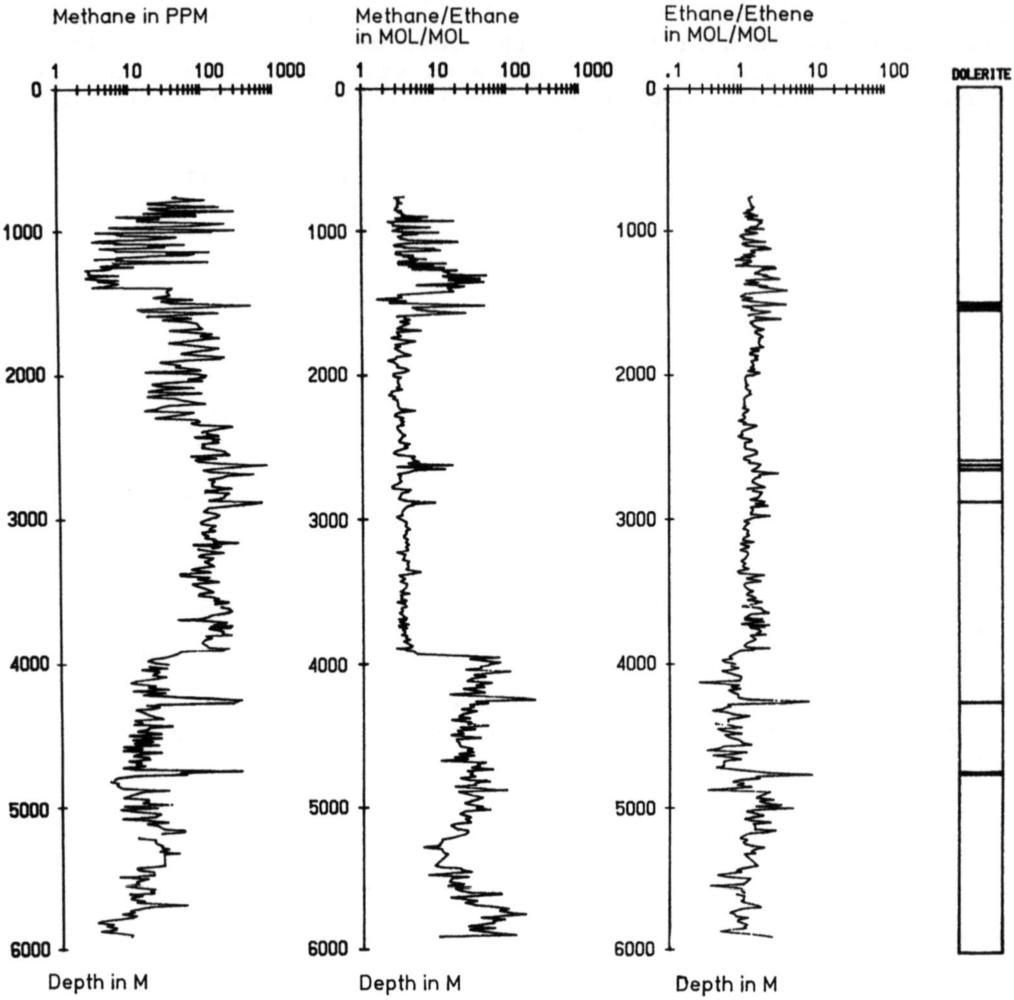

Fig. 1. Plots of head space gas analyses versus depth. Dolerite intrusions from the lithology log are shown on the column to the right

Table 1. Comparison of stable isotope analyses of methane from head space gas and desorbed gas

Depth (m)	3694	3901	4255	4736	5224	5419	5523
$\delta^{13}C$‰, head space:	-33.2[a]	-34.4	-23.0	-15.2	-29.3	-29.1	-28.4
$\delta^{13}C$‰, desorbed :	-34.1	-25.7	-18.0	-10.7	-24.4	-24.4	-23.6

[a] Values are measured relative to PDB standard.

Desorbed Gas Analyses

The canned sample was washed with water on a 0.25 mm sieve. One hundred g of the cuttings were then treated with hot 4 M hydrochloric acid in a closed evacuated system and the gas released was analyzed by gas chromatography. If sufficient methane was present, the gas was flushed into a 100-ml sample loop for stable isotope analysis.

The relative amounts of light hydrocarbons C_1-C_5 were similar for most samples. An example from 4900 m is shown in Table 2.

The relative amounts of C_2 to C_5 are certainly higher than expected for an abiogenic gas from the mantle. They look similar to the light hydrocarbons of petroleum except for the high relative amounts of olefins.

The results of all analyses of desorbed gas are summarized in Fig. 2. The major changes are observed near the dolerite intrusions. The characteristics of the hydrocarbons in the dolerite samples are very close to what might be expected for true magma gases, i.e., almost pure methane with $\delta^{13}C$ around -7‰.

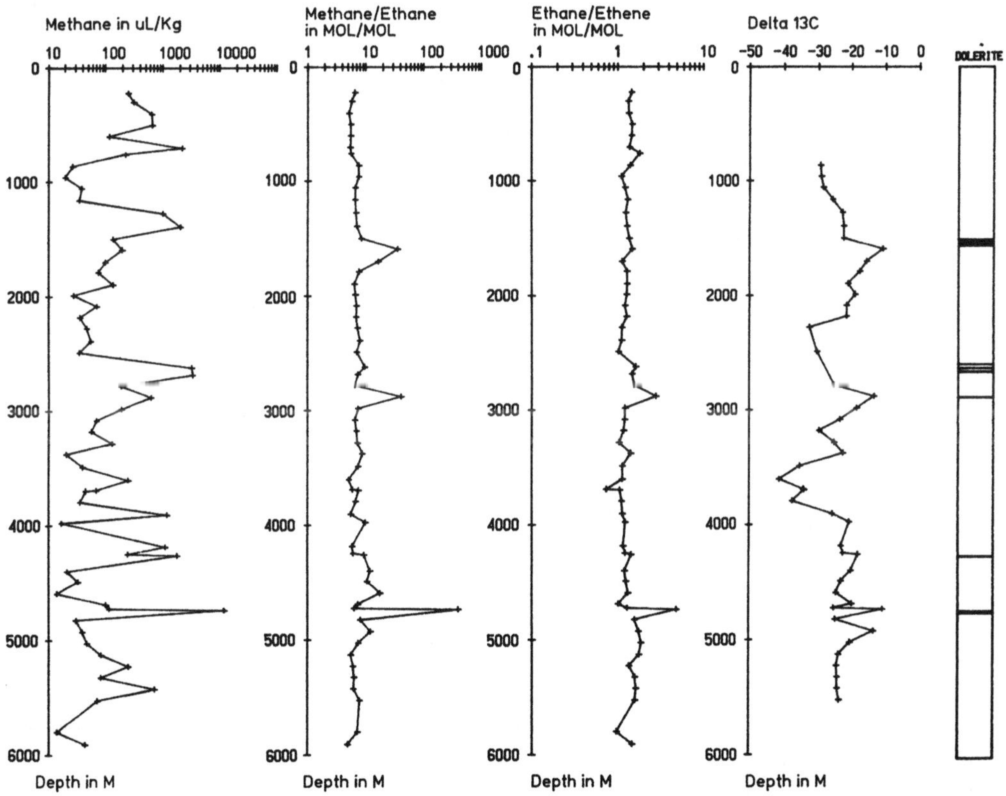

Fig. 2. Plots of desorbed gas analyses versus depth. Dolerite intrusions from the lithology log are shown on the column to the right

Table 2. Relative amounts of light hydrocarbons in granite cuttings from 4900 m

C_1	C_2	C_2=	C_3	C_3=	i-C_4	1-C_4=	n-C_4	2-C_4=	i-C_5	n-C_5
55.5	8.6	8.0	5.4	7.0	0.5	1.6	2.0	10.2	0.5	0.7

Laboratory Experiments

The relative amount of methane in the hydrocarbons of the granites and its stable isotope ratio indicate a gas from thermal cracking of organic matter. If this is the case, the high content of olefins makes it rather unique. In order to find out whether the gases are indigenous to the granite or not, and to elucidate its origin, a number of laboratory experiments were carried out.

Analyses of Cores from Shallow Wells. The core samples were crushed in a mechanical mortar. Desorbed gas analyses were made on the crushed samples by the same method as used for cuttings. The results are summarized in Table 3.

Table 3. Light hydrocarbon analyses of core samples from shallow wells

Well	Depth m	C_1 µl/kg	C_1/C_2 mole/mole	C_2/C_2= mole/mole	$\delta^{13}C_1$ ‰
Gravberg	1.0	408	6.2	1.9	-23.1
	106.2	422	7.8	1.9	-24.0
	220.4	27[a]	8.0	1.8	-16.5
	220.4	4950[b]	6.0	2.1	-21.4
	394.0	405	6.1	2.0	-21.7
	503.9	214	6.9	1.6	-24.7
	582.1	270	6.8	1.8	-23.4
Solberga	472.8	328	6.3	1.7	-24.2
Hättberg	601.0	1000	5.8	1.5	-

[a] Nonmagnetic fraction
[b] Magnetic fraction

The results of core analyses are very similar to those of the cutting analyses.

When comparing core analyses with the cutting analyses it is obvious that the hydrocarbons are not just an artifact of drilling operations. The question still remains whether or not the results of desorbed gas analyses are representative of the rock.

Comparison of Various Methods of Hydrocarbon Analyses. Light hydrocarbons released by crushing or heating of a few rock samples were analyzed and compared with the results of the methods described above.

Fifty g of sample was placed in a vacuum tight ball mill vessel equipped with two valves and septa. The sample was crushed in a ball mill until the hydrocarbon concentration in the vessel remained constant. Ten g of sample was heated to 200°C in a closed 15 ml vial for 1 h. 0.2 ml of the head space gas was withdrawn through the septum of the vial for gas chromatographic analysis. The results are shown in Table 4.

The desorbed gas method gives by far the highest concentrations of all methods. Methane, being the smaller molecule, diffuses more readily out of the rock. This may explain the higher methane to ethane ratio of the ball mill method compared to the desorbed gas method. The results from the heating method differed very much with respect to composition of the hydrocarbons.

Analyses of Magnetic and Non-Magnetic Fractions. The magnetic fraction of cuttings and crushed core samples was removed by a hand magnet and analyzed by the desorbed gas method. It turned out that almost all of the hydrocarbons in the granites are associated with the magnetic fraction. See Tables 3 and 5.

Table 4. Light hydrocarbon analyses by various methods

Sample		Method	C_1	C_1/C_2	C_2/C_2=
			µl/kg	mol/mol	mol/mol
Cuttings	1.274 m	Head space	0.2	21	-
		Ball mill	60	18	1.8
		Heating to 200°C	20	290	0.09
		Acid desorption	1000	6.8	1.2
Core[a]	601 m	Ball mill	55	34	2.7
		Heating to 200°C	40	36	0.07
		Acid desorption	1000	5.8	1.5

[a] From Hättberg shallow well

Table 5. Comparison of light hydrocarbons in magnetic and non-magnetic fractions

Sample	C_1 µl/kg	C_1/C_2 mol/mol	$C_2/C_2=$ mol/mol
Cuttings of granite, Gravberg-1, 1383.8 m	4920[a]	6	1.3
	60[b]	44	.
Cuttings of dolerite, Gravberg-1, 4736.6 m	2340[a]	21	1.8
	12200[b]	407	4.9

[a] Magnetic fraction
[b] Nonmagnetic fraction

Most of the hydrocarbons of the dolerite intrusions are associated with the nonmagnetic fraction of the sample.

Magnetite - Hydrocarbon Reactions. Light hydrocarbons were reacted with synthetic magnetite in a closed vial equipped with a septum.

In the first experiment, 2 g of magnetite was heated to 100°C with a 70 ppm standard of C_1-C_6 n-alkanes. The concentrations of hydrocarbons in the vial decreased with time. After 16 h the concentrations of hydrocarbons varied from 51 ppm to 8 ppm for methane through hexane. The temperature was then raised to 200°C. After 30 min, the concentrations of the n-alkanes varied from 106 ppm to 29 ppm for methane through hexane. In addition to the n-alkanes, 20 ppm of ethylene was found. Several hydrocarbons with more than 6 atoms carbon were observed. After 2 h at 200°C, these higher hydrocarbons had completely disappeared.

In the second experiment, 5 g of magnetite was heated under vacuum at 200°C. After cooling, 100 ppm of the C_2-C_4 olefins was added to the system. The olefins adsorbed quickly on the magnetite and almost all of the C_3-C_4 olefins disappeared from the head space of the vial.

The magnetite was then analyzed by the desorbed gas method. An almost complete recovery of the olefines was achieved and no alkanes were observed.

The magnetite hydrocarbon results may be summarized as follows:

- alkanes polymerize around 100°C in the presence of magnetite.
- thermal cracking takes place around 200°C, and ethylene is produced.
- olefins are not hydrogenated during the desorbed gas method.

Conclusions

Two types of hydrocarbon gas are found in the rocks of the Gravberg-1 well.

One is associated with dolerite intrusions and is clearly of magmatic origin.

The other type is found in the granites. The high percentages of olefins, and the fact that almost all of the gas is associated with the magnetic fraction, have led to the conclusion that the hydrocarbons are formed as a result of a Fischer-Tropsch type reaction Eq.(1).

$$CO + 3 H_2 \xrightarrow{Fe_3O_4} CH_4 + H_2O \qquad (1)$$

The reaction takes place at temperatures from 200°C to 300°C and can produce high amounts of olefins (Biloen and Sachtler 1981).

Equation (1) can lead to kinetic isotopic fractionation (Lancet 1970) which may explain the values observed for methane stable isotope ratios. Some of the variations in stable isotope ratios of methane in the granite may also be due to a certain contribution from mantle gas.

Hydrocarbons containing olefins are reported from an iron ore deposit in the Ukrainian crystalline (Voilov et al. 1975). $^{13}C = -48.6$ ‰ for the hydrocarbons, so they may be of a similar origin as that of the Gravberg gases. The reason why so few examples of this type of gas is reported in the literature may well be that it has not been actually looked for.

Acknowledgments. This work was funded partly by the Danish Ministry of Energy and partly by Vattenfall.
The isotope analyses were carried out by the stable isotope laboratory at the University of Copenhagen.

References

Biolen P, Sachtler WMH (1981) Mechanism of hydrocarbon synthesis over Fischer-Tropsch catalysts. Adv Catal 30:165-216

Lancet MS (1970) Carbon isotope fractionation in the Fischer-Tropsch synthesis and in meteorites. Science 170:980-982

Vattenfall (1985) Deep gas. Swedish Premises: The Siljan Ring structure - investigation and evaluation. Rep No 4 from Vattenfall, Swed State Power Board

Voilov G et al. (1975) Natural gases of the Yuzhno-Belozerskoye iron ore deposit (Ukrainian crystalline shield). Acad Sci USSR Dokl Earth Sci Sect 220:204-206

Trace Elements in Gravberg Drillhole Samples

S. JOVANOVIC and G.W. REED JR[1]

Introduction

Trace elements in Gravberg drillhole samples in granitic rock at the Siljan impact structure in Sweden are being studied. The impact structure is the first nonsedimentary basin-related formation to be used to test the thesis of abiogenic hydrocarbon resources. Trace element data are acquired to correlate with hydrocarbon profiles and other indicators of sources and transport of hydrocarbon-related volatiles such as H_2 and He and with rock porosities. Trace element data are also used to address basic geoscience questions. Volatile trace elements, especially Hg, are being studied. An expectation is that fluids that will transport hydrocarbon compounds and related elements will also transport Hg. At ambient temperatures in the drillhole, Hg will continue to occupy sites after volatiles such as CH_4, H_2 and He have been mobilized. An understanding of the distribution and migration of Hg can provide insights into the behavior of volatiles such as CH_4. Hg is measured in stepwise heating experiments. Labile ($\leqslant$130°C) and retentive (up to 1200°C) sites are identified. The former are related to gases, the latter to less volatile elements and/or compounds. Data on other volatiles such as Se, Sb, Zn, and relatively nonvolatile Fe, Co, Sc, and REE are also utilized.

From an applied perspective, data have been acquired on concentration profiles with depth; the utility of cuttings (sand-sized grains) vis-a-vis core; evidence for permeation of fluids, i.e., in-situ weathering. The latter measurements are also of basic geoscience interest. Also from the basic geoscience perspective, data are obtained on the thermal conditions that existed in the granite and diabase of the formation and evidence or lack thereof is examined for an impact related imprint. Finally, in order to arrive at an understanding of weathering phenomena, results from other Precambrian granitic formations, in particular, Illinois Deep Hole Project core samples and Wolf River, Wisconsin batholite samples are related to Gravberg deep drillhole samples.

---

1 Chemistry Division, Argonne National Laboratory, Argonne, IL 60439, USA

Experimental

Radiochemical and instrumental neutron activation analyses are used to measure trace elements in drillhole samples. Most of the samples are sand-sized cuttings representing ~7 m depth each. From a geochemical point of view, the cuttings probably sample layers of this thickness better than intact cores. Samples selected for study are routinely cleaned ultrasonically in deionized water, rinsed ultrasonically with alcohol and ether and air-dried. The cuttings are sealed in supersil ampules and irradiated at the high flux reactor at Brookhaven National Laboratory. Volatiles, Hg in particular, are extracted by stepwise heating in a vacuum system with a manifold of traps that are sequentially cooled at each temperature step (130°, 250°, 450°, 1200°C) with liquid nitrogen in order to collect the released volatiles. Their radioactivities are measured on a multichannel analyzer.

Results

1. Before examining the data to determine whether there is evidence for volatiles and their migration in the Siljan structure, it is necessary to consider the reliability of results based on submillimeter (sand-sized) grains. Two tests have been made. Silt and coarse sand collected at 1024 m and at 2244 m were compared (Table 1). The 1024 m silt had lower total Hg and the same 130°C Hg as the sand, in spite of a larger surface area. However, the maximum (59%) Hg release was at 130°C for the silt but at 250°C (45%) for the coarser material. A grain-size effect could be a factor. The lower total Hg in the silt suggests that the Hg in the sand-sized samples is not contamination. In contrast to the 1024 m silt, that collected at 2244 m contained much higher total Hg than the sand (13 ppb vs 1 ppb) and almost all (80%) of this is 130°C Hg. This may be evidence for the presence of exogenous Hg. A test is to compare Hg released at ⩽450°C with that released at >450°C. This is a routine procedure which will be discussed in Section 4. The 2244-m sand has a similar $Hg_{\leqslant 450°}/Hg_{>450°}$ ratio to other sand samples. The 2244-m silt has a much higher ratio. If the 130°C Hg fraction is subtracted, the silt ratio compares favorably with sand samples from which 130°C Hg fractions are subtracted.

A further test of whether trace and minor constituents in drillhole cuttings are representative of the rock formation has been carried out using very coarse (~1 cm) chips from 5760 m and sand sized (⩽1 mm) cuttings from 5793 m. Since cutting fluids may not only contaminate but also leach indigenous material, data for elements other than Hg are compared in Table 2. In spite of very different chemistries, the major (Fe) and non-volatile trace elements exhibit a consistent pattern of lower concentration, by an average of 20% in the 5760 m chips than in the 5793 m fines. Cs also behaves in a consistent manner. The difference between the samples may be due to somewhat compositionally different granites separated by 33 m. Since only the surface of the coarse material was exposed to cutting fluids, the element concentrations are those of the bulk fragment. It

Table 1. Stepwise heating release of Hg from Gravberg sand and silt cuttings

Sample	Hg ppb				
	130°C[a]	250°C	450°C	1200°C	Hg_{total}
Gravberg drillhole cuttings					
1024 m granite sand	14	35	20	6.6	75
1024 m " silt	10	5	4.1	0.6±0.2	20
2244 m granite sand	0.2	0.5	0.2	0.1±0.03	1.0
2244 m " silt	11	0.9	1.2	0.5	13

[a] Stepwise heating extraction steps: 130°C for 1 h followed by 2 h; 250°C and 450°C 1 h; 1200°C 30 min. Counting statistical errors are less than 20% or as indicated.

is reasonable to conclude that the trace elements in cuttings are representative of the intact rock.

2. The profile of Hg concentrations down to depths of 6 km is given in Fig. 1. The most interesting feature is the large amount of both 130°C and total Hg at 4622 m. This is also a level of higher porosity and of a steep increase in methane in the granite reported in Vattenfall data. It is possible to ascribe the high 130°C Hg in the interval from 4256 m to about 4726 m (>19 to 81 ppb) and possibly at 5055 m (5 ppb) and 5329 m (3.8 ppb) to an exogenous source. The percentage of 130°C Hg is significantly greater (~50%) in samples from within these depth intervals than the ~15-30% of 130°C Hg in samples from most other intervals. Data on the percent 130°C Hg are included in Table 4. The higher percentage of 130°C Hg coupled with the generally higher, 6 to 100 times, total Hg concentration is additional support for the introduction of Hg. The average 130°C Hg fraction in the granite samples with high total Hg is

Table 2. Element concentration comparisons between coarse (~1 cm) vs. Fine ($\lesssim$1 mm) drillhole cuttings

Sample	Hg_T ppb	$Hg_{>450°C}$ ppb	Fe wt %	Co ppm	Sc ppm	Ce ppm	Eu ppm	Cs ppm
5792 m (0.5-1.0 mm)	2.2	0.61	2.3	6.9	9.1	130	2.3	1.5
5760 m (~1 cm)	3.3	0.68	1.7	4.9	5.8	96	1.4	1.3

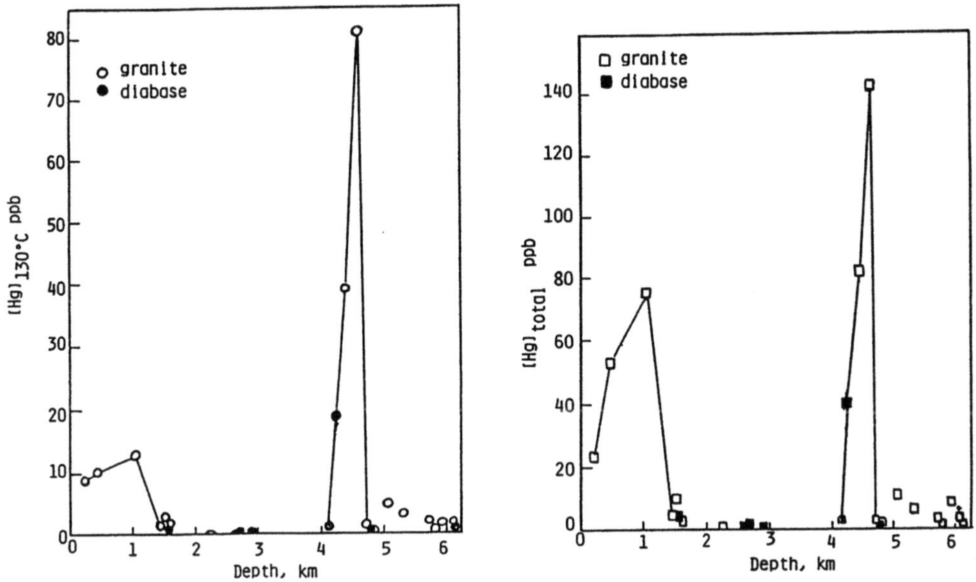

Fig. 1 a, b. Hg concentrations in samples from Gravberg drill-hole. a Hg fraction released at 130°C in stepwise heating experiments; b total Hg in the sample

~47%; that in other granite samples is ~22%. Thus about one half of the 130°C Hg may be exogenous in the samples with high total Hg.

3. The above suggests that Hg has permeated some strata in the formation. The permeation zones may be expected to exhibit alteration (weathering) by the fluids involved. In a study of other Precambrian granitic rock formations, we established that the slope of the chondrite normalized REE trend is related to rock alteration (Jovanovic and Reed, manuscript in preparation). Core samples from the Precambrian basement sampled in the Illinois Deep Hole Project (IDHP) and Wolf River (WR), Wisconsin, batholith quarry samples were measured. The nature of the weathering fluids and the degree of weathering were determined by H_2O and weak acid leaching experiments and by measurement of a secondary mineral, carbonate, deposit. Trace elements, other than REE, were also determined; Cs data, in particular, are relevant to the present considerations.

The chondrite normalized REE trend is less steep in the more weathered IDHP granite; it is identical to that of chlorite, a common alteration product in granite. In the present study detailed REE data could not be obtained in a timely manner; only the longer lived nuclides could be determined. Some curves are given in Fig. 2 for illustrative purposes to suggest trends. The Ce/Tb ratio can be used to assess the degree of weathering. Table 3 summarizes Siljan and Jaerna granite and pertinent IDHP granite data. The "weathered" granite samples identified by

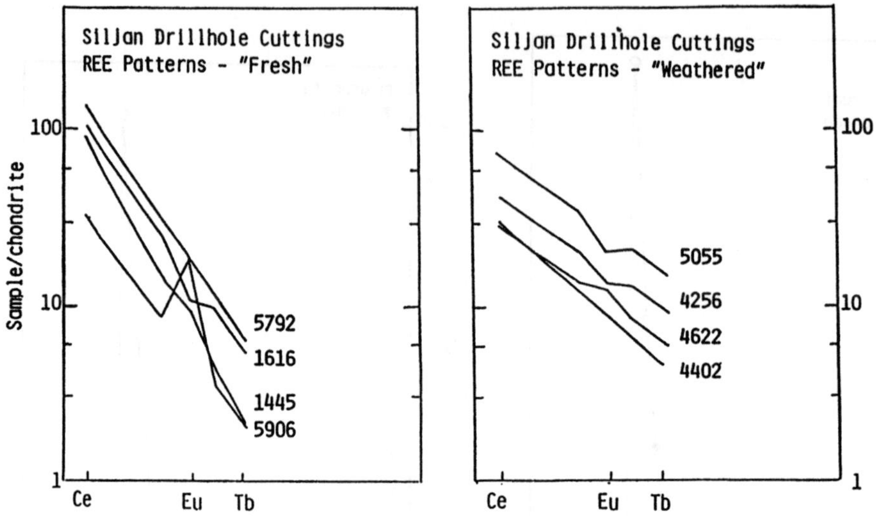

Fig. 2. Chondrite normalized REE patterns for Siljan drillhole cuttings for "fresh" and "weathered" samples

Table 3. Evidence for introduction of Hg related to the degree of in-situ weathering

		Degree of weathering				
		Fe wt %	(REE slope) Ce/Tb		Hg (130°C) ppb	Cs ppm
Siljan samples depth, m						
5055		1.3	74		5.0	7.1
4622		0.84	72		81	8
4402		0.81	192	"weathered"	39	7
5760		1.7	128	"fresh"	1.9	1.9
5906		2.1	113		1.6	1
5792		2.0	167		0.65	1.6
1445		1.5	179		1.2	2.3
IDHP samples depth, m						
1.5	weathered	0.6	72			6.0
31.5	fine grain	0.81	315			3.6
629	coarse grain	1.5	245			3.2

this approach are the ones with the "excess" 130°C Hg. These are also the samples with higher Cs, possibly the result of deposition from the permeating fluids. Trace constituents cannot migrate significant distances from source regions unless carried by an abundant fluid. The excess Hg could be derived from the same source as the enhanced methane reported.

The presence and role of permeating fluids is of basic as well as applied geoscientific interest. Further trace and minor element studies of Siljan and Jaerna granites paralleling those on IDHP and WR granites will provide detailed information.

4. An additional type of basic geoscience information related to the Siljan formation may be obtained from the Hg thermal release data. Hg is released at different temperatures according to whether it is present on surfaces, in grain boundaries and microfissures, dispersed in major mineral phases and/or in accessory minerals such as sulfides. We have proposed a geothermometer based on the thermal equilibration of Hg between low ($\leqslant$450°C) and high (>450°C to 1200°C) temperature sites (Jovanovic and Reed 1985 and included references). The low to high temperature Hg ratios correspond to Hg equilibration during prograde metamorphism or during magma cooling. Regionally metamorphosed pelite schists and quenched magmas are used to establish a temperature scale, Fig. 3. The scale is independent of silicate rock type. Hg equilibration temperatures for Siljan samples are listed in Table 4. There is no difference between granite and diabase except in cases of granite with apparent excess 130°C Hg. For granite and diabase with lower 130°C Hg, the equilibration (closure) temperature is ~900°C. If the 130°C Hg is not included in the ratio, an approximate 44°-70°C (867 → 911°C and 887 → 957°C) increase in apparent equilibration temperature arises for the granite and diabase, respectively. The weathered granites with high percentages of 130°C Hg give lower equilibration temperatures of ~700°C (or ~800°C if the 130°C fraction is not included). Correction for only that part of the 130°C fraction suggested to be exogenous (Sect. 3) would give a Hg equilibration temperature for weathered rock between 700° and 800°C. Some inferences and/or options concerning the thermal history of the Siljan formation may be drawn:

a) The Hg equilibration temperatures are higher than those normally associated with regional metamorphism, $\lesssim$700°C. The general uniformity of equilibration temperatures, however, makes this mechanism attractive.

b) A high temperature heating event is suggested by the almost identical Hg equilibration temperatures for the granite and diabase. This must have occurred as post-diabase intrusion or the diabase and granite emplacement was essentially simultaneous and the temperatures are the Hg **closure** temperatures for an originally hot pluton.

c) The ~700°C temperature of the samples with high 130°C Hg may represent a thermal event occuring after the initial cool-

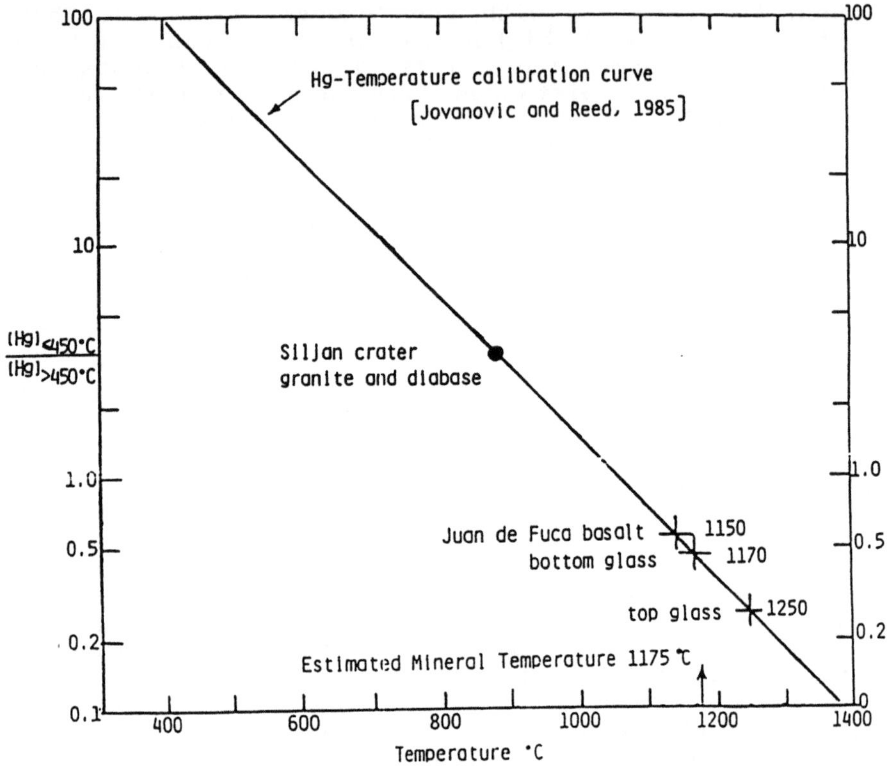

Fig. 3. Thermal equilibration temperatures for granite and diabase Gravberg drillhole samples are estimated using Hg-temperature calibration curve. The calibration curve is based on the ratios of low (>450°C) to high (≤450°C) temperature Hg concentrations released from metamorphic rocks with known temperatures of metamorphism. Data on Juan de Fuca submarine sheet flow basalts and glass support the proposed geothermometer

ing of the formation. The fluids that caused increases in Hg, whatever their origin, could be associated with this event.

d) There is no obvious gradient in Hg equilibration temperatures with depth in the drillhole. It, therefore, is improbable that impact heating is being monitored.

Table 4. Effects of percent Hg released at 130°C on estimates of Hg thermal equilibration temperatures

Siljan samples depth, m	$Hg_{\leq 450°C} / Hg_{>450°C}$ T°C	% 130°C Hg	$Hg_{(\leq 450°C-130°C)} / Hg_{>450°C}$ T°C	Hg_T ppb
Granite				
231	640	35	700	23
1616	700	38	763	3.8
4402	710	48	805	82
4622	645	57	875	141
5055	745	45	840	11
5329	695	60	820	6.3
5760	900	57	1060	3.3
	719		838	
463	968	19	1010	53
1024	770	18	797	75
1445	930	26	990	4.4
2244	785	20	820	1.0
5792	950	30	1017	2.2
5906	800	19	834	8.5
	867		911	
Diabase				
1500	680	35	755	9.1
1561	870	15	900	4.6
2646	960	18	1003	1.0
2659	880	32	950	1.3
2872	804	37	880	0.9
4731	1131	14	1290	2.1
	887		957	

Acknowledgement. This work was supported by the Gas Research Institute

Reference

Jovanovic S, Reed G.W. Jr (1985) The thermal release of Hg from chondrites and their thermal histories. Geochim Cosmochim Acta 49:1743-1751

Calcite Crack Cements in Granite from the Siljan Ring, Sweden: Stable Isotopic Results

J. W. VALLEY[1], S. C. KOMOR[1], K. BAKER[1], A. W. A. JEFFREY[2], I. R. KAPLAN[2]

and A. RÅHEIM[3]

Abstract

Petrographic and stable isotopic analyses have been made for surface samples, and drill cuttings and cores from six shallow (<700 m) and one deep hole (>6 km) in the Siljan Ring meteorite crater, Central Sweden. Most granite samples have fine-scale, cross-cutting fractures that are healed dominantly by calcite. The amounts of calcite in granite range from 0 to 2.4 wt.% above 1.5 km, but are uniformly below 0.2 wt.% at greater depths. In the deep hole at Gravberg, values of $\delta^{18}O$ in calcite range from 7.3 to 22.3 SMOW and show a steady decrease with depth, consistent with precipitation from modern-day meteoric water at variable temperature, from 10°C (near surface) to 70-100°C (6 km depth). Values of $\delta^{13}C$ correlate with the position in the crater, but not with depth. Calcites from two holes nearest the center of the ring are dominantly $\delta^{13}C = -19.8$ to -10 PDB, while those closer to the perimeter range from -19.4 to +8.4.

All evidence suggests that deep circulation of groundwater was the dominant process that resulted in the formation of these carbonates. However, some carbonates are clearly of higher temperature origin and the source of carbon is controversial. The present results suggest that carbon may be derived from two surface sources (organic decay and limestone dissolution) and is carried downward with variable mixing by groundwater. Low $\delta^{13}C$ values can result from organic decay throughout the region while high $\delta^{13}C$ values are found primarily near the perimeter of the ring in proximity to Paleozoic limestones.

Introduction

Nearly all samples of Proterozoic Siljan and Järna granite examined from the Siljan Ring (Fig. 1) contain measurable amounts of carbonate ($\geq$0.01 wt.%). Such carbonate is not unique to Siljan and is also found in other crystalline rocks through-

---

1 Department of Geology and Geophysics, University of Wisconsin, Madison, WI 53706, USA
2 Global Geochemistry Corp., 6919 Eton Ave., Canoga Park, CA 91303-2194, USA
3 Institute for Energy Technology, Box 40, N2007, Kjeller, Norway

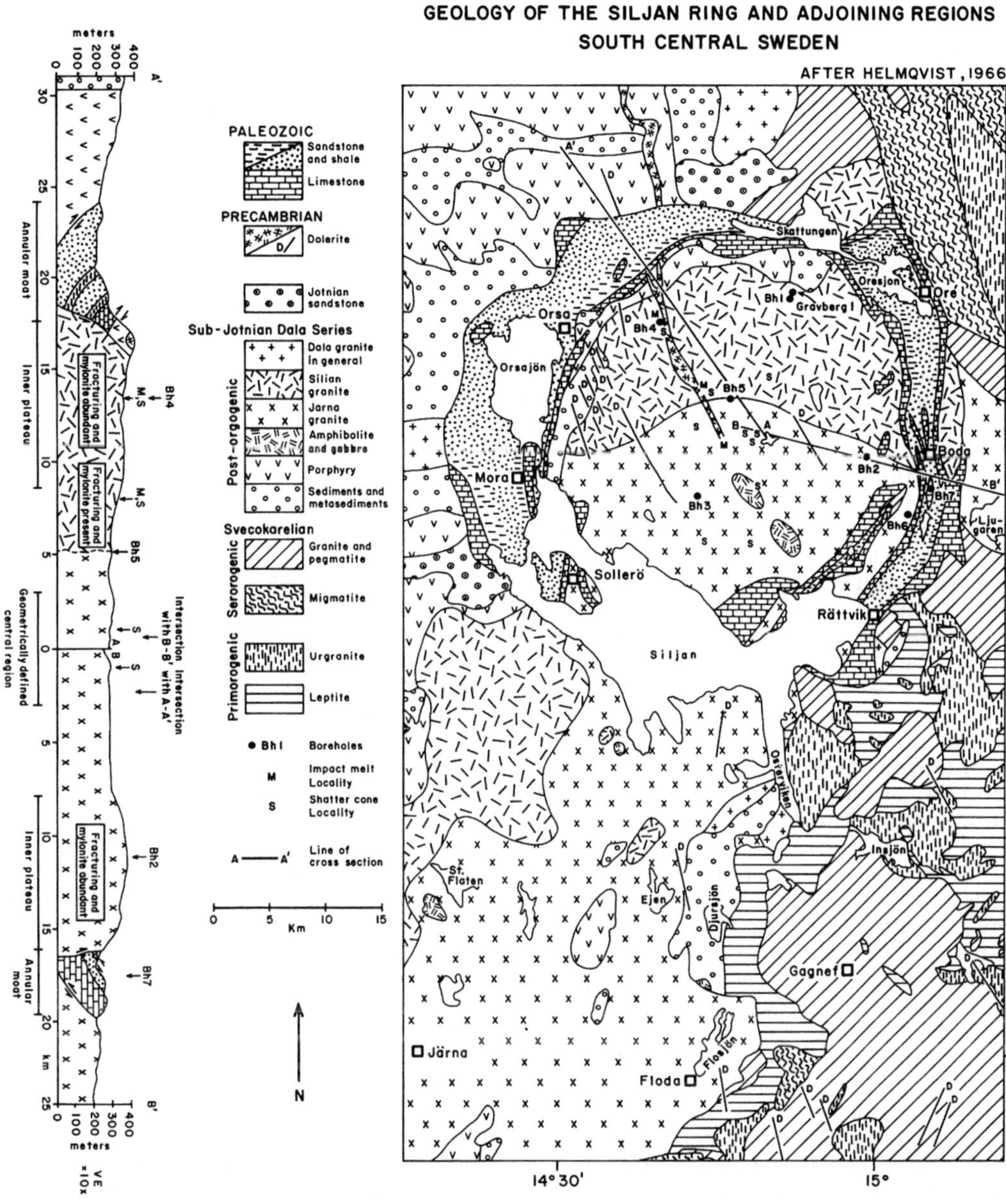

Fig. 1. Geologic map of the Siljan Ring, central Sweden. The locations of seven shallow drill holes (<700 m, Nos. 1-7) and one deep drill hole (>6 km, No. 1) are shown. Paleozoic limestones are restricted to chaotic, down-dropped blocks in the annular ring

out Sweden (see Tullborg 1986). Examination of 155 samples in thin section reveals that calcite at Siljan typically heals late-stage, brittle fractures which cross-cut mineral grains and fabric. The dominant cause of fracturing is related to the formation of the Siljan Ring Structure which represents a meteorite crater (Grieve 1987) formed at 362 my (Bottomly et al. 1978). Thus, the presence of calcite crack cements proves that carbon-bearing fluids have permeated granites in the upper crust underlying approximately 1000 km^2 of the Scandinavian Shield in Post-Devonian times.

The composition, source, and quantity of fluids at Siljan have previously been uncertain as have the nature of the fluid-hosted events. One hypothesis is that the presence of calcite is directly related to the outgassing of methane from the Earth's mantle (Gold and Soter 1980). In this model, mantle-derived carbon, oxidized in the near-surface environment, combines with calcium derived from granite to form calcite (Gold 1985, 1987). Alternatively, a more conventional hypothesis explains the source of calcite cements as resulting totally from processes operative in the upper crust.

The oxidation/reduction state of the mantle and deep crust provide important constraints on theories of mantle outgassing, abiogenic methane, and methane-derived calcite. Studies of lavas and xenoliths from the mantle are controversial in this regard (see Mattioli et al. 1987), but metamorphic rocks from the deep crust show that contrary to Gold and Soter (1980), methane cannot be a **pervasive** fluid in tectonically active areas (Lamb and Valley 1985, Valley and Lamb 1986). However, it is presently controversial to what extent abiogenic, mantle-derived methane rises along **localized** faults and shear zones. The deep fracturing at Siljan, resulting from meteorite impact, has prompted the suggestion that this area represents a methane-rich zone and the Swedish Power Board's deep drilling program is intended to test this possibility.

The ratios of carbon and oxygen isotopes provide important evidence as to the nature of calcite-forming fluids (Valley 1986, Kyser 1986). Thus far, over 400 samples of calcite have been analyzed for carbon and oxygen isotopic ratios, in order to evaluate the competing hypotheses for the nature of fluid-hosted events at Siljan. Analysis of samples from shallow (<700 m) pilot holes at Siljan (numbers 1-7 in Fig. 1) yield values of $\delta^{13}C$ = -19.8 to +8.4. Samples with $\delta^{13}C$ <-10 have been called "anomalous calcite" and the presence of such low $\delta^{13}C$ has been cited as proof that the source of carbon in these rocks is mantle-derived methane (Gold 1985, 1987). The presence of anomalous calcites provided one of the early justifications for the drilling of a deep hole at Gravberg along the north margin of the Siljan structure. This chapter summarizes the preliminary isotopic results and examines the competing hypotheses as to the source of carbon in these calcite crack cements (upper crust vs mantle).

Sample Analysis

Calcites have been analyzed from both Siljan and Järna granites which form the majority of drill core and outcrop samples within the crater as well as from dolerite sills and dikes. Samples have been examined optically after staining for calcite (alizarin red S in dilute HCl) and by cathode luminescence. In the more calcite-rich samples, small veinlets are readily seen, but in many samples calcite occurs along hairline cracks and its presence is only detected by luminescence.

Calcite has been converted to CO_2 for isotopic analysis by the standard phosphoric acid technique (25°C, >8 h, McCrea 1950, Sharma and Clayton 1965). Samples have been extracted from whole-rock powders weighing 1-10 g. Isotopic ratios of oxygen and carbon were measured by mass-spectrometer and are reported in parts per thousand (per mil) relative to the SMOW (oxygen) and PDB (Carbon) standards (Tables 1 and 2). The weight percentage of calcite in each sample is calculated based on the manometrically measured amount of evolved CO_2.

Table 1. Isotopic analyses of calcite in samples of drill core

Sample[a]	Depth (m)	$\delta^{13}C$ (PDB)	$\delta^{18}O$ (SMOW)	$CaCO_3$ %	Rock type[c]	Lab[d]
			Gravberg-1[b]			
GR1-1-16	1006.4	-5.6	12.7	0.03	G	1
GR1-1-16	1006.4	-5.5	12.9	0.03	G	1
GR1-1-25	1009.3	-5.7	14.9	0.09	G	1
GR1-1-25	1009.3	-5.6	18.9	0.08	G	1
GR1-2-1	1496.0	-6.6	9.5	0.85	G	1
GR1-2-1	1496.0	-6.6	9.4	0.12	G	1
GR1-2-3	1497.7	-6.5	9.4	0.03	G	1
GR1-2-3	1497.7	-6.5	9.1	0.07	G	1
GR1-2-9	1498.0	-6.8	9.7	0.75	G	1
GR1-2-9	1498.0	-6.8	9.7	0.76	G	1
GR1-2-11	1498.2	-4.5	9.5	0.08	D	1
GR1-2-14	1498.6	-4.5	14.2	0.10	D	1
GR1-2-14	1498.6	-4.1	10.0	0.08	D	1
GR1-2-26	1500.0	-6.0	22.8	0.12	D	1
GR1-2-27	1501.5	-0.6	12.8	0.03	D	1
GR1-2-27	1501.5	-0.7	15.5	0.03	D	1
GR1-2-38I	1502.7	-9.5	15.2	0.05	D	1
GR1-2-38I	1502.7	-9.7	14.0	0.05	D	1
GR1-2-38I	1502.9	-16.1	20.5	0.02	D	1
GR1-2-38I	1502.9	-16.8	17.8	0.02	D	1
GR1-2-40	1503.6	-17.0	19.8	0.03	D	1
GR1-2-44	1504.2	-10.7	16.3	0.05	D	1
GR1-2-44	1504.2	-11.3	14.3	0.05	D	1
GR1-2-51	1505.2	-21.1	19.9	0.02	D	1
GR1-2-56	1506.4	1.9	9.6	0.03	D	1
GR1-2-56	1506.4	2.6	10.5	0.03	D	1
GR1-3-3	2131.5	-10.1	14.6	0.01	G	1
GR1-3-3	2131.5	-11.6	21.1	0.01	G	1
GR1-3-3	2131.5	-10.6	14.2	0.01	G	1
GR1-4-7	2804.3	-10.2	12.6	0.01	G	1
GR1-4-7	2804.3	-9.1	9.0	0.01	G	1
GR1-5-2	3376.3	-8.2	9.7	0.08	G	1
GR1-6-3	4206.2	-5.1	9.6	0.01	G	1

Table 1. (cont.)

Sample[a]	Depth (m)	$\delta^{13}C$ (PDB)	$\delta^{18}O$ (SMOW)	$CaCO_3$ %	Rock type[c]	Lab[d]
			Gravberg-1[b]			
GR1-6-3	4206.2	-4.9	8.7	0.01	G	1
GR1-6-4	4206.7	-7.1	9.4	0.02	G	1
GR1-6-4	4206.7	-7.0	9.1	0.02	G	1
GR1-6-5	4207.3	-5.8	8.3	0.03	G	1
GR1-6-5	4207.3	-5.6	8.3	0.03	G	1
GR1-6-6	4209.0	-5.7	8.9	0.07	G	1
GR1-6-6	4209.0	-4.7	8.9	0.03	G	1
GR1-6-9	4210.8	-9.4	13.8	0.03	D	1
GR1-6-23	4215.0	-4.0	12.2	0.07	D	1
GR1-6-23	4215.0	-5.8	10.6	0.07	D	1
GR1-5751	5751.0	-3.7	7.2	0.26	G	1
GR1-5751	5751.0	-4.2	6.4	0.25	G	1
GR1-5760	5760.0	-3.7	7.2	0.22	G	1
GR1-5760	5760.0	-3.6	7.2	0.20	G	1
GR1-5760	5760.0	-3.6	6.4	0.16	G	1
GR1-5760A	5760.0	-3.4	8.3	0.15	G	1
GR1-5760R	5760.0	-4.6	7.6	0.18	G	1
GR1-5827F	5827.5	-4.2	8.1	0.06	G	1
GR1-5827R	5827.5	-7.5	11.3	0.04	G	1
GR1-5905	5905.0	-4.7	7.7	0.07	G	1
Cement cuttings	1268.0	-18.6	11.8	14.50		1
Cement cuttings	3944.1	-13.7	14.9	17.50		1
Cement cuttings	4206.2	-19.1	11.9	13.40		1
Cement cuttings	5583.9	-24.7	16.0	8.00		1
		Shallow boreholes				
Shallow borehole 1 - Gravberg						
GA 1:2	52.2	-2.5	19.7	0.20	G	2
GA 2:2	77.9	-7.1	14.9	0.80	G	2
GA 3:2	108.8	-0.7	18.3	0.90	G	2
GA 4:2	157.6	-13.5	16.8	0.60	G	2
GA 5:1	207.1	-8.1	11.1	0.30	G	2
GA 6:1	264.5	-7.4	14.9	0.20	G	2
GA 7:1	328.2	-1.5	15.2	2.00	G	2
GA 9:1	391.0	-5.7		0.05	G	2
GA 10:1	407.9	-7.9	10.9	0.20	G	2
GA 11:1	477.3	-13.9	13.9	0.01	G	2
GA 12:1	513.0	-5.9	8.2	0.01	G	2
GA 13:1	564.0			0.01	G	2
GA 14:1	586.1	-12.0	14.8	0.01	G	2
GC 3	82.8	-12.3	18.1		G	3
GC 4	110.9	-7.9			G	3
GC 5	132.1	-2.6	16.2		G	3
GC 6	160.7	-11.8			G	3
GC 7	181.8	-4.2			G	3
GC 8	208.6	-5.5			G	3
GC 9	234.7	-8.3			G	3

Table 1. (cont.)

Sample[a]	Depth (m)	$\delta^{13}C$ (PDB)	$\delta^{18}O$ (SMOW)	$CaCO_3$ %	Rock type[c]	Lab[d]
GC 10	260.6	-4.7			G	3
GC 11	284.0	-3.2			G	3
GC 12	309.4	-9.8			G	3
GC 13	332.3	-11.8			G	3
GC 14	358.2	-7.7	15.0		G	3
GC 15	383.8	-10.3			G	3
GC 16	405.0	-3.7			G	3
GC 17	434.3	-3.8			G	3
GC 18	463.6	-15.0			G	3
GC 19	464.7	-11.0			G	3
GC 20	505.7	-13.7	15.8		G	3
GC 21	535.0	-12.5			G	3
GC 22	560.5	-9.1			G	3
GC 23	585.0	-15.8			G	3
Shallow borehole 2 - Östbjörka						
OA 1:1	28.9	8.4	22.6	0.20	G	2
OA 2:1	38.8			0.30	G	2
OA 3:1	93.9	2.5	19.6	0.30	G	2
OA 3:4	94.2			0.30	G	2
OA 4:1	149.7			0.20	G	2
OA 5:1	200.2	-13.3	14.1	0.05	G	2
OA 6:1	250.9	-10.6	17.7	2.90	G	2
OA 7:1	299.2	-1.8	10.8	0.40	D	2
OA 8:1	366.0	-8.8	14.2	0.20	G	2
OA 9:1	408.5	-6.6	16.0	0.20	G	2
OA 10:1	451.1	-6.9	16.9	0.30	G	2
OC 7	180.4	-15.7	18.6		G	3
OC 8	202.2	-12.6			G	3
OC 9	227.5	-13.6			G	3
OC 10	248.8	-8.5			G	3
OC 11	278.3	-3.3			G	3
OC 12	302.5	-12.8	17.0		D	3
OC 13	329.0	-3.6			G	3
OC 14	344.4	-5.5			G	3
OC 15	370.9	-15.8			G	3
OC 16	392.2	-15.3	17.8		G	3
OC 17	427.5	-13.2			G	3
OC 18	448.7	-14.2			G	3
Shallow borehole 3 - Fåsås						
3-195.7	195.7	-8.0	11.6	0.04	G	1
3-215.3	215.3	-8.3	14.8	0.11	G	1
3-248.4	248.4	-17.6	19.2	0.03	G	1
3-307.4	307.4	-4.7	9.9	0.03	G	1
3-308.6	308.6	-6.6	13.4	0.01	G	1
FA 1:1	57.9	-17.2	17.2	52.80	G	2
FA 2:1	107.5	-19.5	17.2	0.10	G	2
FA 3:1	217.3	-19.0	15.5	0.05	G	2

Table 1. (cont.)

Sample[a]	Depth (m)	$\delta^{13}C$ (PDB)	$\delta^{18}O$ (SMOW)	$CaCO_3$ %	Rock type[c]	Lab[d]
FA 4:1	268.2	-15.7	15.3	0.05	G	2
FA 5:1	297.6	-11.8	16.0	0.20	G	2
FC 1	32.5	-14.0	19.4		G	3
FC 2	53.3	-14.2			G	3
FC 3	78.9	-15.5			D	3
FC 4	96.7	-12.4			D	3
FC 5	130.2	-7.6			G	3
FC 6	157.3	-16.7			G	3
FC 7	178.1	-16.6	18.2		G	3
FC 9	226.3	-15.4			G	3
FC 10	257.6	-11.6			G	3
FC 11	283.3	-16.5			G	3
FC 12	298.9	-13.9	13.1		G	3
Shallow borehole 4 - Stajsås						
4-167.7	167.7	-5.8	6.0	<0.01	G	1
4-232.8	232.8	-14.7	10.8	0.27	G	1
4-298.4	298.4	-11.1	14.6	0.06	G	1
4-301.1	301.1	-12.5	15.9	0.13	G	1
SA 1:1	31.7	-14.7	14.2	0.10	G	2
SA 2:1	55.0	0.0	18.0	0.05	G	2
SA 3:1	127.0	-16.3	16.2	0.70	G	2
SA 4:1	166.5	-19.4	17.5	2.00	G	2
SA 5:1	230.7	-19.6	17.3	0.50	G	2
SA 6:1	298.8	-19.8	18.2	1.50	G	2
SA 7:1	412.6	-12.0	14.2	0.20	G	2
SA 8:1	436.8	-16.3	13.1	0.01	G	2
SA 9:1	495.9			0.01	G	2
SA 10:1	527.1	-15.7	15.5	0.10	G	2
SA 11:1	642.8	-11.2	15.0	0.01	G	2
SC 1	31.7	-14.7			G	3
SC 2	45.3	-17.9			G	3
SC 3	75.4	-15.3			G	3
SC 4	97.5	-15.6			G	3
SC 5	129.1	-17.0	20.6		G	3
SC 6	155.6	-17.2			G	3
SC 7	173.9	-16.4			G	3
SC 8	200.9	-14.1			G	3
SC 9	226.4	-14.6			G	3
SC 10	253.5	-16.3			G	3
SC 11	275.1	-15.1			G	3
SC 12	301.2	-10.0	11.5		G	3
SC 15	369.4	-19.5			G	3
SC 16	408.0	-16.2			G	3
SC 17	423.0	-15.0			G	3
SC 18	451.8	-16.6			G	3
SC 19	461.4	-16.1	18.6		G	3
SC 22	552.2	-4.2			G	3
SC 23	564.6	-13.3			G	3

Table 1. (cont.)

Sample[a]	Depth (m)	$\delta^{13}C$ (PDB)	$\delta^{18}O$ (SMOW)	$CaCO_3$ %	Rock type[c]	Lab[d]
SC 24	600.7	-18.4			G	3
SC 27	653.6	-9.1	19.7		G	3
Shallow borehole 5 - Hättberg						
5-131.5	131.5	-9.8	19.3	0.02	G	1
5-145.4	145.4	-17.1	18.6	0.01	G	1
5-198.8	198.8	-9.5	16.8	0.01	G	1
5-323.3	323.3	-13.2	17.3	0.13	G	1
5-374.8	374.8	-6.2	20.8	0.02	G	1
HA 1:1	58.6			0.05	G	2
HA 2:1	148.4	-16.6	10.2	0.05	G	2
HA 3:1	222.3	-12.7	19.8	2.00	G	2
HA 4:1	302.0	-14.2	15.2	0.05	G	2
HA 5:1	347.8	-15.9	15.7	0.05	G	2
HA 6:1	426.5	-12.8	18.6	0.05	G	2
HA 7:1	453.7	-5.3	13.4	0.05	G	2
HA 8:1	495.4			0.05	G	2
HA 9:1	547.1	-11.8	19.7	0.05	G	2
HA 10:1	583.8	-14.2	18.2	0.05	G	2
HC1	34.8	-16.4			G	3
HC3	82.0	-13.3			G	3
HC4	95.1	-16.0			G	3
HC5	131.2	-14.0			G	3
HC6	149.9	-10.8			G	3
HC7	175.8	-14.4			G	3
HC8	190.2	-14.9			G	3
HC9	223.6	-14.3			G	3
HC10	244.6	-15.2			G	3
HC18	454.5	-18.9	19.1		G	3
Shallow borehole 7 - Solberga						
SoA 1:1	42.8	2.2	25.5	53.10	L	2
SoA 2:1	141.7	1.3	26.2	53.40	L	2
SoA 3:1	207.6	0.7	24.0	48.70	L	2
SoA 4:1	263.0	-8.7	17.8	0.05	G	2
SoA 5:1	310.9	-17.9	17.5	0.10	G	2
SoA 6:1	426.1	-19.4	13.3	0.05	G	2
SoA 7:1	462.2	-6.3	10.3	0.05	G	2
SoA 8:1	499.9	-13.9	11.3	0.05	G	2

[a] Samples GR1-5751 to GR1-5905 are coarse cuttings (1 cm).
[b] Analyses of replicates are shown for some Gravberg-1 samples.
[c] Rock type: G, granite; D, dolerite; L, limestone.
[d] Lab: 1 University of Wisconsin, USA; 2 Institute for Energy Technology, Oslo, Norway; 3 Global Geochemistry, California, USA.

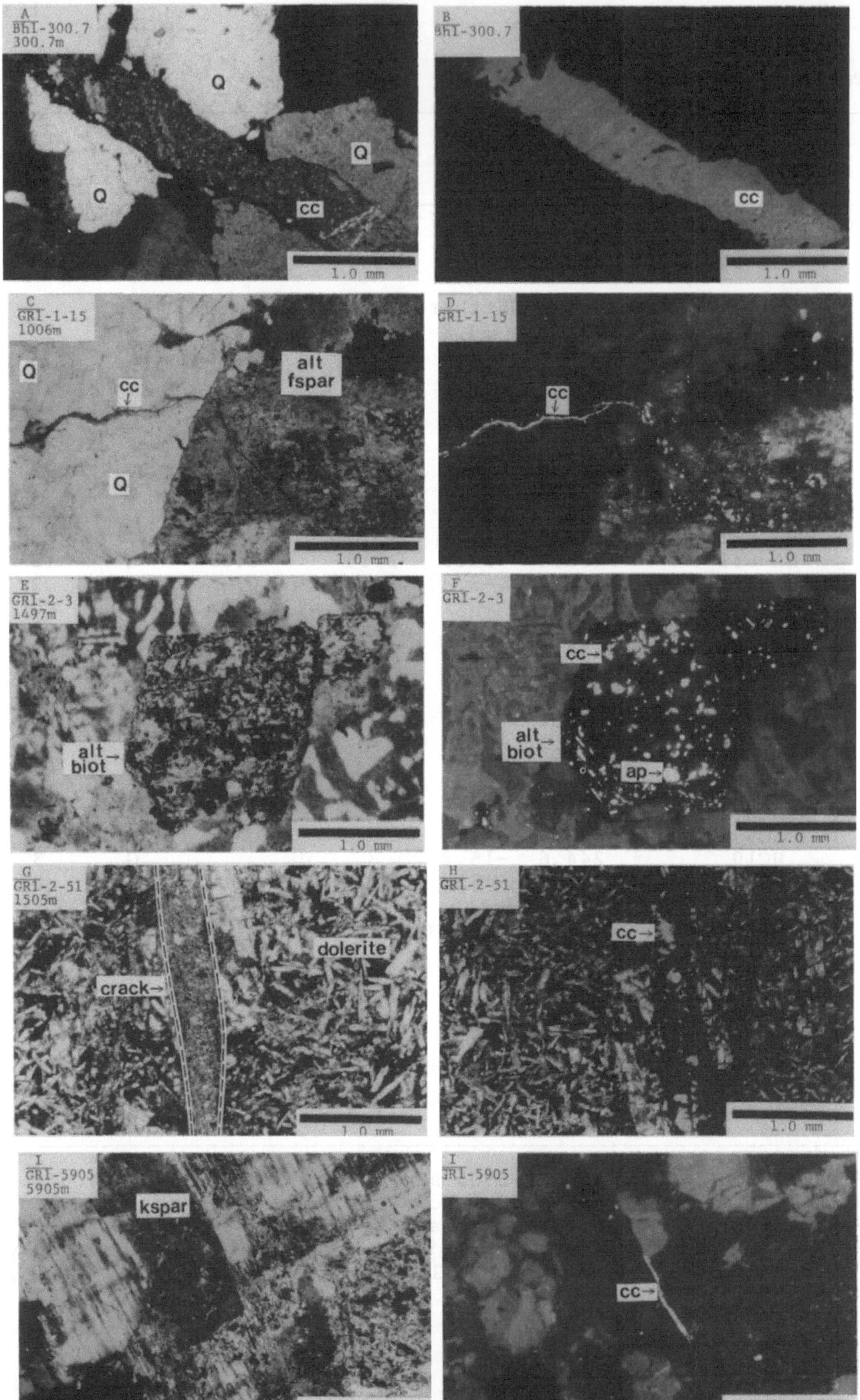

A
BhI-300.7
300.7m
Q
Q
Q
Q
cc
1.0 mm
B
BhI-300.7
cc
1.0 mm
C
GRI-1-15
1006m
Q
cc
alt fspar
Q
1.0 mm
D
GRI-1-15
cc
1.0 mm
E
GRI-2-3
1497m
alt biot
1.0 mm
F
GRI-2-3
cc
alt biot
ap
1.0 mm
G
GRI-2-51
1505m
crack
dolerite
1.0 mm
H
GRI-2-51
cc
1.0 mm
I
GRI-5905
5905m
kspar
0.5 mm
I
GRI-5905
cc
0.5 mm

Petrography of Calcite Occurrences

Core and surface samples of granite and dolerite contain calcite-healed cracks and patches of calcite replacing primary phases. Figure 2 shows some examples of calcite occurrences in samples recovered from Gravberg-1. Patchy calcite replacing primary biotite and amphibole is always accompanied by chlorite. Calcite replacing feldspar occurs with fine-grained mixtures of sericite and albite. Patches of calcite replacing primary phases are fed by networks of crack-healing calcite. Where calcite-healed cracks traverse quartz, the boundaries between calcite and quartz are sharp. Upon entering grains of feldspar of mafic minerals, crack boundaries usually become more vague and eventually the crack may terminate in a patch of replacement calcite. This suggests that the texture of calcite may be controlled by the reactivity of the host phase. Where carbonate-bearing solutions flow through unreactive quartz, calcite precipitates only within the boundaries of the crack. Where the solutions encounter more reactive feldspars and mafic minerals, portions of the host phases are removed in solution while calcite and other secondary minerals are deposited as patchy replacement phases.

Measurements of CO_2 liberated by acid treatment show that the abundance of calcite decreases with depth in Gravberg-1 (Fig. 3). In thin section, the decrease of calcite abundance with depth is reflected by decreases in: the length and width of cracks healed by calcite; the average density of cracks; the average number of grains replaced by calcite in a thin section; and the average size of calcite patches in altered minerals.

Sample Selection and Contaminants

Many of the early carbonate analyses from the deep drill hole at Gravberg were from silt-size cuttings. These cuttings are comparatively free of carbon-bearing impurity at depths of 0 to 4.2 km when the drilling fluid was water and few additives were used (Jeffrey et al. 1987). However, occasional cementing

---

Fig. 2 a-i. Calcite occurrences in drill core from Gravberg-1 and the shallow Gravberg pilot hole. Photos in **left-hand column are taken under plain or polarized light. Photos in right-hand column** are taken under cathode luminescence. Sample names and depths are shown in the **upper left-hand corner** of each photo. Note the general decrease in abundance of calcite (cc) with depth. **a**, b Calcite-healed crack in quartz (Q) from granite. c, d Calcite-healed crack in quartz from granite feeding small patches of calcite in altered feldspar (**alt fspar**). **e**, f Altered biotite (**alt biot**) in graphic textured granite. Biotite is replaced by chlorite, calcite, and opaques. Apatite (**ap**) is a primary phase included in nearly all Gravberg-1 biotites. **g**, **h** Dolerite cut by a crack filled with granulated mineral fragments and patches of calcite. Sample is from the Rl seismic reflector. **i-left, i-right:** Altered potassium feldspar containing a hairline crack healed by calcite

Table 2. Isotopic analyses of calcite in surface samples (Tab. 1)

Sample	^{13}C (PDB)	^{18}O (SMOW)	Wt% $CaCO_3$	Radial Dist (km)	Lithology and comments
87SJN2a	-3.3	14.4	1.71	17.7	Digerberg sandstone
87SJN3a	-3.0	17.5	0.18	16.7	Fresh Dala granite
87SJN4b	-2.8	12.9	0.12	13.1	Aplitic dike in Siljan granite
87SJN5a	-0.2	18.9	<0.01	12.6	Quartz breccia
87SJN7b	-1.4	19.2	5.15	16.7	Boda Pb-Zn ore
87SJN8b	-3.1	22.6	0.03	6.4	Shattercone in Järna granite
87SJN9c	-14.2	14.1	0.01	9.6	Shattercone in Järna granite
87SJN10a	-5.5	19.2	0.04	30.3	Siljan granite near Siljan/Järna contact
87SJN11	-4.1	16.1	1.03	20.6	Järna granite in annular moat
87SJN12a	-2.7	18.6	0.12	21.5	Järna granite in annular moat
87SJN14b	-1.0	15.4	0.40	11.6	Siljan granite near dolerite/granite contact
87SJN15b	4.8	25.6	0.15	5.1	Granite near pseudotachylite
87SJN15f	-5.7	11.4	9.41	5.1	Pseudotachylite
87SJN19a	-11.5	24.1	0.04	12.3	Järna granite near limestone block
87SJN20a	2.8	25.5	0.19	12.8	Migmatite
87SJN21c	-4.3	11.0	0.03	13.4	Järna granite
87SJN22e	-5.9	13.7	0.06	13.7	Siljan granite with vugs
87SJN24	-3.0	19.9	0.01	23.2	Dala granite N of Ring
87SJN25a	-2.0	16.9	0.04	24	Dala granite N of Ring
87SJN26	2.0	23.8	0.02	24.7	Dala granite N of Ring
87SJN27	-8.7	24.1	<0.01	14.5	Siljan granite near annulus
87SJN28	-9.5	3.0	0.01	14.5	Siljan granite near annulus
87SJN29	-6.7	19.5	0.01	13.2	Siljan granite near annulus
87SJN31a	2.2	24.5	0.15	15.9	Porphyry
87SJN32a	-7.2	16.3	0.07	22.2	Red porphyry at Storstupet
87SJN33b	-11.6	25.9	0.07	22.4	Red porphyry at Storstupet
87SJN35b	-8.0	17.4	0.03	30.3	Dala granite N of Ring
87SJN36b	-3.0	22.5	0.03	20	Dala granite E of Ring
87SJN36e	-7.5	15.0	0.44	20	Dala granite E of Ring
87SJN37b	-10.4	28.1	0.09	2.3	Dolerite near pseudotachylite
87SJN37f	-22.8	10.8	0.01	2.3	Granite near same pseudotachylite
87SJN38	-4.0	9.5	1.19	29.4	Dala granite W of Ring
87SJN39a	-5.2	14.5	0.01	15.4	Granite near granite/sediment contact at Stygforssen
87SJN39m	-11.2	9.4	0.02	15.4	Granite cut by quartz veins at Stygforssen
87SJN40b	-1.0	17.1	0.02	67	Järna granite S of Ring
87SJN41b	-3.5	13.3	0.08	67	Järna granite S of Ring
87SJN42e	-5.5	8.8	2.35	64.8	Järna granite S of Ring
87SJN43a	-6.1	10.5	0.05	60.1	Järna granite S of Ring

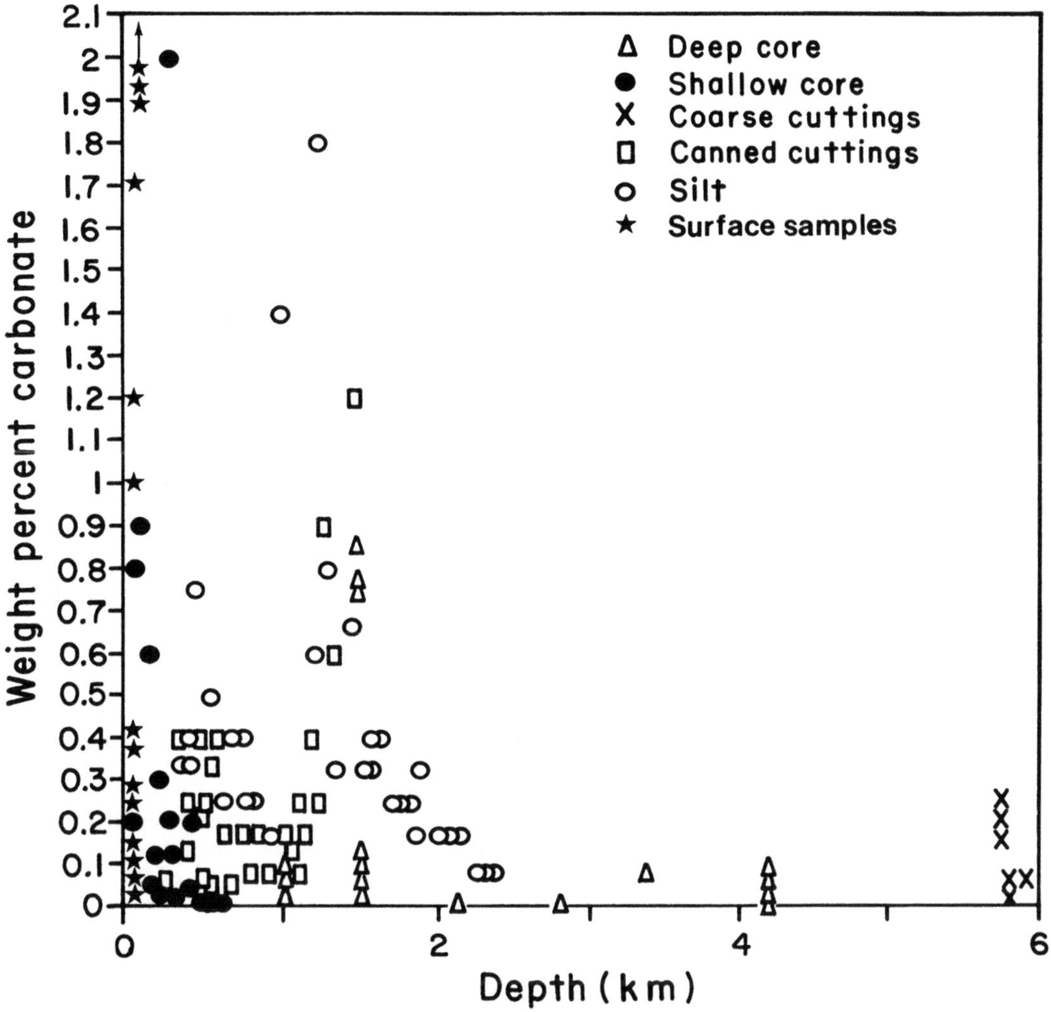

Fig. 3. The percentage of calcite in granite and dolerite samples from the surface and the pilot and deep holes. One surface granite sample from outside the Ring plots off the vertical axis at 2.4 wt.% calcite, as do two unusual lithologies (Pb-Zn ore (5.2 wt.%) and pseudotachylite (9.4 wt.%)). Many surface samples plotting at <0.1% carbonate are omitted for clarity. The amount of calcite drops to below 0.2 wt.% at 1.5 km, approximately the depth of the first dolerite sill (60 m thick, seismic reflector #1)

operations result in visible contamination of some samples and all samples contain large amounts of metal filings from the drill. It is estimated that 20,000 lbs of metal were abraded from the drill string per month of drilling with water (T. Brittenham, pers. comm., 1987). In the laboratory, reaction of iron with phosphoric acid produces large amounts of hydrogen gas at 25°C. The presence of impurities compromises the silt data even when attempts are made to avoid or remove impurities.

Accordingly, present and future analyses are only being made on coarse (>1 cm) cuttings, drill core, and surface samples.

The presence in drill cuttings of large amounts of fine metal particles with large surface area is also significant in light of the trace levels of H_2 and CH_4 that have been found in drilling fluids at Siljan. At elevated down-hole temperatures, it is clear that iron reacts with drilling fluid and H_2 gas may result. The amounts of carbon found in calcite and metal are more than sufficient to explain the measured CH_4 concentrations as well, but the importance of this process is not known (eg. Gold 1987, Jeffrey et al. 1987).

Calcite Concentration

Figure 3 shows the concentration of calcite in surface samples, and in sand or silt-sized cuttings and drill cores, from both the shallow and deep holes at Gravberg (Gravberg-1 and Bhl in Fig. 1). Granite samples from 0 to 1500 m contain from 0 to 2.4 wt.% calcite, but concentrations drop off sharply to below 0.2 wt.% in the deeper samples. (Two unusual lithologies (Pb-Zn ore and pseudotachylite) have calcite concentrations as high as 9.4 wt.%). This transition roughly corresponds to a 60-m-thick dolerite sill at 1500 m (seismic reflector #1) and is slightly below the contact of Siljan and Järna granites (~1150 m). It appears that these features have influenced the post-impact hydrologic system that was responsible for precipitation of calcite.

A total of 8.5 m of 8 cm diameter drill core was retrieved from six intervals in the deep hole at Gravberg. These intervals are shown by X's in Fig. 3. Limited deep samples have also been obtained at seven intervals by side-wall coring, but quantities have not permitted isotopic analysis. The results in Fig. 3 suggest that cuttings are on average more calcite-rich than cores, raising the possibility that coring preferentially recovers samples of less fractured rock.

Isotopic Compositions

Surface Samples

Isotopic data for surface samples must be interpreted cautiously because these samples are the most likely to have reacted with modern-day surface water. Surface samples have $\delta^{13}C$=-22.8 to 4.8, and $\delta^{18}O$=+3 to 28.1 (Fig. 4). Anomalous calcite (with $\delta^{13}C$ <-10; see discussion below) is present in six samples inside the Ring, but the two most anomalous values (-22.8 and -14.2) we found within 10 km of the geometric center.

Silt vs Core

Figure 5 shows values of $\delta^{18}O$ and $\delta^{13}C$ for samples from Siljan and Fig. 6 shows values only for the drill core. Values of

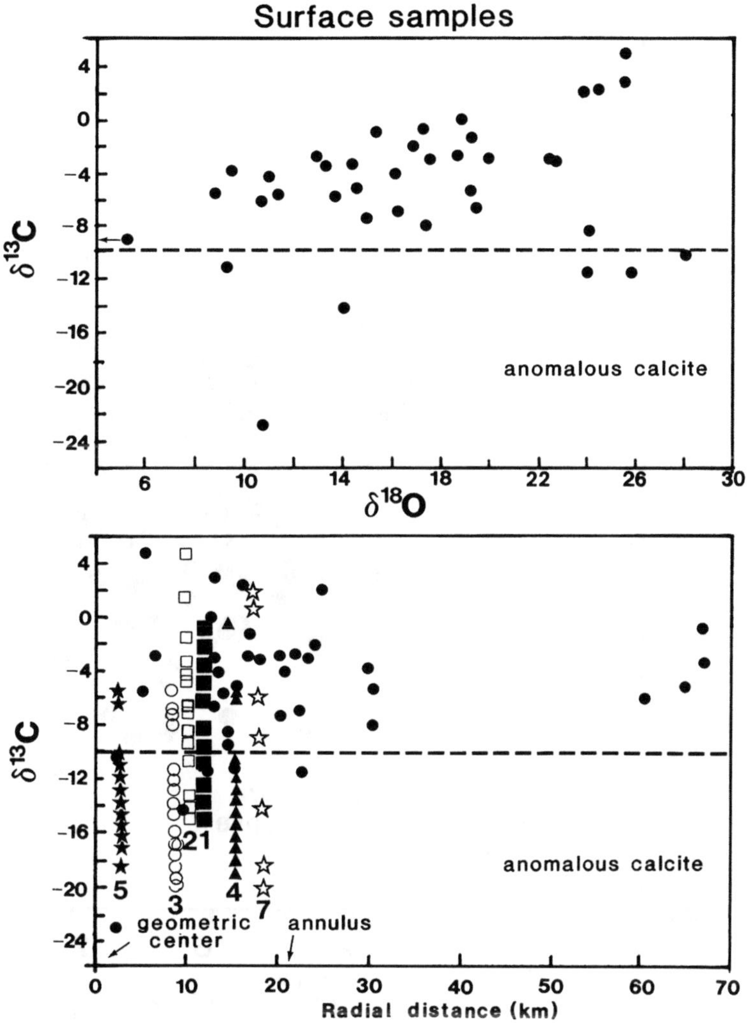

Fig. 4. $\delta^{13}C$ vs $\delta^{18}O$ in calcite (top) and radial distance from the geometric center of the Siljan Ring (bottom). Also shown in the lower plot are $\delta^{13}C$ values in calcites from the shallow boreholes. The numbers of the shallow boreholes (1-5 and 7) are shown at the base of the corresponding data columns. The radial distance between shallow boreholes 1 and 4 is exaggerated for clarity. The annular moat encircles the Ring at a distance of 20-25 km

$\delta^{18}O$ and $\delta^{13}C$ are higher in silt, on average, than in core. It is not known if these differences reflect contamination of silt or real differences due to preferential sampling of heterogeneous rocks by the drilling process. In the future, analysis will be made of individual veinlets to evaluate the possibility of the many different calcite generations. The present discussion will concentrate on core samples so as to avoid ambiguity.

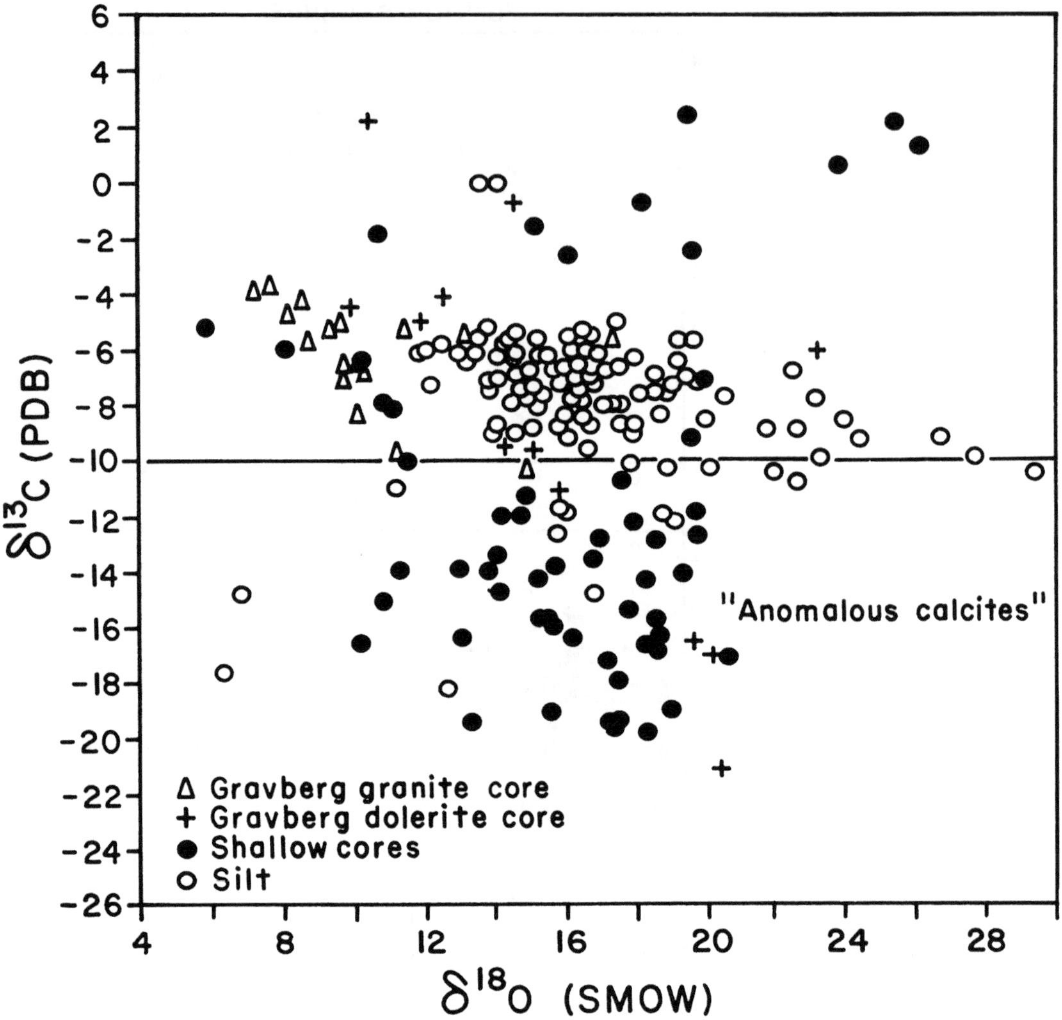

Fig. 5. Values of $\delta^{13}C$ vs $\delta^{18}O$ for carbonates in all Siljan samples. Samples with $\delta^{13}C$ <-10 have been proposed to be anomalous and to represent oxidation of mantle-derived methane. Some overlapping data points have been omitted for clarity

Anomalous Carbonates

Values of $\delta^{13}C$ vs $\delta^{18}O$ are plotted in Fig. 7 for samples from six of the shallow pilot holes. In four of the cores, carbon isotopic ratios span the full range from $\delta^{13}C$ = -19.4 to +8.4, but in the other two, values are dominantly in the "anomalous" range of <-10‰. It can be seen in Fig. 1 that the two drill holes with "anomalous" values, numbers 3 and 5, are also the two that are nearest the center of the crater. In fact, Grieve (1987) concludes that the diameter of the excavated crater was only 26 km, approximately one-half of the diameter of the ring.

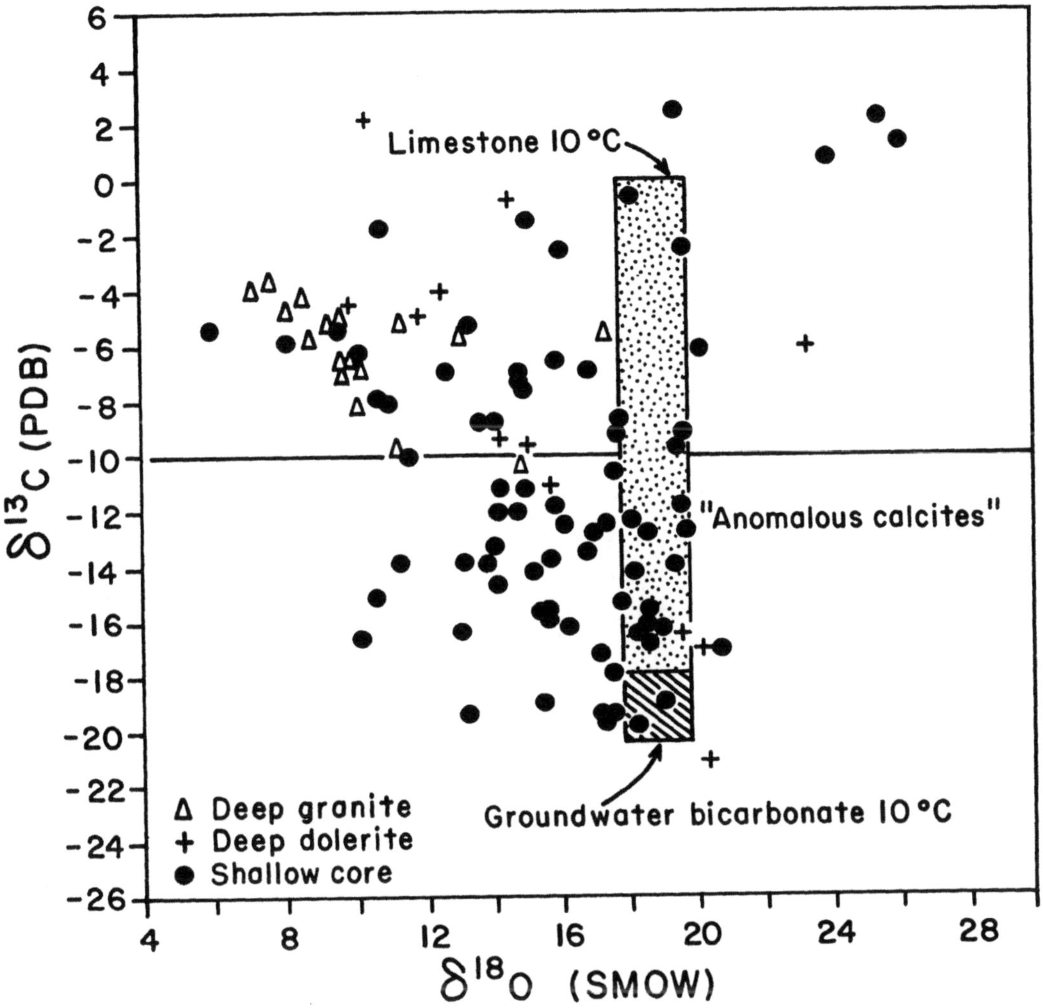

Fig. 6. Values of $\delta^{13}C$ and $\delta^{18}O$ for carbonates in drill cores and coarse cuttings at Siljan. Silt has been omitted as explained in text. The majority of "anomalous calcites" are from shallow drill cores and very few are from Gravberg. The "groundwater bicarbonate" box represents the composition of calcite that would be in equilibrium at 10°C with organically-derived groundwater samples such as were collected at 0-684 m (depth) in the Taavinunnanen gabbro, northern Sweden (Larsen and Tullborg 1984). The values near $\delta^{13}C = 0$, in the "limestone" box represent calcites precipitated at 10°C by groundwater that has exchanged with limestone. Thus, the full range of $\delta^{13}C$ values is explicable in terms of groundwater mixing from two surface reservoirs; organic decay and limestone dissolution

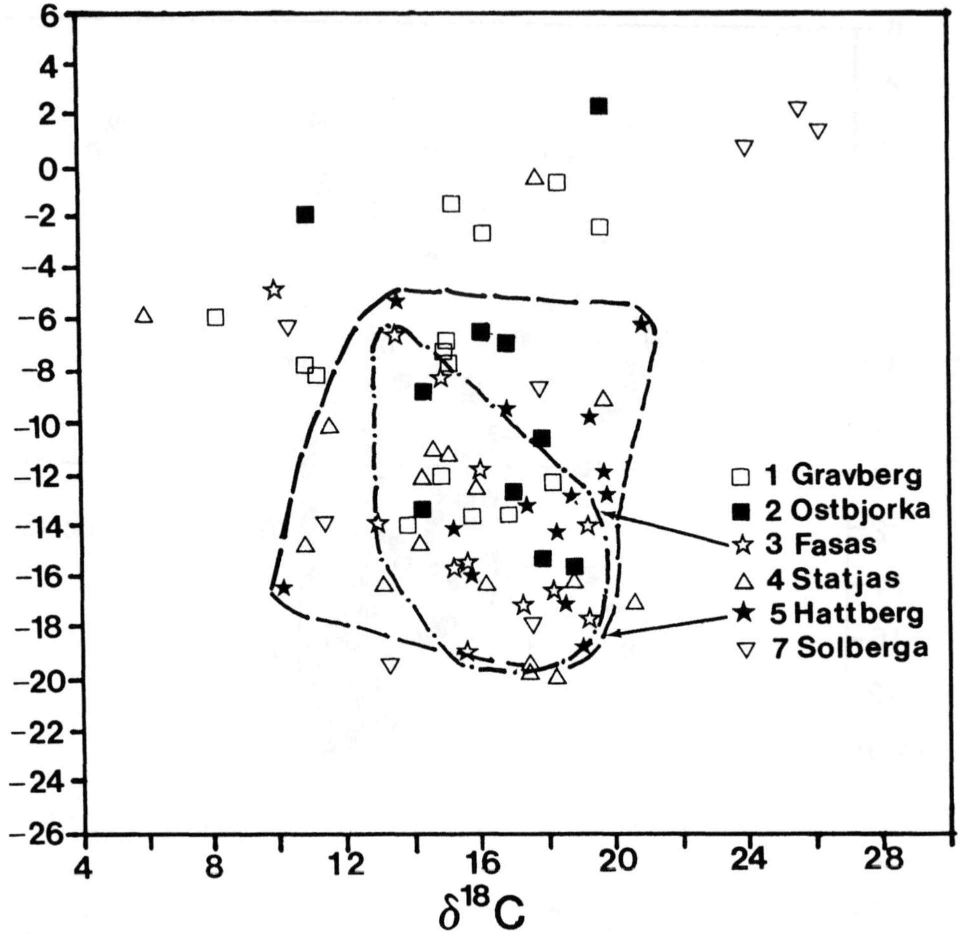

Fig. 7. Values of $\delta^{13}C$ and $\delta^{18}O$ for calcite from shallow drill cores at Siljan. Deep samples are omitted. Calcites from holes 3 and 5 (Fåsås, Hättberg, respectively) are almost exclusively in the "anomalous" range, whereas holes 1, 2, 4, and 7 (Gravberg, Östbjörka, Stajsås, Solberga) include all values. The low $\delta^{13}C$ samples correlate with the center of the Siljan ring (Fig. 1) suggesting that groundwater is dominated by organic decay and samples from the periphery of the Ring have both high and low $\delta^{13}C$, consistent with limestone dissolution and groundwater mixing

Thus, a striking correlation exists; "anomalous carbonates" are largely found in samples from beneath the original area of excavation.

Two hypotheses may explain the correlation of $\delta^{13}C$ and position within the crater: (1) If the center of the Ring is most thoroughly fractured by impact and most permeable, then the best access is provided for deep-seated fluids. Conversely, the center may actually be more compacted and less permeable (Grieve 1987). (2) The center of the ring is the area furthest from Paleozoic limestones which are restricted to the annulus of the ring (Fig. 1). If the source of carbon is exclusively shallow, then around the crater margin groundwater will contain a mixture of carbon from organic decay and carbon from limestone dissolution with the result that $\delta^{13}C$ values will be highly variable from -20 (organic) to 0 (limestone). Samples from the center are furthest from limestone and would thus be expected to have the lowest values of $\delta^{13}C$.

Oxygen Isotopes

Any consideration of the origin of these isotopic values requires knowledge of the temperature of formation. Figures 4-6 show many values of $\delta^{18}O$ that are >+15. These values can only form at low temperatures where isotopic fractionations are large.

Values of $\delta^{18}O$ are plotted versus depth for calcite from drill core samples in granite (Fig. 8). The general trend is decreasing $\delta^{18}O$ with depth. The scatter in values is not surprising considering the complex nature of the samples. The "meteoric water" bar shows the calculated $\delta^{18}O$ of calcite in equilibrium with modern-day meteoric water for the Siljan area (-12 to -14) with a temperature of 10°C at 0 km increasing to 70°C at 6 km. This geothermal gradient (10°C km^{-1}) is cooler than that measured at Gravberg (17°C km^{-1}). This hotter geotherm results in T = 100°C at 6 km and calculations at this temperature yield $\delta^{18}O(Cc)$ = 3 to 5. Thus, the $\delta^{18}O$ values in calcite can be explained by this range in temperature. Slight isotopic shifting by water/rock interaction may account for some scatter in values.

It is also possible that the $\delta^{18}O$ values near 10 (deeper than 1.5 km) may result from higher temperature hydrothermal processes such as are proposed by Komor et al. (1987), based on fluid inclusion evidence. Such fluids would likely be in equilibrium with silicate minerals in the granite. Five samples of granite have wholerock $\delta^{18}O$ values that are 7.5 to 9.1‰ (Table 3) suggesting that the meteoric water/rock ratio was not high. Thus, the upper 3 km at Gravberg shows low temperature calcites, consistent with precipitation from modern meteoric water at ambient temperatures (10-100°C). The deeper samples may have a similar origin or may be hydrothermal (200-400°C).

The deep penetration of oxidizing groundwater at Siljan is also suggested by the color of granite in drill cuttings. Shallow samples are deep red due to the presence of hematite in feldspar and along cracks. The red color gradually changes to gray at depth. Occasional zones of red between 4.5 and 6 km suggest that groundwater oxidation may occur in restricted areas.

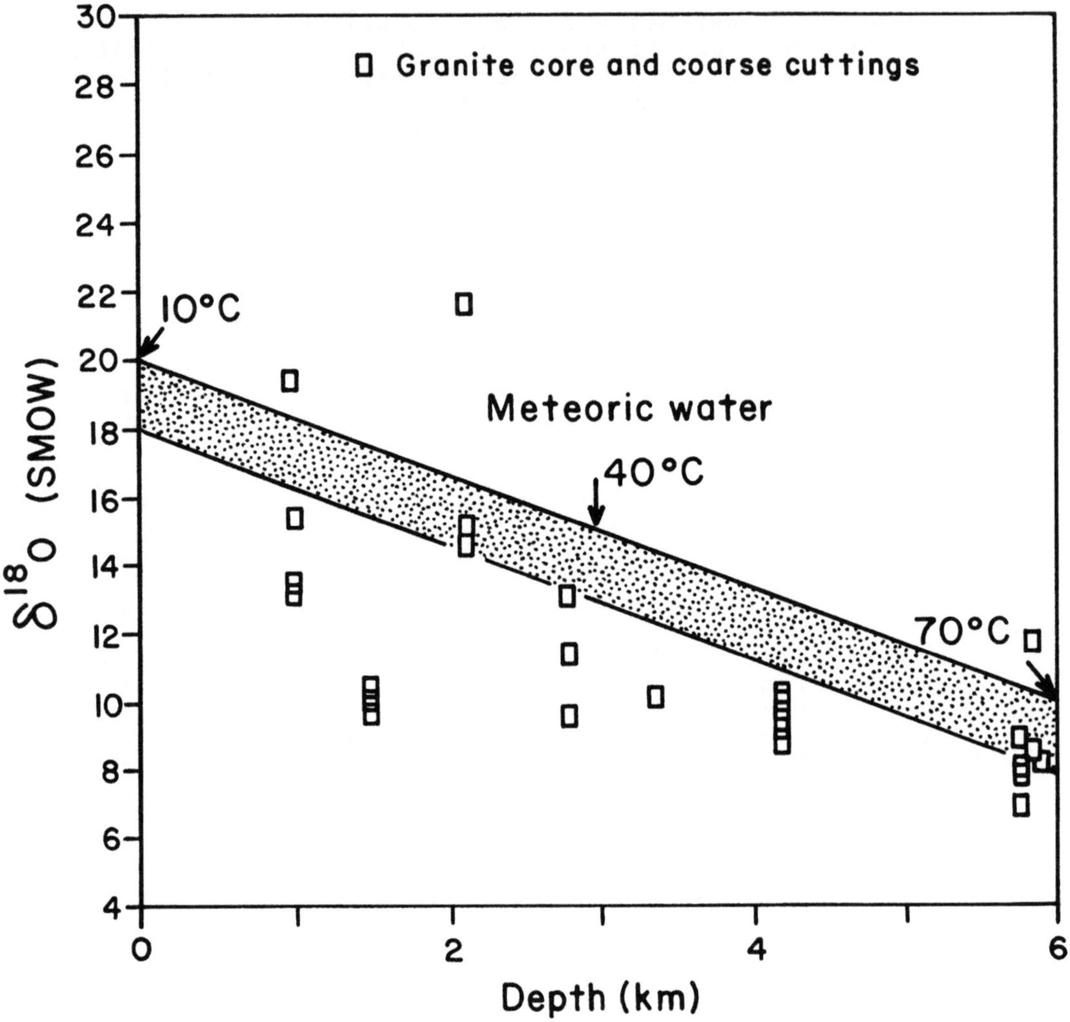

Fig. 8. Values of $\delta^{18}O$ decrease with depth for calcite from granite samples in the deep drill hole at Gravberg. The "meteoric water" bar represents the composition of calcite precipitated by modern meteoric water in central Sweden assuming 10°C at the surface and 70°C at 6 km depth. This geotherm, 10°C km^{-1}, is slightly lower than that measured down-hole at Gravberg (~17°C km^{-1}). If T = 100°C at 6 km, then the calculated $\delta^{18}O(Cc)$ = 3 to 5 and the "meteoric water" bar has a steeper negative slope. Samples from 0-3 km, with $\delta^{18}O$ >15, must be of low temperature origin, however, deep samples with $\delta^{18}O$ ~10 may have precipitated either from meteoric water at 50-100°C or from hydrothermal fluids at higher temperature

Table 3. Oxygen isotopic values of silicates

Samples	Depth (m)	Material Analyzed	$\delta^{18}O$ (SMOW)	Comments
GR1-1-16	1006	feldspar	9.1	60% feldspar/35% quartz/5% biotite; chlorite-filled cracks; cloudy feldspar.
GR1-2-9	1498	feldspar	8.6	55% feldspar/15% quartz/ 28% biotite/2% opaques & apatite; chlorite replacing biotite; cloudy feldspar.
GR1-4-7	2804	feldspar	7.7	62% feldspar/35% quartz/3% mafics; parallel cross-cutting fractures filled with granulated minerals; chlorite replacing mafics.
GR1-5-2	3376	feldspar	7.7	50% feldspar/45% quartz/5% mafics; fractured rock with grain boundary granulation; quartz and epidote-filled veins; cloudy feldspar.
GR1-6-1	4206	whole rock	7.5	62% feldspar/30% quartz/8% mafics & accessories; fresh, undeformed.

Carbon Isotopes

Samples in Fig. 6 include many in the "anomalous" range with $\delta^{13}C$ below -10. It is important to understand the genesis of these values and also why the majority of low $\delta^{13}C$ samples are from shallow pilot holes and no granites from the deep Gravberg hole have $\delta^{13}C$ (Cc) <-14.

The box labeled "groundwater bicarbonate" in Fig. 6 represents the composition of calcite that would be in equilibrium at 10°C with samples of groundwater from a drill hole in the Taavinunnanen gabbro from near Kiruna in northern Sweden. These waters were sampled at depths of 0 to 684 m and contain dissolved bicarbonate with $\delta^{13}C$ = -18.7 to -21.2 (Larson and Tullborg 1984). The low $\delta^{13}C$ values reflect a carbon source that is dominated by surface-derived organic matter (Salomons and Mook 1986). The top of the box labeled "limestone" in Fig. 6 represents equilibrium at 10°C of present-day meteoric water with the Paleozoic limestones from the annulus of the Siljan Ring. Thus, the full range of isotopic values in calcites at Siljan can be explained by mixing of groundwater at low temperature and it is not necessary to invoke deep-seated or anomalous sources of carbon.

In order to evaluate further the hypothesis that low $\delta^{13}C$ calcites reflect precipitation from mantle-derived methane, it is instructive to consider what the composition of mantle methane should be. The average $\delta^{13}C$ of carbon in the mantle is well known to be near -5 from studies of diamonds, kimberlites, and igneous rocks (Kyser, 1986). Methane, at mantle temperatures of >1000°C, will have $\delta^{13}C \approx -7$. It is important not to confuse such mantle methane ($\delta^{13}C \approx -7$) with thermogenic methane ($\delta^{13}C$ = -30 to -50) which is largely derived by thermal breakdown of organic matter. Mantle methane at $\delta^{13}C \approx -7$ will not form carbonate cements with $\delta^{13}C$ = -10 to -20 except in unusual and rare circumstances.

The fact that most metamorphic rocks from the deep crust have not equilibrated with a methane-rich fluid (they are too oxidized) indicates that large amounts of deep earth gas does not rise **pervasively** from the mantle (Valley and Lamb, 1986). Thus, methane can only be proposed as prevalent along **localized** zones in the deep crust; any migration routes would be methane-dominated; and deep-seated isotopic modification by rock interaction cannot cause significant shifts in $\delta^{13}C(CH_4)$ for the large methane flux proposed by Gold (1987). In the near-surface environment, if methane becomes oxidized, the equilibrium precipitation of calcite will thus yield $\delta^{13}C$ values in calcite of over +40. This process will drive $\delta^{13}C$ of CH_4 to lower values as proposed by Gold (1985). If calcite is precipitated gradually then isotopic values can be modeled as a Rayleigh process (Valley, 1986) and it is possible that a large range of $\delta^{13}C$ values will result with the extreme values of $\delta^{13}C$ (calcite) extending down to negative values. However, the law of mass balance provides a critical constraint on this process. The majority of calcite precipitated in equilibrium with mantle methane at 10-100°C will have $\delta^{13}C \geq 0$. Low values of $\delta^{13}C$ <-10 would thus be rare and can only result after most methane has been precipitated. Such low values of $\delta^{13}C$ would only be produced in rock-dominated systems where the quantitites of CH_4 were small.

An alternate mechanism of methane oxidation by microbial metabolism might be hypothesized in a high porosity, low temperature environment. In this model, the fractionation of carbon isotopes in calcite may show an isotopic reversal, may not attain equilibrium, and $\delta^{13}C$ (calcite) may be lower than that of methane. However, mass balance must also apply and $\delta^{13}C$ of methane will be driven toward high values (>-7). It is significant that all methane sampled in granite at Siljan has $\delta^{13}C$ = -38 to -16 (Jeffrey et al. 1987, this Vol.) indicating either that the methane is not mantle-derived or that the disequilibrium precipitation of low $\delta^{13}C$ calcite is not an important process at Siljan. In any circumstance, there is no simple process whereby the measured range of carbon isotope values at Siljan can be derived from mantle methane. Furthermore, the prediction of Gold (1985, p. 105), that "isotopically heavier ($^{13}C/^{12}C$) carbonates will be found at deeper levels" in the Gravberg-1 well, has not been verified.

Hydrothermal Fluid Convection

Homogenization temperatures of 135-333°C have been measured at the Siljan Ring in low salinity, aqueous fluid inclusions from quartz in granites (Komor et al., 1987) and in quartz-cemented breccia (Lindblom and Wickman, 1985). These high temperatures are believed to have existed after the time of meteorite impact, in which case the source of heat is likely to be impact-related as no Phanerozoic igneous activity is known in this part of the Scandinavian Shield and pre-impact limestones are petroliferous and unmetamorphosed. It is well documented that hydrothermal convection can result from thermal gradients such as are indicated by the fluid inclusion temperatures (Criss and Taylor 1986) and hydrothermal alteration is described in other meteorite impact craters (Newsom et al. 1986).

Post-impact erosion at Siljan is believed to have removed about 1 km of rock (Grieve 1987) and thus, any post-impact fluids would have been shallow and were likely dominated by meteoric water. The effects of heated meteoric water on hydrogen and oxygen isotopic ratios in granite are amply documented (see Criss and Taylor 1986) and are dependent on both temperature and latitude. However, the mass transport effects of convection are less well known, yet are of more importance to the goals of the Swedish Drilling program. Large-scale, shallow convection would be accompanied by hot spring activity and surface recharge. The possibility exists that near-surface material, including carbon, would have been carried to depth within the granite, accounting for some of the unusual aspects of the area.

This chapter represents an interim report and work is in progress to investigate further the complexities of fluid history in these rocks. Surface samples have been collected, both inside and up to 60 km outside of the Ring to evaluate effects that are not impact-related in comparison to those that show a spatial relation to the crater. Microsampling of calcite veins is proceeding by dental drill and laser beam to document the various generations of calcite veins. It is hoped that strontium isotopic analysis of calcite will test the model for groundwater mixing at Siljan (Smalley et al., 1988). Oxygen and hydrogen isotopic ratios in silicates from igneous rocks and veins will restrict the temperature of hydrothermal exchange, quantities of water/rock, and the geometry of the plumbing system.

Acknowledgments. This research has been supported primarily by the Gas Research Institute (5086-260-1425). Additional support has come from the U.S. National Science Foundation (EAR 85-08102). We thank John Castano, Bengt Collini, Ferol Fish, Tony Gorody, Dick Karlson, Tord Lindbo, and Paul Westcott for their assistance and thought provoking conversations. Helpful reviews of this manuscript were made by Mary Anderson, Jean Bahr, and Tony Gorody.

References

Bottomly RJ, York D, Grieve RAF (1978) $^{40}Ar-^{39}Ar$ ages Scandinavian impact structures: I, Mien and Siljan. Contri Mineral Petrol 68:79-84

Criss RE, Taylor HP (1986) Meteoric-hydrothermal systems. In: Valley JW, Taylor HP, O'Neil JR (eds) Stable isotopes in high temperature geological processes, M.S.A. Reviews 16, 373-424

Gold T (1985) The Siljan deep gas project: evaluation of results to date. Vattenfall Gas Project G.2 Rept No. 4, January 1985, 100-112

Gold T (1987) Power from the Earth, JM Dent and Sons, 208 p

Gold T, Soter S (1980) The deep-earth gas hypothesis. Sci Am, 154-161

Grieve RAF (1987) The formation of large impact structures and constraints on the Siljan impact structure. This volume

Jeffrey AWA, Kaplan IR, Castano JR (1987) Analysis of gases in the Gravberg-1 well, this Vol

Komor S, Valley JW, Brown PE (1987) Petrologic results from deep drilling in the Siljan Ring structure, Sweden: petrography and fluid inclusions, this Vol

Kyser TK (1986) Stable isotope variations in the mantle. In: Valley JW, Taylor HP, O'Neil JR (eds) Stable isotopes in high temperature geological processes, M.S.A. Reviews 16, 141-164

Lamb WM and Valley JW (1985) C-O-H calculations and granulite genesis. In: Tobi AC, Touret JLR (eds) The deep proterozoic crust in the N. Atlantic provinces, Reidel, 119-131

Larson SA, Tullborg EL (1984) Fracture fillings in the gabbro massif of Taavinunnanen, northern Sweden. Svensk Kärnbränslehantering AB Tech Rept 84-08, 38 p

Lindblom S, Wickman FE (1985) Fluid inclusions in quartz from a quartz breccia in the Siljan Ring structure, central Sweden. Geologiska Föreningens i Stockholm Förhandlingar 107:53-58

Mattioli GS, Wood BJ, Carmichael ISE (1987) Ternary-spinel volumes in the system $MgAl_2O_4$ - Fe_3O_4 - $Fe_{8/3}O_4$: implications for the effect of P on intrinsic fO_2 measurements of mantle-xenolith spinels. Am Mineral 72:468-480

McCrea JM (1950) On the isotope chemistry of carbonates and a paleotemperature scale. Jour Chem Physics 16:849-857

Newsom HE, Graup G, Sewards T, Keil K (1986) Fluidization and hydrothermal alteration of the suevite deposit at Ries Crater, W. Germany, and implications for Mars. J Geoph Res 91, no. B13, E239-251

Salomons W, Mook WG (1986) Isotope geochemistry of carbonates in the weathering zone. In: Fritz P, Fontes JCh (eds) Handbook of environmental isotope geochemistry 2:239-269

Sharma T, Clayton RN (1965) Measurement of $^{18}O/^{16}O$ ratios of total oxygen of carbonates. Geochim Cosmochim Acta 29:1347-1353

Smalley PC, Blomqvist R, Råheim A (1988) Sr isotopic evidence for discrete saline components in stratified groundwaters from crystalline bedrocks, Outokumpu, Finland. Geology 16, 354-357

Tullborg EL (1986) Fissure fillings from the Klippearas study site. Svensk Kärnbränslehantering AB Tech Rept 86-10, 59 p

Valley JW (1986) Stable isotope geochemistry of metamorphic rocks. In: Valley JW, Taylor HP, O'Neil JR (eds) Stable isotopes in high temperature geological processes, M.S.A. Reviews 16 455-490

Valley JW, Lamb WM (1986) The deep earth gas hypothesis: a metamorphic test. Gas Res Inst Report 86/0234, 221 p

Fluid Inclusions in Granite from the Siljan Ring Impact Structure and Surrounding Regions

S. C. Komor[1], J. W. Valley[1], P. E. Brown[1] and B. Collini[2]

Introduction

Meteorite impacts impart large amounts of energy to target rocks (Dence et al. 1977). Temperature and pressure distributions in impact craters can be mapped based on the distribution of melted and shocked rocks. Another line of evidence which potentially can be used to characterize the thermal effects of terrestrial bolide impacts are the properties of fluid inclusions contained within target rocks. Trapping temperatures of fluid inclusions may record evidence for a thermal aureole generated by the impact. The Siljan Ring structure is a meteorite impact crater formed in late Devonian time (362 Ma; Bottomley et al. 1978). Recent drilling in the granitic basement that forms the crater floor has made available samples from depths to 6000 m. These samples, supplemented by surface samples from inside and outside the crater, make possible three-dimensional studies of impact crater characteristics. This report documents the occurrence of fluid inclusions in granites from the Siljan Ring and adjoining regions. Trapping temperatures of fluid inclusions from shallow depths within the impact crater provide evidence for an impact-generated thermal aureole. The dimensions of the impact-generated aureole are large, extending to depths of more than 1 km in basement, and to the edge of the 45-km-wide crater. Outside the crater and at depths below the impact-generated aureole, fluid inclusion trapping temperatures appear to show evidence for hydrothermal activity related to Precambrian granitic plutonism.

Regional Precambrian Geology

The geology of the Siljan Ring and adjoining regions is shown in Fig. 1 (Helmqvist 1966). The oldest rocks in this part of south-central Sweden date from the 1750-2000 Ma Svecokarelian orogeny. Recent workers have attempted to fit the various features of this orogeny into a plate tectonic framework involving generation of volcanics and granites above a subduction zone

---

1 Department of Geology and Geophysics, University of Wisconsin, Madison, WI 53706, USA
2 Department of Mineralogy and Petrology, Uppsala University, Box 555, 751 22 Uppsala, Sweden

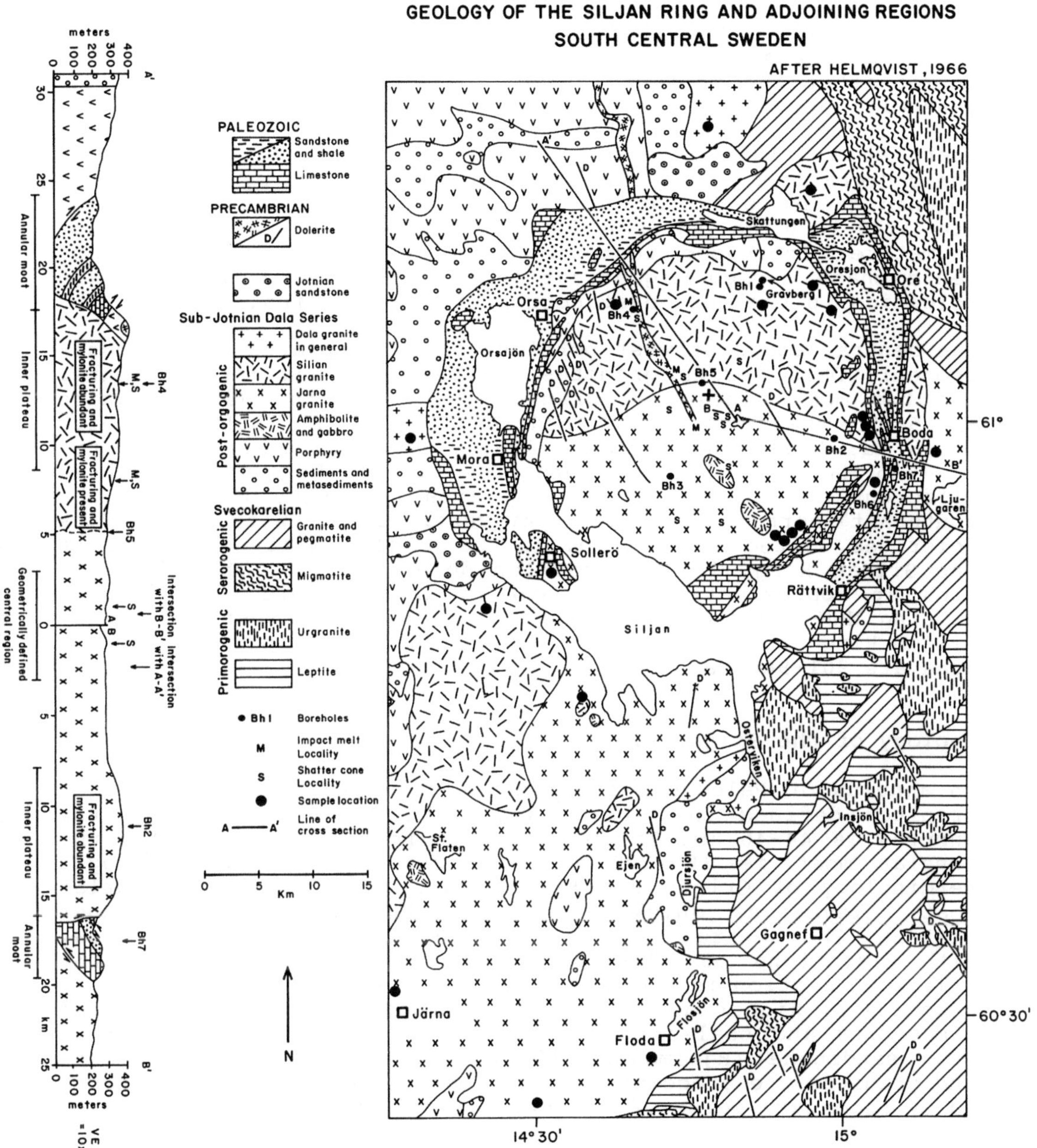

Fig. 1. Geology of the Siljan Ring and vicinity, simplified after Hjelmqvist (1966). The locations of shallow boreholes (Bh1-Bh7) Gravberg 1, and surface samples (dots) are shown. The geometric center of the Ring is shown by a "+"

(e.g. Lofgren 1979, Lundqvist 1979, Patchett et al. 1987). Rocks formed in the early or late stages of the orogeny are termed primorogenic or serorogenic, respectively, whereas those formed after the orogenic episode are post-orogenic. Rock ages generally decrease from southeast to northwest across the map area. In the southeast, 1900-2000 Ma leptites (meta-rhyolites; Åberg et al. 1984, Kresten 1986) are intruded by the primorogenic Urgranites (1845-1925 Ma; Åberg 1978), and by minor primorogenic gabbro and diorite. Serorogenic granites in the region are 1730-1800 Ma (Wilson 1980), and include large areas of pegmatite. Increased tectonic activity attending the serorogenic phase caused migmatization of primorogenic granites and volcanics in the northeast corner of the map area.

The central and southwest portions of the map area expose post-orogenic Dala series granites, porphyries, and metasediments. The two most widespread granite types are the Järna granite (1675-1734 Ma; Åberg and Fredrikson 1984) and Siljan granite (1646 Ma; Åberg and Bollmark 1983). Dala porphyries are dated at 1635 Ma by Wehlin and Lundqvist (1970). Dolerites of three different ages occur in the map area (Patchett 1978). In the southeast corner are the near-vertical NE-SW-trending Tuna dolerites, dated at 1371 Ma. In the north, partly contained in the Siljan Ring, is the Emadalen dolerite sill. This dolerite dips at approximately 11°, and is 1223 Ma old. The steeply dipping NNW-trending dikes that cut across the west rim of the Siljan Ring and continue into the south-central part of the map area are 901-966 Ma old.

Geology of the Siljan Ring

An annulus of Paleozoic sediments surrounding Järna and Siljan granites defines the Siljan Ring. The outer diameter of the annulus is 44-46 km. The water level in the lakes forming the annular moat is 160-200 m above sea level (geomorphologic terminology of Grieve and Head 1983). Inward of the annular moat, elevations quickly rise to 300-400 m on the inner plateau. The central region of the Ring is a topographic low, with elevations of 280-320 m (Fig. 1). At the base of the Paleozoic section is the Ordovician Obulus conglomerate. Stratigraphically higher are a series of Ordovician through Silurian limestones and shales; the sequence is capped by the Silurian Orsa sandstone (Hjelmqvist 1966). Contacts between sediments and granites are brecciated fault zones. These often are cemented by secondary veins of quartz, calcite, chlorite and epidote.

Samples discussed in this report are from outcrops, shallow boreholes drilled within the Ring, and the deep drill hole Gravberg 1. Outcrop samples of granite were collected both within and outside the Ring (Fig. 1). Core recovery from the shallow boreholes is almost 100%, whereas only ~9 m of core was recovered from Gravberg 1. Sample depths are quoted with respect to sea level, with positive (+) or negative (-) signs to denote elevations above or below sea level.

Shallow boreholes 2, 3, 5 and 6 began in Järna granite, whereas boreholes 1 and 4 were spudded into Siljan granite. Cores recovered from these boreholes consist mainly of granite; other less common lithologies include pegmatite, dolerite and impact melt. In boreholes 1 and 2 near the perimeter of the Ring, granites are commonly fractured, brecciated, and mylonitized. Fracture surfaces sometimes exhibit slickensides. Brecciated zones and fractures are cemented with calcite, quartz, or chlorite fillings. Also observed are fine-grained microbreccias consisting of granulated fragments of granitic minerals. A similar style of deformation is observed in boreholes 3, 4, and 5 nearer to the geometric center of the Ring, but it is somewhat less pervasive and intense than in boreholes 1 and 2. Laumontite-healed fractures in granite are observed in borehole 4. Borehole 6 was spudded into granite, and drilled at an inclination of 60° to the southeast. After 68 m of drilling in granite, borehole 6 penetrated gray sandstone at +174 m. The 2-m-wide contact zone consists of quartz, brecciated granite, and thin clay layers. Bedding in the sandstone below the contact dips at ~50 degrees (the direction of dip is uncertain due to rotation of the core). Borehole 7 was spudded into limestone and drilled through limestone and thin marly shale layers for 217 m. At this depth (+3 m), the Obulus conglomerate is in fault contact with the underlying Siljan granite. Quartz- and epidote-healed brecciated zones are present near the contact.

Gravberg-1 is projected to reach a total depth of -7200 m (the Gravberg-1 site is at +300 m). In this report we present data for samples to -5711 m. Prior to drilling, seismic imaging in the Siljan Ring defined subhorizontal reflectors in five depth intervals (Juhlin and Pedersen 1988). The shallowest reflector, R1, is strong and coherent under the Gravberg-1 site but becomes segmented and disrupted toward the center of the Ring. Reflectors deeper than R1 are continuous for successively greater distances toward the center of the Ring. Drilling in Gravberg-1 demonstrates that the uppermost three reflectors are dolerite sills. The R1 reflector is a 60-m-thick dolerite between -1200 and -1260 m, dated at 906 Ma by K-Ar techniques (G. Åberg unpublished data). R2 is a composite reflection from three dolerites up to 9 m thick between -2289 and -2343 m, and R3 is a 21-m-thick dolerite sill first penetrated at -4427 m. The dolerites have chilled margins, showing that the surrounding granites were cool at the time of dolerite intrusion.

R4 and R5, at approximately -7200 m and -8000 m respectively, are below the present depth of Gravberg-1 but are presumed to be dolerites. Granites encountered in Gravberg-1 include the Siljan and Järna variants, and Dala granites transitional between these types.

Evidence for an Impact Origin

Complex terrestrial impact craters have a structurally uplifted central region, a peripheral depression, and a faulted annular rim (Grieve 1987). The structural form of Siljan Ring generally fits this description, except that its central region is a to-

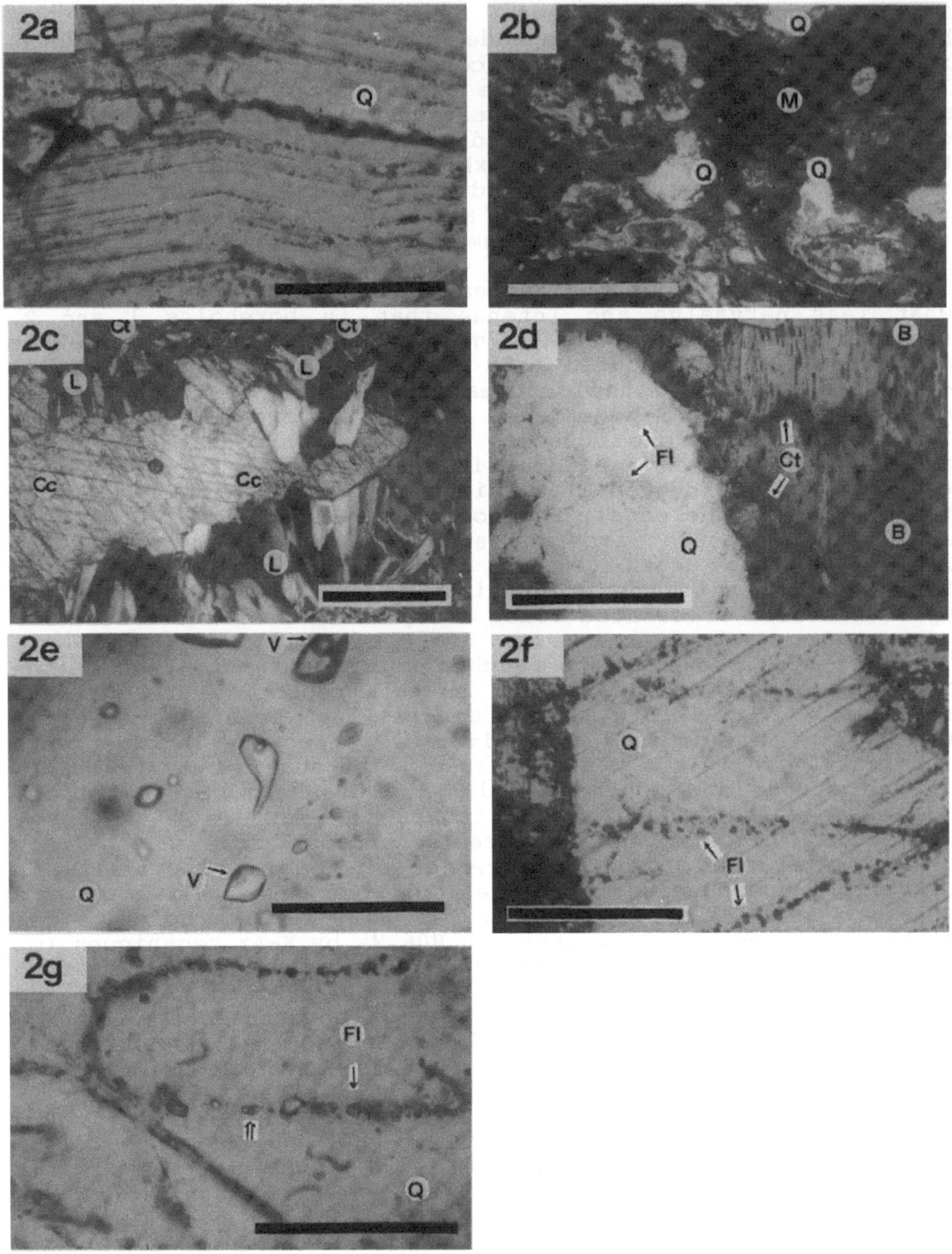
2a
Q
2b
Q
M
Q
Q
2c
Ct
Ct
L
L
Cc
Cc
L
2d
B
Fl
Ct
Q
B
2e
V
V
Q
2f
Q
Fl
2g
Fl
Q

pographic low. This might be due to extensive post-impact faulting and erosion of the central part of the crater. Alternatively, the impact center of the Siljan Ring may be offset from the geometric center (cf. Wickman 1980). Other evidence of an impact origin for the Siljan Ring includes the following features:

1. The sediments around the Ring comprise a megabreccia that appears to have slumped off the crater rim. The contacts between sedimentary blocks, and between sediments and granite, are interpreted to be listric faults that flatten into the crater floor (see cross-section in Fig. 1). This pattern of faulting has been documented in lunar craters (Settle and Head 1987) and in other terrestrial impact structures (e.g. the Manicouagan crater, Canada; Grieve and Head 1983). The presence of sedimentary megabreccia in the annular moat suggests that Paleozoic sediments outcropped at the time of impact.

Both the Siljan Ring and the Manicouagan impact structure have annular moats with associated blocks of sedimentary megabreccia. At the better-studied Manicouagan crater (Grieve and Head 1983), the diameter (D) of the annular moat is 65 km, impact related fracturing extends beyond the annular moat to D = 100 km, and an outer disturbed zone continues to D = 150 km. The ratios relating the annular moat D to fracturing and disturbed zone D's are 1.5 and 2.3, respectively. Applying these ratios to the Siljan Ring suggests that fracturing extends to D = 68 km, and impact-related disturbances continue to D = 104 km.

Fig. 2 a-g. a Planar elements in quartz from a granite. Planar elements are decorated with tiny inclusions of diaplectic glass. Sample 5-400.4, -109 m in borehole 5. (From depth of 400.4 m in borehole 5. Elevation with respect to sea level is -109 m.) Scale bar = 0.05 mm. **b** Impact melt from 3.5 km south of borehole 5. Partially melted quartz clasts (Q) and streaks of quartz-rich melt set in a matrix of glassy melt (M) with doleritic composition. Sample 87SJN-37F. Scale bar = 1.0 mm. **c** Crack healing mineral assemblage in Rl dolerite. Chlorite **(CT)** lines crack walls, laumontite **(L)** grows inward from chlorite substrate, and calcite **(CC)** fills crack center. Sample GRl-240, -1204 m in Gravberg-1 well. Scale bar = 0.5 mm. **d** Chlorite-filled **(CT)** crack in biotite **(B)** coplanar with fluid inclusion **(FI)** trail in adjacent quartz **(Q)** grain. Sample 4-275.1, +63 m in borehole 4. Scale bar = 0.2 mm. **e** Typical two-phase fluid inclusions hosted by quartz **(Q)**. **V** vapor. Vapor bubble in center inclusion is not marked. Sample 2-27.1, +347 m in borehole 2. Scale bar = 0.3 mm. **f** Fluid inclusion trails **(FI)** in quartz **(Q)** at an oblique angle to NE-SW trending planar elements. Sample 5-400.4, -109 m in borehole 5. Scale bar = 0.10 mm. **g** Fluid inclusion trail **(FI)** in quartz clast from pseudotachylite at Hattberg nature preserve, ~2.5 km west of borehole 5. Pseudotachylite occurs at granite-dolerite contact. The fluid inclusion marked with the double arrow homogenized at 296°C and decrepitated at 380°C. Sample 87SJN15h. Scale bar = 0.02 mm

2. Quartz grains in granites from the central part of the Ring contain sets of planar elements decorated with tiny glass inclusions (Fig. 2 a). These have been found to a depth of 400 m in borehole 5. Planar elements have been documented in quartz and other minerals from other impact craters (e.g. von Engelhardt and Bertsch 1969, Stoffler 1972), and have been reproduced in experimentally shocked materials (Borg 1972). Planar elements in granites collected 2 km northeast of the Ring's geographic center result from an estimated mean shock pressure of 16 GPa (Tamminen and Wickman 1980).

3. Whole or partial shattercones of granite are observed at many outcrops in the Siljan Ring (Fig. 1; Svensson 1973, Wickman 1980). The cones are usually a few tens of centimeters to a meter or two in length, and are covered with horsetail arrangements of striae. Similar shattercones are present in other impact craters (e.g., the Sudbury structure, Ontario; French 1972; the Charlevoix crater, Quebec; Robertson 1975), and Roddy and Davis (1977) describe shattercones formed by experimental cratering explosions.

4. Glassy to cyptocrystalline, purplish-red to black impact melts are found in a patchy distribution of outcrops within the Ring (Fig. 1). Melt rocks are also observed in cores from shallow boreholes 4 and 5, at depths down to 600 m. In the cores, melts occur as veins and pods up to 1.5 m thick with orientations from sub-horizontal to vertical. Clay minerals replace portions of the glassy melt rocks in some samples. Impact melts commonly are clast-bearing, with the clast content varying from <1% to ~30%. Both pseudotachylites bearing angular clasts and true impact melts with rounded clasts and flow structures (Fig. 2 b) are present in the Ring (see Reimold et al. 1987, for criteria to distinguish pseudotachylites from true impact melts). Some clasts have granitic compositions, but most consist of quartz or feldspar alone. Dolerite clasts are also observed and tend to be larger than granitic clasts in the same melts. Some clasts are cross-cut by channels of melt. Certain melt rocks collected near the annuals have compositions too silicic to be mixtures of granite and dolerite; a component of sedimentary material (sandstone) is required to produce these compositions. These unusual melt compositions provide further evidence that sediments outcropped or were near the surface at the time of impact.

Reimold et al. (1987) estimated that pseudotachylites in the Rochechouart crater, France formed at 700-750°C. Minimum melting temperatures in the model granite system quartz-albite-orthoclase are also near 750°C at 1000 bar water pressure (Tuttle and Bowen 1958). An approximate formation temperature of 750°C appears likely for pseudotachylites and granitic impact melts within the Siljan Ring.

Petrography

Järna granite contains subhedral plagioclase and microcline, intergranular quartz, biotite and hornblende amounting to 10-20% of the mode, and accessory sphene, zircon, allanite, rutile, apatite and magnetite. Siljan granite is more leucocratic, and biotite is the only common mafic mineral (Table 1). Small amounts of muscovite (<3%) are present in some granites below 5 km in Gravberg-1. In many samples, chlorite partially to completely replaces the mafic minerals and feldspars are altered to sericite. Granite coloration changes with increasing depth from red through reddish gray to gray. This trend reflects decreasing oxidation of iron-bearing phases with depth. Dolerites contain plagioclase, titaniferous augite, olivine, and magnetite in an ophitic texture. Chlorite and minor serpentine partially replace the mafic phases.

Fractures and shear zones are found in the surface granites, and at all levels in the drill cores. Crack-widths reach several mm in shallow samples, but in deeper samples most cracks are thin, hairline fractures. The frequency of fracturing in the granites wanes gradually with depth, and appears to decrease abruptly across the interval -1200 to -1260 m where the R1 dolerite is found. Almost all cracks are filled, and little porosity is evident. Electric logging in Gravberg-1 indicates porosities of <1% over most of the hole (D. Karlsson unpublished report 1987). Hydraulic leak-off tests in Gravberg-1 show very little pressure leakage, providing evidence for very low permeabilities (G. Gustafson unpublished report 1987). In samples from shallow depths (above -1200 m), fractures may be filled by granulated fragments of the granitic minerals or by crack-healing minerals, including calcite, chlorite, quartz, and epidote. Granites from deeper levels have most fractures filled only by crack-healing minerals. Above -1200 m, calcite is an

Table 1. Representative granite compositions. (Hjelmqvist 1966, Kresten 1986)

	Siljan granite	Järna Granite
Quartz	34	17
K-feldspar	44	17
Plagioclase	19	48
Biotite	2	11
Hornblende	-	5
Accessories	1	2
SiO_2	72.94	67.45
TiO_2	0.35	0.54
Al_2O_3	13.14	14.83
Fe_2O_3	4.06	3.91
MnO	0.06	0.08
MgO	0.42	1.04
CaO	1.03	2.65
Na_2O	3.33	3.31
K_2O	4.82	4.79
P_2O_5	0.02	0.19
Total	100.17	98.79

abundant crack-healing mineral; below that depth chlorite and epidote become more common, while the abundance of calcite diminishes (Valley et al. this Vol.). Some cracks in the R1 dolerite contain a regular zonation of secondary minerals. Chlorite lines the crack walls and forms a substrate for prismatic laumontite, which projects into the center of the crack; calcite occupies the center of the crack (Fig. 2 c).

Fluid Inclusions

Fluid inclusions are present within the matrix quartz of granites, and in crack-healing quartz veins. Despite intensive searching, no fluid inclusions were found in crack healing minerals other than quartz. Most fluid inclusion data discussed in this report are for secondary inclusions within granitic matrix quartz. Also presented are inclusion data from four samples of vein quartz from Stygforssen near Boda on the east side of the Ring (Fig. 1). These veins occur near or at granite/sedimentary-megabreccia contacts in many locations around the periphery of the Ring.

Secondary fluid inclusions occur in linear trains, planar arrays, and dispersed apparently at random throughout quartz grains. In some instances, the trails of fluid inclusions are contiguous with veins of alteration minerals in adjacent phases (Fig. 2 d). Most studied inclusions are 5-10 microns across, and contain two phases: a bubble of water vapor and a low-salinity aqueous fluid (Fig. 2 e).

A minority of inclusions contain anisotropic daughter minerals 1-2 microns across. One of these daughters in a granite from -706 m in Gravberg-1 was identified as calcite by laser Raman analysis (Pasteris et al. 1986, personal communication 1987). In addition to this sample, laser Raman analysis was applied to samples from -1196 m, -3907 m, and -4051 m in Gravberg-1. The analyzed fluid inclusions contain no detectable H_2, H_2S, CO_2, CH_4, or N_2 (detection limit = 3-5 bar partial pressure).

The timing of fluid inclusion formation with respect to the meteorite impact that formed the Siljan crater may be documented in certain shallow samples by examining textural relationships between inclusions and impact-generated features. Planar elements in quartz sometimes are cross-cut by trails of fluid inclusions (Fig. 2 f). Partially melted quartz clasts in pseudotachylites (Fig. 2 b) also contain fluid inclusion trails (Fig. 2 g). These inclusion trails merge with channels of altered melt that cut the clasts. Two phase inclusions (water plus vapor bubble) comprise only a fraction (<10%) of the fluid inclusions that cross-cut planar elements or that occur in melt clasts. The majority of fluid inclusions associated with impact features contain only one phase, or are empty and appear to have leaked.

Analytical Methods

Homogenization (T_h) and final melting (T_m) temperatures of fluid inclusions were determined on a Fluid Inc. heating/freezing stage calibrated at 0 and -56.6°C. Inclusions were cooled with liquid nitrogen until completely frozen, which required temperatures of -30 to -60°C. After freezing, a rapid heating run served to roughly locate T_m. In a second run, a heating rate of 2°C/min or slower was used to accurately determine T_m. A similar two-step method was used to determine T_h. Measurements of T_h are reproducible to within ±3°C, whereas T_m measurements are reproducible to within ±0.2°C. The equivalent weight percent NaCl ($NaCl_{eq}$) in the inclusions was calculated from the equations of Potter et al. (1978). Pressure corrections (Potter 1977) were made to average T_h values assuming a hydrostatic pressure gradient existed at the time of inclusion trapping, and that one km of overburden has been eroded from the crater since the impact. Pressure corrected homogenization temperatures are hereafter termed trapping temperatures (T_t). Fluid inclusion data is presented in Table 2.

The concentration of water soluble chlorine (Peters 1986) was measured in 25 granite samples from Gravberg-1. Sample chips were ultrasonically cleaned and rinsed in 17 megaohm/cm distilled water, then crushed to an average particle diameter of <5 microns as measured in oil at 40 x magnification. Fifty ml of distilled water were added to about 5 g of powder, and the slurry was agitated at room temperature in a shaker bath for 72 h. Cl^- concentrations in the filtered solutions were measured with a Dionex ion chromatograph (Table 3). Accuracy and precision of the measurements are approximately 3% relative.

Geochemical Results

Fluid salinities in most inclusions are less than 2 wt.% NaCl equivalent (Fig. 3). Much less common are isolated inclusions containing water with up to 23.2 wt.% NaCl equivalent. No trend with depth is defined by the inclusion salinity data, except that the highest average salinities (≥4.2 wt.% $NaCl_{eq}$) are only present in samples below -2504 m (and in one sample from far outside the crater; 87SJN-40b, Table 2). Whole rock NaCl equivalents, calculated from Cl^- measuremnts, decrease from 100 ppm in near surface samples to 20 ppm at -1200 m (Table 3, Fig. 4). Below this depth, whole-rock salinities increase to 70-170 ppm near 6 km depth. Both whole rock and fluid inclusion salinity data are available for 8 samples. The number of inclusions per cubic centimeter of quartz may be calculated for these samples (Table 4). In samples above -3070 m, inclusion densities vary around 5×10^7 inclusions/cc of quartz. Below that depth, inclusion densities range over 3 orders of magnitude, from 5.2×10^6 to 1.3×10^9 inclusions/cc of quartz. For comparison, the Stripa granite (Sweden) contains 5.0×10^7 to 3.1×10^8 inclusions/cc of quartz (average = 1.74×10^8; Lindblom 1984).

Table 2. Fluid inclusion data

Sample	Type of: Sample	Granite	Depth (m)	Elev. WRT SL (m)	Radial Dist. (km)	n	Avg. Incl. Size (um)	Vol. % Vapor (Range)	Temperature (C): Homogenization Avg.	sigma	Min	Max	P-Crxn	Trapping Temp.	n	Final melting Temp. Avg.	Avg. % NaCl Equiv.
14F 5b:4	H	S	0	350	10.5	10	6	1-5	225	23	192	267	4	229	5	0.0	0.0
14F 5c:9	H	S	0	275	14.6	10	6	2-50	313	20	283	344	9	332	3	- 1.6	2.7
14F 6c:2	H	S	0	285	15.1	10	5	2-30	286	21	240	323	9	295	-	-	
14F 1b:40a	H	J	0	265	13.1	10	8	2-30	321	33	262	355	10	331	6	0.0	0.0
14F 1b:41	H	J	0	275	13.5	12	10	5-20	257	55	185	343	8	265	6	- 0.6	1.0
14F 2d:20	H	J	0	295	15.6	10	9	1-20	273	41	232	341	9	282	11	- 0.7	1.2
87SJN-10a	H	S	0	205	30.3	12	8	2-20	148	35	81	207	9	157	-	-	
87SJn-12a	H	J	0	190	21.5	5	6	1-5	135	6	131	142	12	147	-	-	
87SJN-21cq	H	vein in J	0	255	13.4	16	8	5-40	261	53	206	345	9	270	9	- 0.1	
87SJN-21cg	H	J	0	255	13.4	5	9	5-10	279	40	247	335	9	288	8	- 1.3	
87SJN-24	H	D	0	320	23.2	10	8	5-20	225	19	194	246	8	233	4	- 0.8	1.4
87SJN-29	H	S	0	328	13.2	10	6	2-20	254	38	209	310	8	262	-	-	
87SJN-35b	H	D	0	390	30.3	10	5	2-15	183	42	132	229	10	293	-	-	
87SJN-36b	H	S	0	240	20.0	10	6	1-15	180	56	124	262	9	189	-	-	
87SJN-38	H	D	0	200	29.4	10	6	2-20	265	32	202	311	9	274	-	-	
87SJN-39k	H	vein in S	0	240	18.8	13	5	5-20	235	12	217	258	9	244	10	- 0.6	1.0
87SJN-391	H	vein in S	0	240	18.8	10	7	2-20	222	21	187	265	9	231	-	-	
87SJN-39m	H	vein in S	0	240	18.8	10	4	1-5	246	9	226	259	9	255	-	-	
87SJN-40b	H	J	0	230	67.0	11	10	5-25	269	13	262	289	5	274	10	-10.7	14.7
87SJN-41b	H	J	0	170	67.0	11	11	3-25	251	23	218	287	9	260	3	- 0.1	0.2
87SJN-43a	H	J	0	170	60.1	16	12	2-25	216	37	168	277	9	225	8	- 1.1	1.9
87SJN-44	H	S	0	510	28.0	11	10	5-20	207	37	151	259	8	215	5	- 0.2	0.4
Bh1-300.7	C	S	300	0	11.7	33	5	1-40	250	35	172	319	9	259	5	- 0.1	0.2
Bh2-27.1	C	J	27	247	9.8	9	7	2-20	213	42	132	274	10	223	8	- 0.9	1.6
Bh2-427.5	C	J	428	-54	9.8	10	10	1-25	259	50	187	341	9	268	-	-	
Bh3-226.3	C	J	226	-6	7.8	9	5	1-50	300	30	268	362	10	310	10	- 0.1	0.2
Bh4-275.1	C	S	275	63	11.8	6	8	2-15	317	13	298	330	10	327	-	-	
Bh4-369.3	C	S	369	- 31	11.8	1	7	15	332				10	342	-	-	
Bh4-628.2	C	S	628	- 290	11.8	10	7	1-15	287	24	246	323	9	296	-	-	
Bh5-400.4	C	S	400	- 109	3.2	6	9	2-10	257	19	219	279	9	266	-	-	
GR1-1-16	C	S	1006	- 706	12.2	12	8	5-30	302	34	261	370	15	317	3	- 1.3	2.2

Table 2. (continued)

Sample	Type of: Sample	Type of: Granite	Depth (m)	Elev. WRT SL (m)	Radial Dist. (km)	n	Avg. Incl. Size (um)	Vol. % Vapor (Range)	Temperature (C) Homogenization Avg.	sigma	Min	Max	P-Crxn	Trapping Temp.	n	Final melting Temp. Avg.	Avg. % NaCl Equiv.
GR1-1-25	C	S	1009	- 709	12.2	5	9	1-20	273	5	267	280	15	288	2	- 1.0	1.7
GR1-2-9	C	D	1498	-1198	12.2	11	7	1-10	284	19	249	307	20	304	10	- 0.4	0.7
GR1-3-3	C	D	2132	-1832	12.2	11	7	1-5	218	16	199	244	25	243	10	- 1.6	2.7
GR1-4-7	C	D	2804	-2504	12.2	11	5	10	283	25	239	319	27	310	8	- 2.5	4.2
GR1-5-1	C	S	3370	-3070	12.2	10	6	0.1-2	250	21	214	282	36	286	3	- 0.9	1.6
GR1-4023	N	D	4023	-3723	12.2	10	5	1-10	296	27	252	340	41	337	9	- 1.1	1.9
GR1-4206	N	D	4206	-3906	12.2	10	5	1-10	262	28	231	311	41	303	4	- 0.9	1.4
GR1-6-3	C	D	4206	-3906	12.2	10	11	1-10	198	40	145	262	47	245	8	- 5.7	8.1
GR1-6-4	C	D	4207	-3907	12.2	10	10	5-20	314	24	278	348	50	364	6	- 0.2	0.4
GR1-6-5	C	D	4207	-3907	12.2	10	13	2-10	310	19	278	334	49	359	-	-	
GR1-4351	C	D	4351	-4051	12.2	9	19	1-10	276	17	252	302	42	318	11	-11.3	15.3
GR1-5135	C	J	5135	-4835	12.2	10	11	2-15	324	19	295	357	55	379	-	-	
GR1-5751F	X	J	5751	-5461	12.2	2	7	2-5	290	5	285	294	60	350	-	-	
GR1-5760F	X	J	5760	-5460	12.2	3	6	2-5	306	25	271	329	63	369	-	-	
GR1-5827A	X	J	5827	-5527	12.2	9	8	2-10	322	18	294	341	67	389	-	-	
GR1-5827R	X	J	5827	-5527	12.2	11	10	1-10	292	12	279	318	64	356	5	- 0.5	0.9
GR1-6011	X	J	6011	-5711	12.2	10	6	1-5	309	10	295	329	70	379	4	- 0.1	0.2

Explanation

Type of: Sample - H, surface outcrop sample; C, core sample from borehole; X, cuttings sample from borehole.

Granite - S, Siljan granite; J, Jarna granite; D, other Dala series granites.
All analyzed fluid inclusions are in granitic matrix quartz, except for four samples of vein quartz.

Depth - Depth (m) below surface.

Elev. W.R.T. SL - Elevation with respect to sea level.

Radial Dist. - Distance (km) from geographic center of crater.

n - number of samples. n in column 7 refers to data through column 10. n in column 16 refers to melting temperature data only.

Inclusion Size - Length (microns) of longest dimension of fluid inclusion.

Vol. % Vapor - Volume percent of inclusion occupied by vapor inclusion.

P-crxn - Pressure correction made using Potter's (1977) curves, and assuming that 1 km of overburden has been removed since the meteorite impact.

Trapping temp. - Measured homogenization T plus pressure correction.

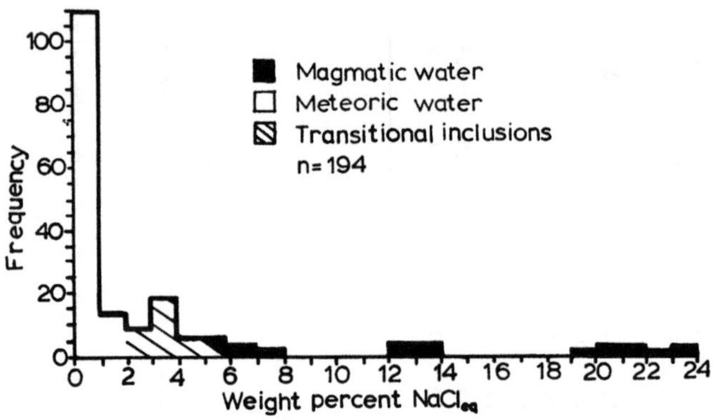

Fig. 3. Weight percent NaCl equivalents in fluid inclusions determined from freezing point depression measurements and equations of Potter et al. (1978). 194 individual measurements are plotted. High salinity inclusions freeze at lower temperature (50 to 60°C), tend to be larger, are texturally distinct, and contain fluids with a greenish tinge in contrast to the clear fluids in low salinity inclusions. These inclusions are thought to be fluids released by crystallizing granitic magmas. Salinities of inclusions labeled meteoric water are typical of those found in hydrothermal systems hosted by silicic rocks and recharged by meteoric water. (Henley 1985)

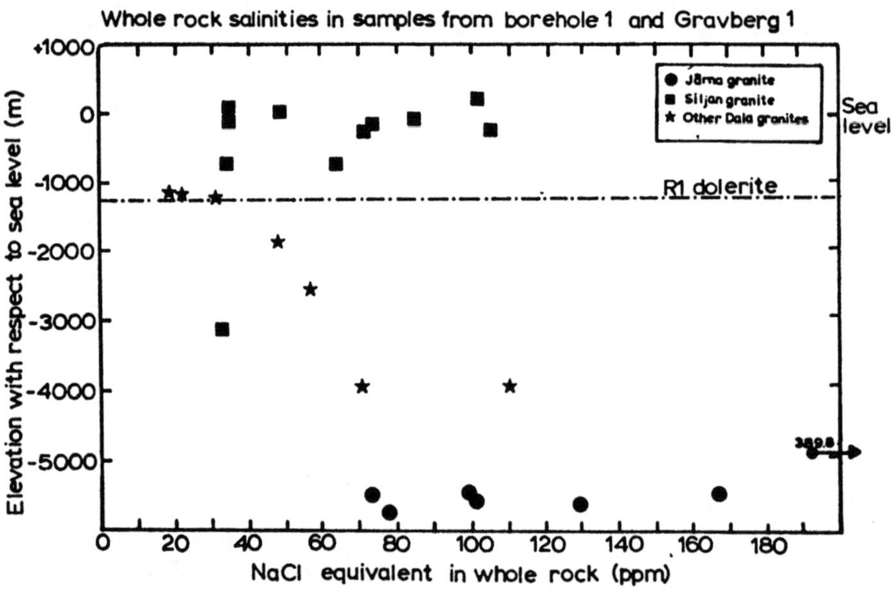

Fig. 4. Parts per million NaCl equivalents in bulk granite samples. Calculated from measured Cl^- concentrations in room temperature aqueous leachates (Table 3)

Table 3. Salinities of whole rock aqueous leachates

Sample	Depth (m)	Elev. WRT SL (m)	Measured [Cl−] in granite (ppm)	Calculated [$NaCl_{eq}$] in granite (ppm)
1- 46.7	47	253	62.0	102.3
1-251.9	252	48	29.6	48.8
1-275.3	275	25	21.2	34.9
1-323.7	324	-24	52.1	85.9
1-375.1	375	-75	21.1	34.8
1-425.6	426	-126	44.7	73.6
1-476	476	-176	64.3	106.0
1-526.3	526	-226	43.8	72.3
GR1-1-15	1006	-706	21.1	34.8
GR1-1-16	1006	-706	39.3	64.7
GR1-2-1	1496	-1196	13.3	21.8
GR1-2-3	1497	-1197	18.9	31.1
GR1-2-9	1497	-1197	13.5	22.3
GR1-3-3	2132	-1832	29.2	48.1
GR1-4-7	2804	-2504	34.4	56.6
GR1-5-2	3376	-3076	19.8	32.7
GR1-6-3	4206	-3906	42.7	70.4
GR1-6-6	4209	-3909	66.9	110.3
GR1-5135	5135	-4835	236.5	389.8
GR1-5724	5724	-5424	101.7	167.6
GR1-5751F	5751	-5451	60.3	99.3
GR1-5760	5760	-5460	44.2	72.8
GR1-5827	5827	-5527	61.1	100.8
GR1-5905b	5905	-5605	78.5	129.4
GR1-6011	6011	-5711	47.2	77.8

Fluid Inclusion Results

Vapor-liquid volumetric ratios inclusions vary widely in shallow samples, but have more restricted ranges at depth (Fig. 5). Most inclusions homogenize to liquid, although in a few shallow samples some inclusions homogenize to vapor. Homogenization temperatures versus elevation for samples from radial distances of 12 ± 5 km are plotted in Fig. 6 a. The data from Gravberg-1 (radial distance = 12.2 km) are included in this figure. The vertical scale of Fig. 6 a requires grouping of samples from several depth intervals. Individual histograms for these grouped samples are presented in Fig. 6 b. T_h's for samples not in the radial band (12 ± 5 km) are plotted in Fig. 6 c.

Homogenization temperatures in samples from the radial band (Figs. 6 a, 6 b) mostly are in the range 200-350°C. Median T_h's (midpoints of the histograms) increase gradually from elevations above sea level to -1200 m, just above the R1 dolerite.

Below R1, another trend of increasing average temperature with depth extends from sample GR1-3-3 (-1832 m) to the deepest sample GR1-6011 (-5711 m). T_h's in quartz veins from granite/megabreccia contacts (Fig. 6 c) primarily fall between 200 and 275°C. Other samples in the vicinity of the annular moat have lower homogenization temperatures, mostly between 100 and 250°C. A wide range of T_h's is present in samples in the fractured zone outside the moat. Temperatures near 200°C are most common in these samples. Outside the zone of fracturing, T_h's also span a wide range, with temperatures around 275°C most common. A fluid inclusion within a quartz clast (diameter = 3 mm) in a pseudotachylite (Fig. 2 f) homogenized at 296°C, and decrepitated at 380°C. This is considerably less than the estimated formation temperature of 750°C for pseudotachylites, suggesting that fluid entrapment occured after melt generation.

The wide range of homogenization temperatures and vapor to liquid volumetric ratios (Figs. 5, 6; Table 2) may result from inclusion necking, which can occur during healing of host quartz crystals (Bodnar et al. 1985). Measured homogenization temperatures in necked inclusions can be both higher and lower than in the original inclusion (Fig. 7). The effect on T_h histograms is to spread data in both directions from the original homogenization temperature. Therefore, if one suspects that necking is the cause of large ranges of homogenization temperatures, the best approach is to use mean temperatures for interpretations and modeling. Pressure corrected mean homogenization temperatures (i.e., trapping temperatures) are plotted versus

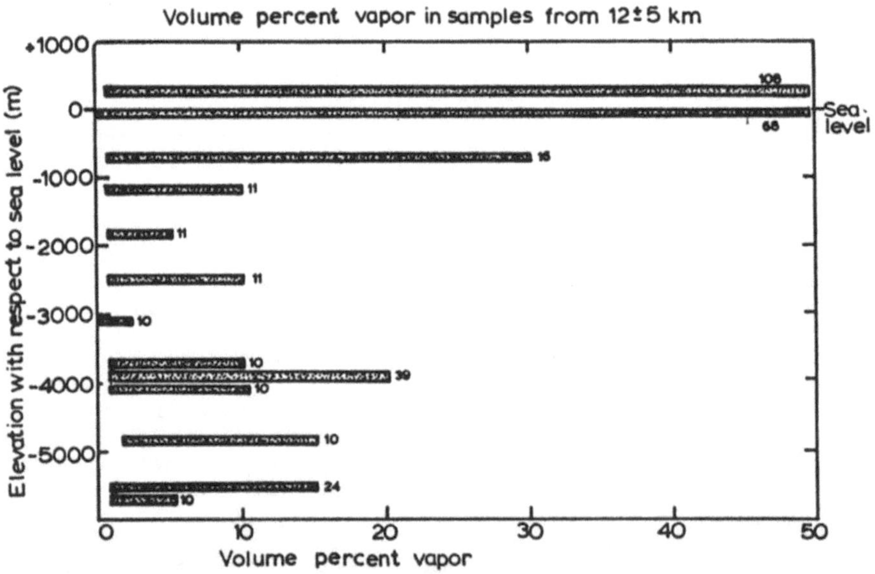

Fig. 5. Volumetric fractions of fluid inclusions occupied by vapor. The number of measurements in each range is shown next to the bars

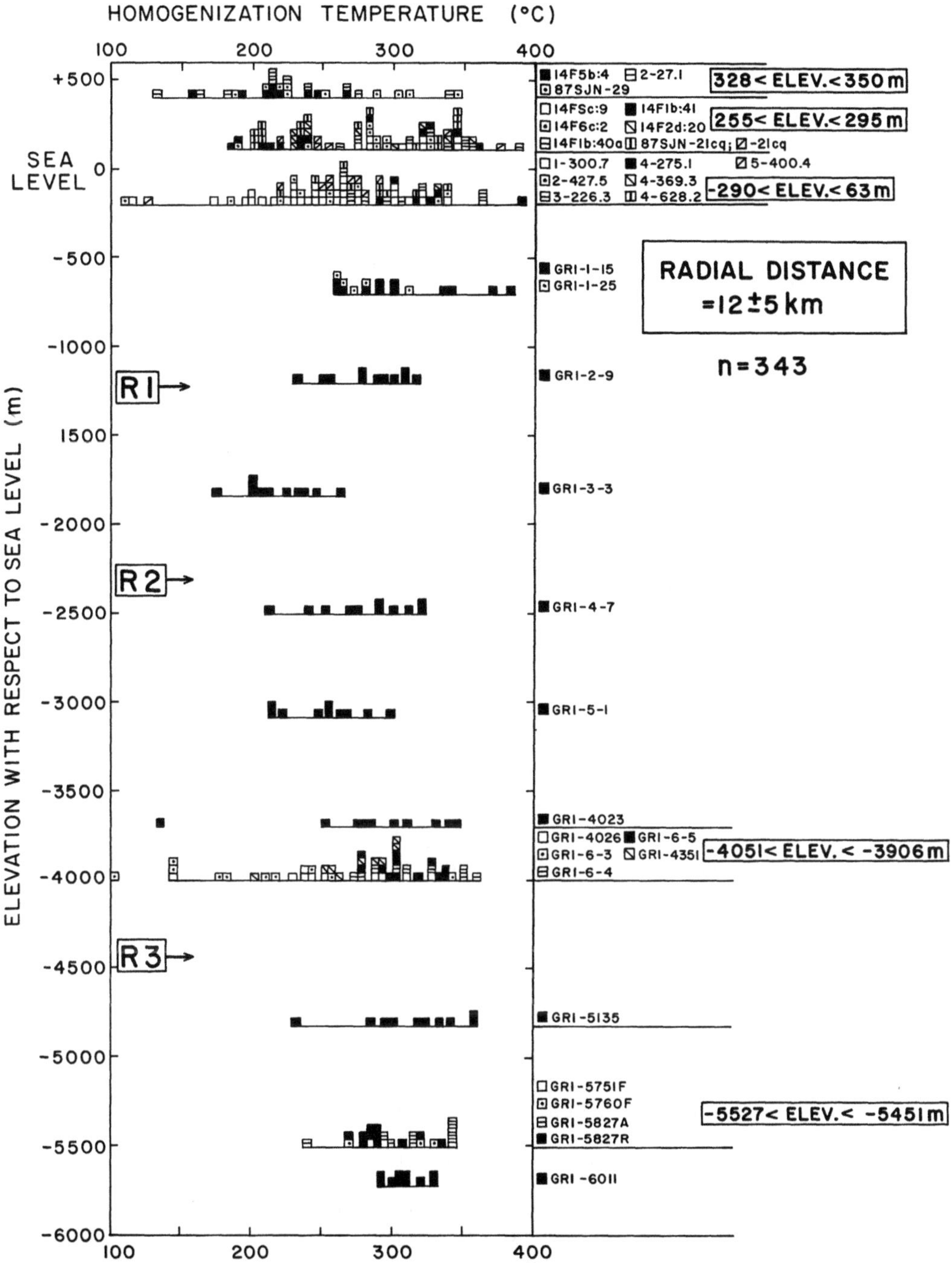

Fig. 6 a. Homogenization temperatures in samples from radial distances of 12 ± 5 km

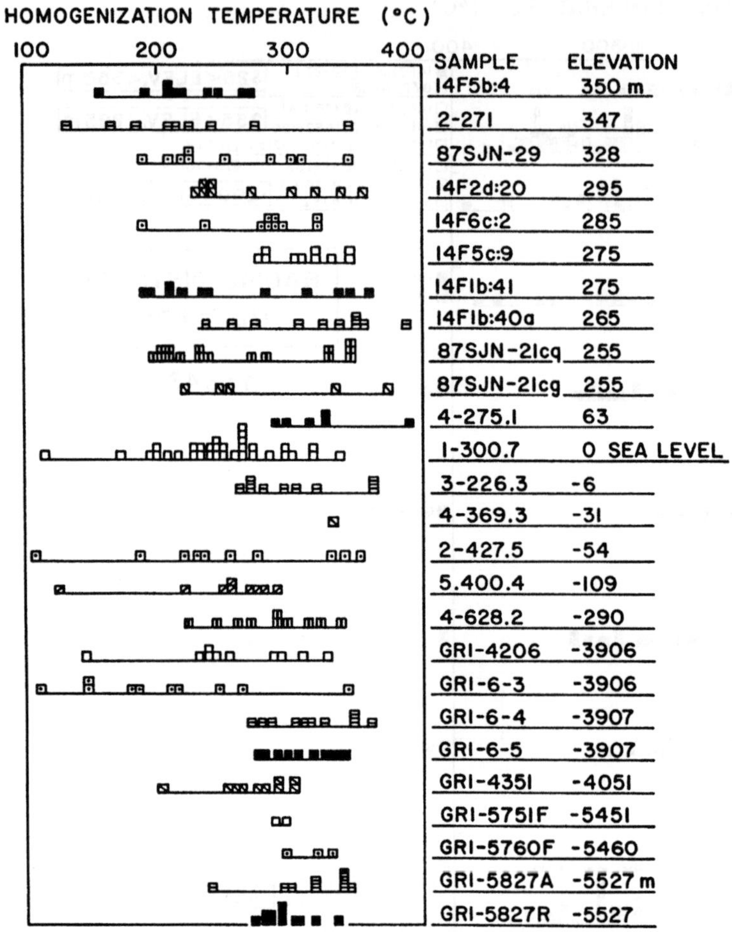

Fig. 6 b. Homogenization temperatures for samples from congested parts of a

radial distance (Fig. 8) and elevation (Fig. 9). In surface samples, peak T_t's occur on the inner plateau, whereas minimum T_t's are found in the vicinity of the annular moat (Fig. 8). In the subsurface, samples below -1200 m define a clear trend of increasing temperature with depth (Fig. 9). The slope of this trend is 29°C/km. Above -1200 m is a separate, less well-defined trend of increasing temperature with depth.

The purpose of plotting data from the radial band within 12 ± 5 km (Figs. 6 a, b, 9) relates to the expected temperature distribution around an impact crater. Impact craters are radially symmetric features, and impact generated isotherms and isobars from radially symmetric patterns centered about the geometric centers of craters (e.g. Grieve et al. 1977). An impact-generated thermal event recorded by fluid inclusions also might create a radially symmetric pattern of isotherms. Consequently, plots of homogenization versus elevation should be restricted

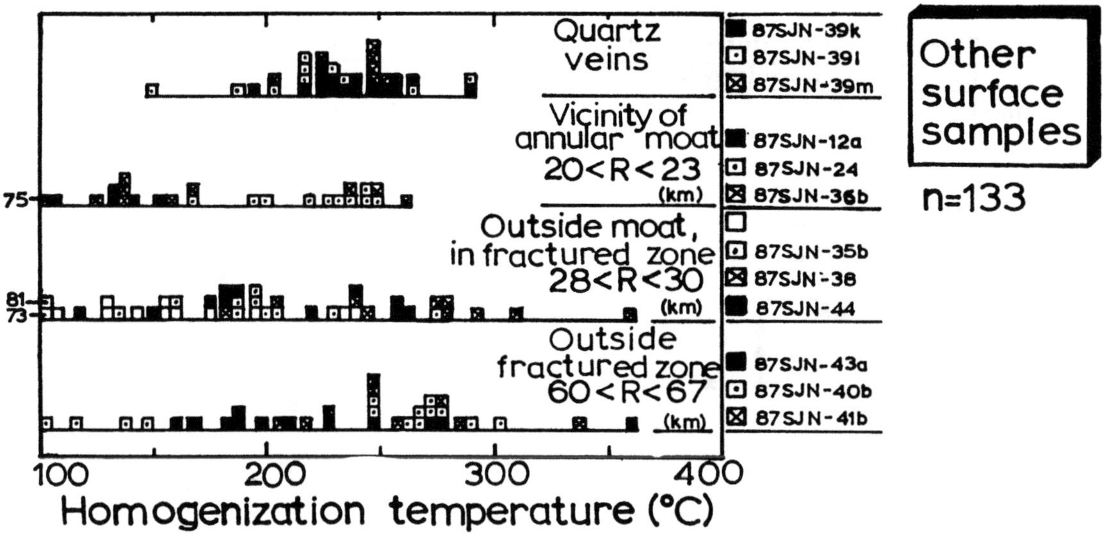

Fig. 6 c. Homogenization temperatures in samples not in the radial band defined by 12 ± 5 km

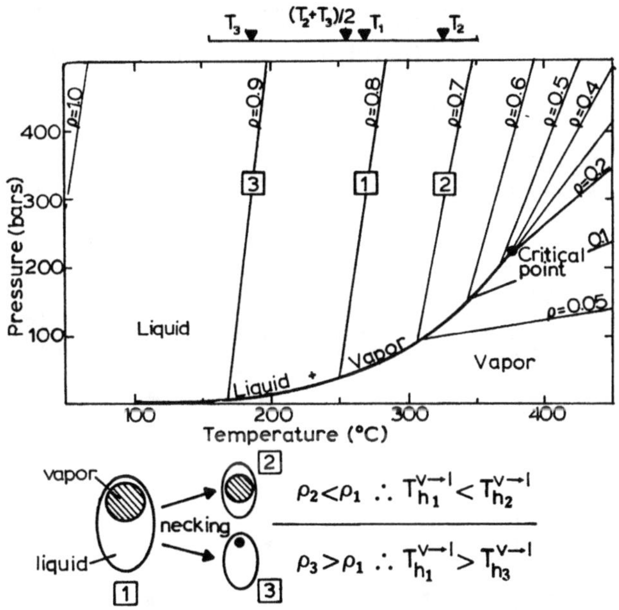

Fig. 7. Schematic illustration of how inclusion necking can affect measured homogenization temperatures. The large inclusion at lower left necks down to from two smaller inclusions with variable vapor/liquid ratios. Compared to inclusion 1, inclusion 2 contains more vapor and therefore has a lower density. The opposite is true of inclusion 3. The phase diagram for water (Fisher 1976) shows that compared to inclusion 1, inclusion 2 will homogenize at a nigher temperature and inclusion 3 at lower temperature. The histogram at the top of the figure shows that T_h of inclusion 1 can be approximated from the average T_h's of inclusions 2 and 3

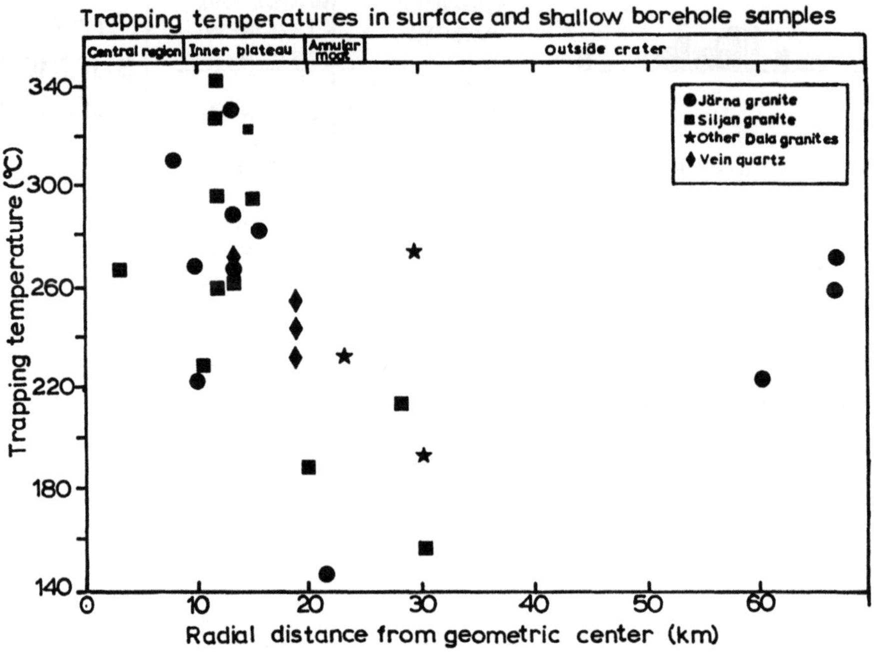

Fig. 8. Average trapping temperatures of fluid inclusions (Table 2) versus radial distance from the geometric center of the Siljan Ring. The data distribution is irrespective of the particular type of granite analyzed (i.e., Järna, Siljan, or Dala)

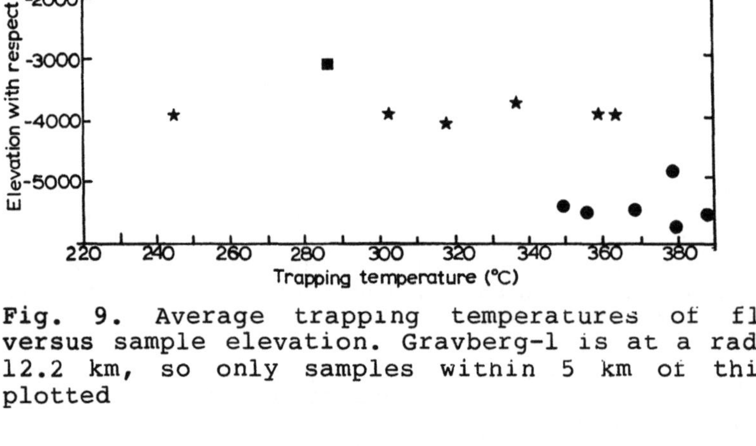

Fig. 9. Average trapping temperatures of fluid inclusions versus sample elevation. Gravberg-1 is at a radial distance of 12.2 km, so only samples within 5 km of this distance are plotted

to subsurface and surface data from approximately the same radial distances. This concentrates the plot on depth-dependent thermal variations, and minimizes the influence of radial distance on the data pattern. Similarly, plots of radial distance versus trapping temperature should include only data for surface and shallow samples. This focuses the plot on the surficial thermal pattern, and minimizes the effect of depth-related variations.

Evidence for Hydrothermal Activity in Impact Craters

Hydrothermal activity driven by the heat of meteorite impacts has been documented at several terrestrial impact craters. It is also suspected to occur on Mars (Newsom 1980). Evidence of intense hydrothermal activity is observed in layers above and below melt sheets. Impact melts and breccias often are altered to mixtures of clays (Fe-chlorites and smectites), zeolites (commonly analcite), potassium feldspar and quartz. These assemblages indicate water temperatures of 100-300°C (Allen et al. 1982). In impact breccias at the West Clearwater crater in Quebec, transformation of low-Ca pyroxene to Mg-rich saponitic clay suggests maximum temperatures of 500°C (Phinney et al. 1978). At the Ries crater in southern Germany, deposits of suevite (impact breccia containing abundant melt particles) are cut by vertical degassing pipes (Newsom et al. 1987). Degassing pipes are interpreted to be escape channels for steam and other volatiles generated by the heat of the impact. Passage of fluids through the pipes began soon after deposition of the suevite, at temperatures in excess of 500°C. Hydrothermal assemblages found in veins with pseudotachylite at the Rouchechouart crater, France, include arsenopyrite, pyrite, quartz, calcite and dolomite. These assemblages suggest fluid temperatures of ~700°C (Reimold et al. 1987).

In the Siljan Ring, evidence of hydrothermal activity is provided by secondary mineral assemblages and fluid inclusions. Crack-healing minerals in granites (i.e., chlorite, calcite, quartz, epidote, ziolites) are similar to hydrothermal assemblages described in other granite-hosted hydrothermal systems (e.g., Stone and Exley 1985). The calcite-chlorite-laumontite vein-healing assemblage found in the R1 dolerite at -1200 m has been experimentally reproduced at 250°C and 500 bars (Savage et al. 1987). At Boda in the eastern part of the Ring (Fig. 1), Pb-Zn ores were deposited by hydrothermal solutions thought to be related to the meteorite impact (Johansson 1984). Thick quartz veins at granite/sedimentary-megabreccia contacts and other locations near the annulus (Lindblom and Wickman 1985) provide evidence for transport and precipitation of quartz by hydrothermal solutions. Other evidence of hydrothermal activity is the partial replacement of glassy melt rocks by clay minerals.

Fluids in inclusions are demonstrably related to certain secondary assemblages. In some melt rocks fluid inclusion trails in quartz clasts can be traced to altered melt channels cutting the clasts. In certain granites, chloritized cracks in biotite are contiguous with fluid inclusion trails (Fig. 2 d). Further-

Table 4. Measured and calculated properties of granites and fluid inclusions

Sample	Elev. WRT SL (m)	Modal QTZ (vol.%)	Granite density (g/cc)	Fluid inclusions per cc QTZ	Fluid concentration in granite (ppm)
GR1-1-16	-706	32	2.61	$5.3x10^7$	1733
GR1-2-9	-1198	13	2.61	$8.4x10^7$	754
GR1-3-3	-1832	27	2.64	$3.0x10^7$	560
GR1-4-7	-2504	33	2.63	$5.0x10^7$	3315
GR1-5-1	-3070	32	2.60	$5.6x10^7$	777
GR1-6-3	-3906	28	2.64	$5.2x10^6$	382
GR1-5827	-5527	20	2.64	$1.3x10^8$	2628
GR1-6011	-5711	33	2.64	$1.3x10^9$	18225

Notes:
1. Elev. WRT SL - Elevation with respect to sea level.
2. Modal quartz from wellsite cuttings analysis.
3. Granite density from wireline electric log.
4. Fluid inclusions/cc QTZ calculated from average data in Table 2 and whole rock and inclusion salinities. Values calculated from [(Whole rock [Cl^-]/modal QTZ)/([Cl^-] in inclusion)]/(Inclusion volume x fluid density)
   Assumptions: Inclusions are spherical.
   Fluid density = 1.0 g/cc
   All inclusions in QTZ. All Cl^- in inclusions occurs as NaCl.
5. Fluid concentration in granite calculated from:
   Inclusion volume x number of inclusions in QTZ x modal QTZ x fluid density/granite density

more, fluid inclusions are abundant in the secondary quartz veins. These observations show that some fluids trapped in inclusions used the same pathways through the rocks as fluids that precipitated secondary chlorite and quartz. The concentrations of included fluids in granites range from 382 to 2628 ppm (excluding 1 sample; Table 4). Since the fluids in inclusions were trapped by sealing of fluid pathways, these values are minimum estimates of fluid concentrations when fluids actually moved through the rocks. The highest fluid concentration (18,225 ppm) occurs in the sample from -5711 m. At this depth in Gravberg-1, drilling rates increased and fracturing and secondary mineralization become more abundant (J. Castano unpublished report). All these observations are consistent with higher permeabilities and fluid concentrations in this zone. Water trapped in quartz-hosted fluid inclusions may have several sources: breakdown of hydrous phases in granites (biotite, amphibole); exsolution from quartz (Roedder 1984); or it may be introduced from external sources, such as a hydrothermal system recharged by surface water. Samples in which biotite is altered to chlorite in varying degrees are found throughout Gravberg-1.

Chlorite contains about 12 wt.% water compared to 2% in biotite, so extensive chloritization of biotite requires introduction of water from external sources. In a typical granite containing 10 modal percent biotite or amphibole with the balance feldspar and quartz, the weight fraction of mafic phases is about 12%. The weight fraction of water in the mafic phases is 2%, so 2400 ppm H_2O is the total water available from mafic mineral breakdown. This value is close to the water concentrations in the granites listed in Table 4. However, this process requires complete breakdown of the mafic minerals into anhydrous phases such as pyroxenes, which is not observed in any sample. Therefore, this process cannot be responsible for all the trapped fluids, and cannot supply enough water for extensive alteration of biotite to chlorite. In studies reviewed by Roedder (1984), the maximum concentration of water exsolved from quartz is 500 ppm. In a granite containing 20% modal quartz, 10% mafic phases and the balance feldspar, this amounts to about 100 ppm H_2O in the bulk rock, significantly less than the concentrations listed in Table 4. Although we have not quantitatively identified the proportions of fluids from different sources, it seems likely that some of the trapped fluids must have been introduced from sources external to the granites.

Fluid inclusions provide evidence for high temperature (250-350°C) aqueous fluids in granites at the Siljan Ring. In contrast to these results, Valley et al. (this Vol.) present isotopic data suggestive of low temperature (<100°C) meteoric water circulation. These low temperature waters are thought to be responsible for much of the calcite precipitation in the granites. Our preliminary optical and x-ray diffraction studies of chlorites in the Siljan Ring provide evidence for several generations of chlorite, at least some of which also are low temperature in origin (type Ib chlorite, T <150°C; Hayes 1970). The fluid inclusion data presented here depict only part of a complex and polygenetic hydrologic regime in the Siljan Ring.

Heat Source for the Siljan Ring Hydrothermal Activity

Three major thermal events are known to have influenced the Siljan Ring: granite intrusion (1650 Ma), dolerite intrusion (901-1371 Ma), and meteorite impact (362 Ma). Certain features of the fluid inclusion data indicate entrapment following granite intrusion and following the meteorite impact. No fluid inclusions studied so far record evidence of dolerite intrusion.

1. Meteorite Impact. The estimated formation temperature of pseudotachylite is 750°C (Reimold et al. 1987), much higher than the decrepitation temperature (380°C) of an inclusion in a clast contained within pseudotachylite melt (Fig. 2 g). Thermal modeling of clast-melt interactions for impact melts (Onorato et al. 1978) suggests that clasts and melt thermally equilibrate in less than a second. Therefore, it is likely that fluid inclusions in melt clasts would decrepitate if they pre-dated the melt. Similar reasoning applies to planar elements in quartz that are cross-cut by fluid inclusion trails. Planar elements form in response to extremely high strain rates, which cause disruption of the quartz crystal lattice by crystal gliding (Stoffler 1972). Any fluid inclusions contained within the crystal lattice would be likely to leak under these conditions. In fact, many fluid inclusions associated with impact features in the Siljan Ring rocks have leaked; these inclusions probably pre-date the impact. However intact two phase inclusions associated with impact features (Fig. 2 f) are likely to post-date the impact. These inclusions may preserve samples of post-impact hydrothermal fluids.

Heating associated with the meteorite impact provides the most likely explanation for higher trapping temperatures inside the Ring compared to outside (Fig. 8). This pattern would be fortuitous if fluid temperatures were not influenced in some way by heat derived from the impact. Quartz veins associated with blocks of sedimentary megabreccia are best explained as precipitates from impact-generated hydrothermal solutions (Lindblom and Wickman 1985):

2. Granite Intrusion. The last column in Table 2 shows that high average inclusion salinities ($\geq$4.2 wt.% $NaCl_{eq}$) are present in samples below -2504 m in Gravberg-1 and in one sample at a radial distance of 67 km. No highly saline inclusions are found in areas profoundly affected by the meteorite impact. The saline fluid inclusions probably contain samples of H_2O released from crystallizing magmas. For a granitic magma containing 5 wt.% water, chloride concentrations in evolved aqueous fluids range from ~6 to 18 wt.% NaCl equivalent (Holland 1972). These values apply under open system conditions in which evolved fluids leak away from the magma body; closed system conditions can generate fluids with even higher salinities. Similar magmatic processes may be responsible for high salinities in certain of the Siljan inclusions (Fig. 3).

3. Dolerite Intrusion. Doleritic magma intruded at 1200°C into cold granite would heat the contact to ~700°C (Table 5, Jaeger 1968). Two sources of water are available for hydrothermal activity in this environment. Fluid inclusions and hydrous phases pre-existing in the granite could evolve fluid. In addition, analyses of dolerites from the area (Hjelmqvist 1966) show that they can contain up to 3 wt.% water. Although the analyzed dolerites may contain some hydrous alteration minerals, their large water contents suggest that the dolerite magma was sufficiently wet to have released volatiles during crystallization. However, there is no evidence of very high trapping temperatures (near 700°C) in inclusions close to the dolerite/granite contact (T_t for sample GR1-2-9 at -1198, 2 m above R1, is 304°C). Nor are unusually high trapping temperatures present in sample GR1-4-7 200 m below R2, or GR1-4351 400 m above R3. Dolerite intrusion undoubtedly heated the surrounding granites, but it does not appear to have had a major influence on the fluid inclusion characteristics of the granites.

Discussion

In a plot of fluid inclusion trapping temperatures versus depth (Fig. 10), the data appear to define two populations, one above and one below -1200m. This depth coincides with the location of the R1 dolerite. The data below R1 can be fit to a line representing a thermal gradient of 29°C/km. Trapping temperatures above R1 are distinct from those below, and they are hotter for a given depth than would be predicted from an extension of the gradient below R1. The extra thermal input required to achieve the higher temperatures in samples above R1 is believed to be derived from the meteorite impact. Petrographic data reviewed earlier provide addi-tional evidence that fluid inclusions in shallow samples trap post-impact fluids.

Although the data provide strong evidence for heating due to the impact, the exact nature of the heating event is not clear. The data above R1 are scattered, but they may be interpreted to

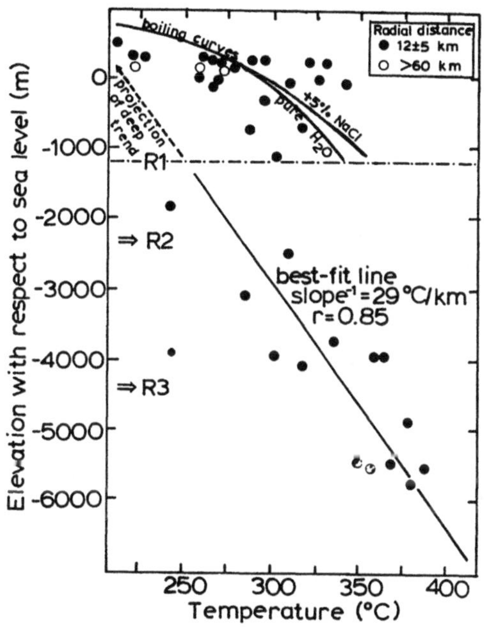

Fig. 10. Trapping temperatures versus sample elevation. One point at -3906 (GR1-6-3) was not regressed to calculate the best fit trend below -1200 m. Exclusion of this sample is justified by the consistently higher trapping temperatures determined for five other samples in the interval between -3723 and -4051 m (Table 2). The boiling curves above -1200 m are from Haas (1971). They are drawn assuming a hydrostatic pressure gradient and that 1 km of overburden has been removed from the crater since it was formed. The dashed line above -1200 m shows the projection of the deep trend to the present-day surface. Open circles are data from samples at radial distances greater than 60 km

follow boiling curves for low salinity fluids (Fig. 9). The boiling curves predict an increase of temperature with depth to ~350°C just above R1. The fluid inclusions do not provide unambiguous textural evidence for boiling, although the wide variation in the volume percent vapor especially prevalent in shallow samples (Fig. 5, Table 2) might be supportive evidence for boiling. The deep gradient of 29°C/km below R1 may represent a hydrothermal convection system recharged by meteoric water within a cooling granite pluton. The low salinities in most of the fluid inclusions are typical of those found in hydrothermal systems hosted by silicic rocks and recharged by meteoric water (Edmunds et al. 1984, Henley 1985). We are presently evaluating various processes that may produce the data trends defined by the shallow and deep fluid inclusion data.

The R1 dolerite occurs at the interface between the shallow, impact-generated hydrothermal system, and the deep hydrothermal activity attributed to plutonic cooling. This dolerite presumably represented a low permeability barrier that isolated the relict thermal structure of the deep system from overprinting by the impact-generated system. At the center of the Siljan Ring where the R1 reflector is disrupted or absent, the aquiclude is probably not present. In the central part of the Ring, the hydrothermal structure is likely to be other than described here.

Concluding Remarks

There are several lines of evidence that fluid inclusions in shallow granites from the Siljan Ring record impact-generated hydrothermal activity. Fluid inclusions are present in thick quartz veins that cement impact-related faults around the annulus. Certain lithologic features of undisputed impact origin (i.e., clasts in pseudotachylites) are cross cut by fluid inclusion trails. Samples within the annular moat to a depth of -1200 m have fluid characteristics distinct from those of samples outside this zone. The available data may indicate that impact-related heating generated boiling fluids in the crater. The increase in temperature of these fluids away from the point of impact is contrary to the expected thermal regime around an impact target. This suggests that patterns of fluid flow are complex in the vicinity of meteorite impacts. The study of crack-filling calcites by Valley et al. (this Vol.) provides evidence for low temperature fluids in granites at the Siljan Ring. The timing of low temperature fluid infiltration is not certain, but it is likely to be post-impact because the cracks filled by calcite are thought to be impact-generated. This low temperature fluid infiltration may be part of a continuum with the higher temperature event recorded by the fluid inclusions. Alternatively, it may represent a discrete episode, unrelated to the hydrothermal activity recorded in the fluid inclusions. Detailed studies of mineral assemblages and stable isotope compositions in the Ring will help clarify the fluid history of the crater.

Acknowledgments. We thank Paul Westcott, Dick Karlsson, John Costano, Ferol Fish, Tord Lindbo and Tony Gorody for help with this project. S.W. Bailey helped with chlorite identification, and P. Dombrowski drafted Fig. 1. This work was supported by the Gas Research Institute (5086-260-1425).

References

Åberg G (1978) Precambrian geochronology of southeastern Sweden. Geol Foeren Stockholm Foerh 100:125-154

Åberg G, Bollmark B (1983) Rb-Sr dating of a postorogenic Svecokarelian composite dike within the Siljan Ring structure, central Sweden. Geol Foeren Stockholm Foerh 105:75-77

Åberg G, Fredrikson G (1984) Radiometric dating of the postorogenic Järna granite, south central Sweden. Geol Foeren Stockholm Foerh 106:171-174

Åberg G, Levi B, Fredrikson G (1984) Zircon ages of metavolcanic and synorogenic granitic rocks from the Svardsjo and Yxsjoberg areas, south central Sweden. Geol Foeren Stockholm Foerh 105:199-203

Allen CG, Gooding JL, Keil K (1982) Hydrothermally altered impact melt rock and breccia: Contributions to the soil of Mars. Geophys Res 87:10083-10101

Borg IY (1972) Some shock effects in granodiorite to 270 kilobars at the Piledriver site. Am Geophys Un, Geophys Monogr 16:293-311

Bodnar RJ, Reynolds TJ, Kuehn CA (1985) Fluid inclusion systematics in epithermal systems. In: Berger BR, Bethke PM (eds) Geology and geochemistry of epithermal systems. Soc Econ Geol Rev Econ Geol 2:73-96

Bottomley RJ, York D, Grieve RAF (1978) ^{40}Ar-^{39}Ar ages of Scandinavian impact structures. I Mien and Siljan. Contrib Mineral Petrol 68:79-84

Dence MR, Grieve RAF, Robertson PB (1977) Terrestrial impact structures: Principal characteristics and energy considerations. In: Roddy DJ, Pepin RO, Merrill RB (eds) Impact and explosion cratering. Pergamon, New York, pp 247-275

Edmunds WM, Andrews JN, Burgess WG, Kay RLF, Lee DJ (1984) The evolution of saline and thermal groundwaters in the Carnmenellis granite. Mineral Mag 48:407-424

Engelhardt W von, Bertsch W (1969) Shock-induced planar deformation structures in quartz from the Ries crater, Germany. Contrib Mineral Petrol 20:203-234

Fisher JR (1976) The volumetric properties of H_2O-graphical portrayal. J Res US Geol Surv 4:189-193

French BM (1972) Shock-metamorphic features in the Sudbury structure, Ontario, a review, Geol Soc Am Spec Pap 10:19-28

Grieve RAF (1987) Terrestrial impact structures. Annu Rev Earth Planet Sci 15:245-270

Grieve RAF, Head JW (1983) The Manicouagan impact structure: An analysis of its original dimensions. J Geophys Res 88:A807-A818

Grieve RAF, Dence MR, Robertson PB (1977) Cratering processes: As interpreted from the occurrence of impact melts. In: Roddy DJ, Pepin RO, Merrill RB (eds) Impact and explosion cratering. Pergamon, New York, pp 791-814

Haas JL (1971) The effect of salinity on the maximum thermal gradients of a hydrothermal system at hydrostatic pressure. Econ Geol 66:940-946

Hayes JB (1970) Polytypism of chlorite in sedimentary rocks. Clays Clay Minerals 18:285-306

Henley RW (1985) The geothermal framework of epithermal deposits. In: Berger BR, Bethke PM (eds) geology and geochemistry of epithermal systems. Soc Econ Geol Rev Econ Geol 2:1-24

Hjelmqvist S (1966) Summary description to map of the pre-Quarternary rocks of the Kopparberg county, central Sweden (in Swedish with English summary) with map (1:200,000). Sver Geol Unders C40:1-217

Holland HD (1972) Granites, solutions, and base metal deposits. Econ Geol 67:281-301

Jaeger JC (1968) Cooling and solidification of igneous rocks. In: Hess HH, Poldervaart A (Eds) Basalts, Vol. 2. Wiley Interscience, New York, pp 503-556

Johansson A (1984) Geochemical studies on the Boda Pb-Zn deposit in the Siljan astrobleme, central Sweden. Geol Foeren Stockholm Foerh 106:15-25

Juhlin C, Pedersen LB (1988) Reflection seismic investigations of the Siljan impact structure. J Geophys Res 92:14113-14122

Kresten P (1986) Geochemistry and tectonic setting of metavolcanics and granitoids from the Falun area, south central Sweden. Geol Foren Stockholm Foerh 107:275-285

Lindblom S (1984) Hydrogeological and hydrogeochemical investigations in boreholes-fluid inclusion studies in the Stripa granite. Stripa Proj Intern Rep 84-07:34

Lindblom S, Wickman FE (1985) Fluid inclusions in quartz from a quartz breccia in the Siljan Ring structure, central Sweden. Geol Foeren Stockholm Foerh 107:53-58

Lofgren C (1979) Do leptites represent Precambrian island arc rocks? Lithos 12:159-165

Lundqvist T (1979) The Precambrian of Sweden. Sver Geol Unders C768:1-87

Newsom HE (1980) Hydrothermal alteration of impact melt sheets with implications for Mars. Icarus 44:207-216

Newsom HE, Graup G, Sewards T, Keil K (1987) Fluidization and hydrothermal alteration of the suevite deposit at the Ries crater, West Germany, and implications for Mars. J Geophys Res 91:E239-E251

Onorato PIK, Uhlman DR, Simonds CH (1978) The thermal history of the Manicouagan impact melt sheet, Quebec. J Geophys Res 83:2789-2798

Pasteris JD, Kuehn CA, Bodnar RJ (1986) Applications of the laser Raman microprobe RAMANOR U-1000 to hydrothermal ore deposits, Carlin as an example. Econ Geol 81:915-930

Patchett PJ (1978) Rb/Sr ages of Precambrian dolerites and syenites in southern and central Sweden. Sver Geol Unders C747:1-63

Patchett PJ, Todt W, Gorbatschev R (1987) Origin of continental crust of 1.9-1.7 Ga age: Nd isotopes in the Svecofennian orogenic terrains of Sweden. Precamb Res 35:145-160

Phinney WC, Simonds CH, Cochran A, McGee PE (1978) West Clearwater, Quebec impact structure, part II. Petrology: Proc 9th Lunar Planet Sci Conf 2:2659-2694

Peters T (1986) Structurally incorporated and water extractable chlorine in the Boettstein granite (N. Switzerland). Contrib Mineral Petrol 94:272-273

Potter RW (1977) Pressure corrections for fluid-inclusion homogenization temperatures based on the volumetric properties of the system NaCl-H2O. U S Geol Surv J Res 5:603-607

Potter RW, Clynne MA, Brown DL (1978) Freezing point depression of aqueous sodium chloride solutions. Econ geol 73:284-285

Reimold WU, Oskierski W, Huth J (1987) The pseudotachylite from Champagnac in the Rochechouart meteorite crater, France. J Geophys Res 92:E737-E748

Robertson PB (1975) Zones of shock metamorphism at the Charlevoix impact structure, Quebec. Geol Soc Am Bull 86:1630-1638

Roddy DJ, Davis KL (1977) Shatter cones formed in large-scale experimental explosion craters. In: Roddy DJ, Pepin RO, Merrill/RB (eds) Impact and explosion cratering. Pergamon, New York, pp 715-750

Roedder E (1984) Fluid inclusions. Mineral Soc Am Rev Mineral 12:644

Savage D, Cave MR, Milodowski AE, George I (1987) Hydrothermal alteration of granite by meteoric fluid, an example from the Carnmenellis granite. United Kingdom. Contrib Mineral Petrol 96:391-405

Settle M, Head JW (1987) The role of rim slumping in the modification of lunar impact craters. J Geophys Res 84:3081-3096

Stoffler D (1972) Deformation and transformation of rockforming minerals by natural and experimental shock processes. Fortschr Mineral 49:50-113

Stone M, Exley CS (1985) High heat production granites of southwest England and their associated mineralization, a review, in High heat production (HHP) Granites, hydrothermal circulation and ore genesis. Inst Min Metall, London, pp 571-593

Svensson NB (1973) Shatter cones from the Siljan structure, central Sweden. Geol Foeren Stockholm Foerh 95:139-143

Tamminen J, Wickman FE (1980) Shock effects in pegmatitic quartz from the Siljan ring structure, central Sweden. Geol Foeren Stockholm Foerh 102:275-278

Tuttle OF, Bowen NL (1958) Origin of granite in the light of experimental studies in the system $NaAlSi_3O_8$-$KAlSi_3O_8$-SiO_2-H_2O. Geol Soc Am Mem 74

Wehlin E, Lundqvist T (1970) New Rb-Sr age data for the Sub-Jotnian volcanics (Dala porphyries) in the Los-Harma region, central Sweden. Geol Foeren Stockholm Foerh 92:35-39

Wickman FE (1980) The impact center of the Siljan structure, Central Sweden. Geol Foren Stockholm Foerh 102:105-109

Wilson MR (1980) Granite types in Sweden. Geol Foeren Stockholm Foerh 102:167-176

Low Temperature Alterations in Granitic Rocks from the Siljan Ring Structure, Central Sweden

A. A. ALDAHAN[1], K. RAMSEYER[2], S. MORAD[1] and B. COLLINI[1]

Introduction

This chapter presents results on hydrothermal alterations in granitic rocks (1,635-1,735 Ma old; Lundqvist 1979; Åberg and Fredriksson 1984) obtained from the Gravberg-1 well (six cores) and three outcrops in the Siljan Ring Structure (Fig. 1). This structure represents a remnant of a meteorite impact (e.g., Fredriksson and Wickman 1963; Svensson 1973), which is about 362 Ma old (Bottomley et al. 1978).

The rocks investigated are red to greyish red, medium to coarsegrained (sometimes porphyritic) and show allotriomorphic to hypidiomorphic textures. Microperthitic and antiperthitic feldspars as well as granophyric texture occur in most of the rocks. Subhedral crystals of plagioclase showing zonal variation from oligoclase cores to albite rims, occur in some

Table 1. Modal composition of the samples investigated

Samples	Quartz	K-feldspar	Plagioclase	Chlorite	Biotite	Opaques	Epidote, sphene hornblende	Sericite
outcrops								
1	37	40	20	0.5	0.5	<0.5	<0.5	<0.5
2	43	33	17	3	1	1	1	1
3	36	54	10	2	1	<0.5	<0.5	<0.5
Borehole								
4 (1002 m)	40	40	15	2	0.5	0.5	<0.5	1
5 (1496 m)	21	43	20	5	2	2	4	3
6 (2135 m)	21	44	22	5	3	1	2	2
7 (4205 m)	34	36	24	2	2	0.5	1	0.5

1 Department of Mineralogy and Petrology, Institute of Geology, Uppsala University, Box 555, 751 22 Uppsala, Sweden
2 Department of Geology, University of Bern, 3012 Bern, Switzerland

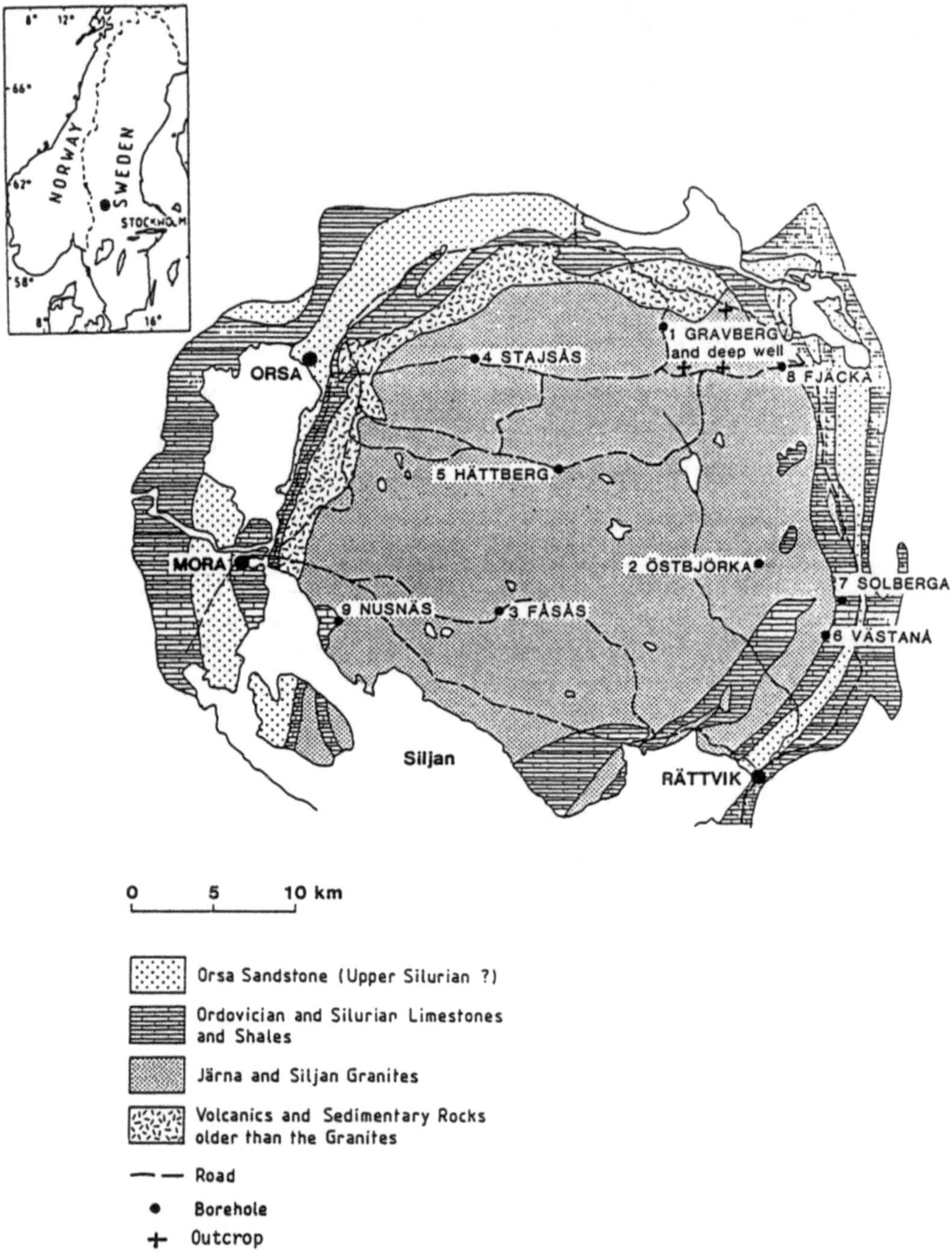

Fig. 1. A simplified geologic map of the Siljan Ring Structure. (After Hjelmqvist 1966)

samples. The mineralogical composition (Table 1) indicates a two-feldspar granite to adamellite (Hughes 1982) with minor amounts of calcite, zircon, apatite, and fluorite. The opaque minerals are mainly hematite, Ti-oxides, and Fe-Ti oxides (mainly ilmenite) in samples 1-5 and magnetite and ilmenite in samples 6 and 7.

In addition to light microscopy, a JEOL SEM instrument (equipped with EDS) and a cathodoluminescence microscope (e.g., Matter and Ramseyer 1985) were used in examinations of the investigated rocks.

Mineralogical and Textural Alterations

The most common feature of alteration in the rocks investigated is the formation of albite and K-feldspar (adularia), which are rather pure ($NaAlSi_3O_8$ and $KAlSi_3O_8$, respectively). The albite occurs as replacement of primary feldspars (microcline, oligoclase, and andesine) and fracture fillings. The K-feldspar is found as fracture fillings too. The cathodoluminescence color of the secondary K-feldspar is pale brown when compared to the bright blue color of the primary microcline. The albite shows dark brownish colors compared to the bright yellow and pink colors of primary plagioclases (oligoclase and andesine) and albite (in perthitic and antiperthitic textures), respectively. Some apparently deformed K-feldspar and albite crystals (Fig. 2 a and b) also show cathodoluminescence colors similar to those exhibited by secondary feldspars above. The origin of these secondary crystals is not clear, but their texture indicates flow under high pressure, as suggested by sutured boundaries with adjacent quartz.

The amounts of secondary (altered) feldspars generally decrease with depth and an estimation of about 50-60% of total feldspar is secondary in the outcrop samples and the 1002 m-sample of the borehole and about 20-30% in sample 7 (4205 m depth). Samples 5 and 6 (1496 and 2135 m) occur in between the above ranges. In most of the albitized feldspars there is commonly a mixture of sericite and/or smectite ± hematite ± chlorite ± epidote ± sphene (Fig. 3 a and b). The occurrence of only smectite appearing as honeycomb texture, is found in some altered plagioclases (Fig. 4 a).

Another extensively altered mineral in the rocks studied is biotite. The amount of biotite (particularly fresh grains) increases with depth. The main alterations of biotite are the occurrence of chlorite and/or phengite as interstratification together with hematite and Ti-oxides. The chlorite is usually pale green in color and Fe-rich (up to 25 wt % FeO). In some biotite grains, the alteration assemblage is epidote + sphene + violet-colored chlorite, and less commonly pumpellyite ± prehnite ± chlorite ± sphene (sample 7; 4205 m).

The Fe-Ti oxides (ilmenite and titanomagnetite) and magnetite were altered into hematite and Ti-oxides. The extent of this

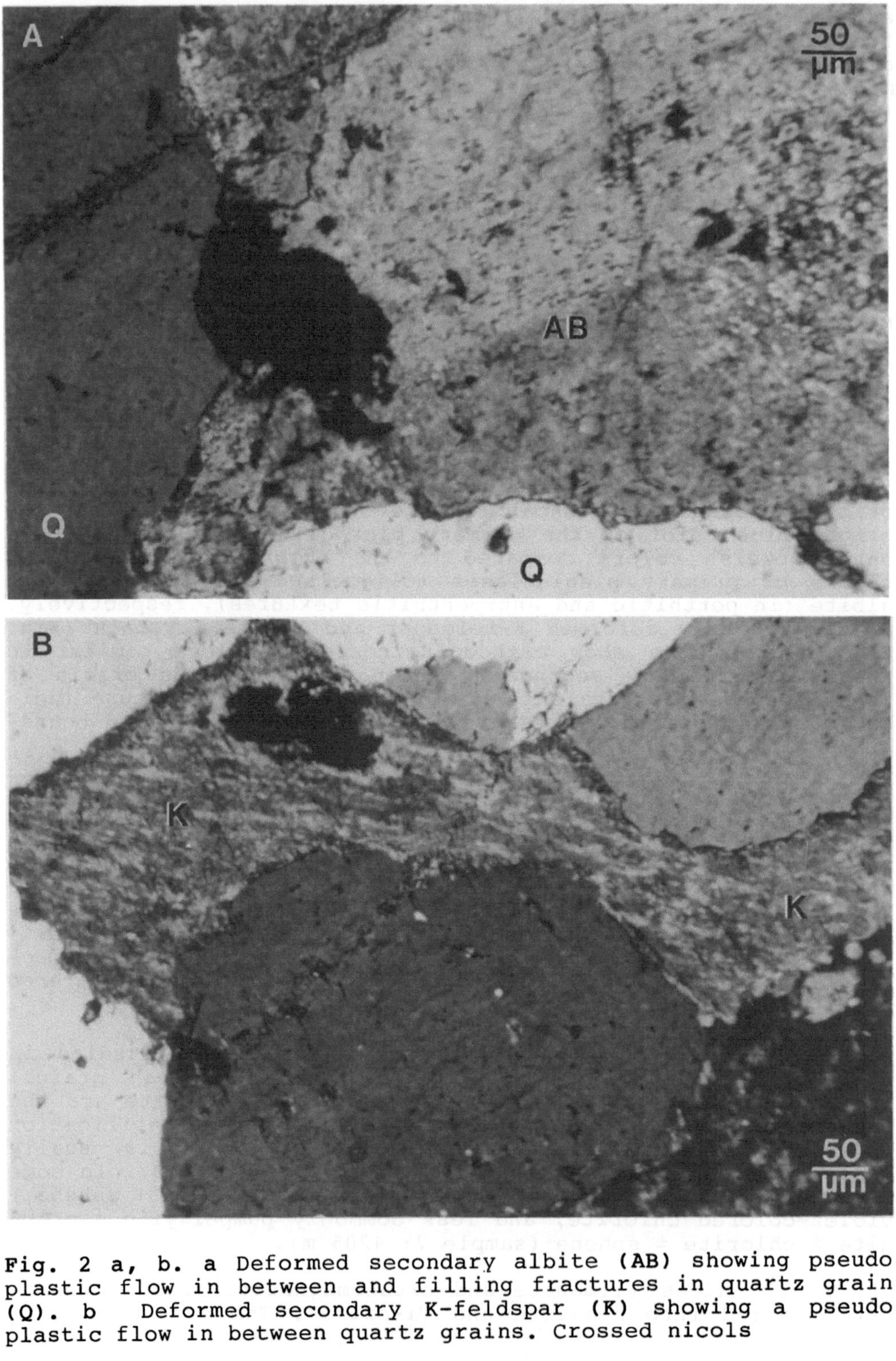

Fig. 2 a, b. a Deformed secondary albite (AB) showing pseudo plastic flow in between and filling fractures in quartz grain (Q). b Deformed secondary K-feldspar (K) showing a pseudo plastic flow in between quartz grains. Crossed nicols

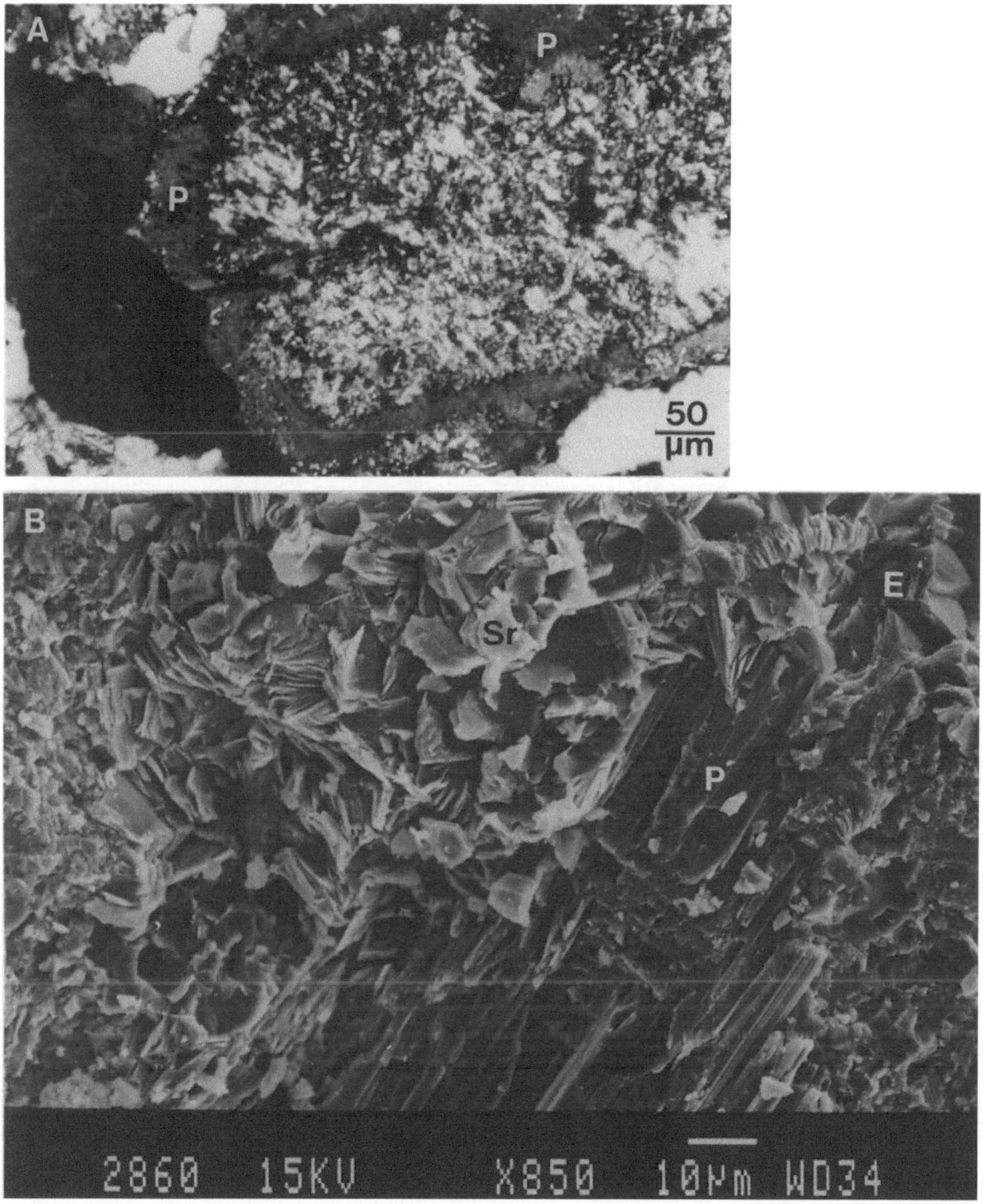

Fig. 3 a, b. a A mixture of sericite, smectite, and epidote replacing the central part of a plagioclase (P). Crossed nicols. b Sericite (Sr) and epidote (E) replacing a plagioclase (P). SEM photograph

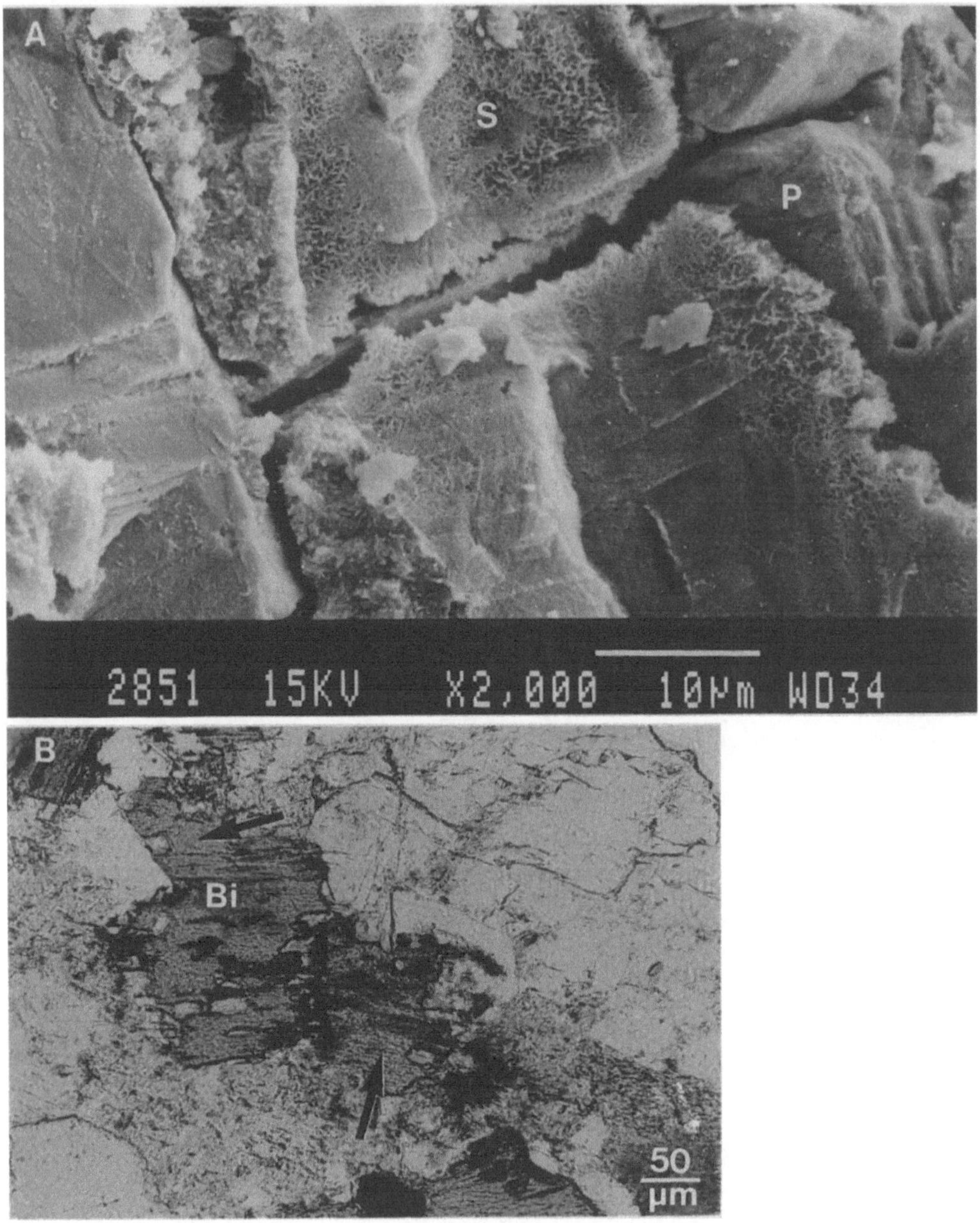

Fig. 4 a, b. a Smectite (S) replacing a plagioclase (P). SEM photograph. b Biotite (Bi) showing alteration into chlorite (arrows), hematite and Ti-oxide lumps (dark spots). Plane polarized light

alteration decreases with depth and in samples 6 and 7 (2135 and 4205 m) fresh grains of the Fe-Ti oxides and magnetite are common.

Hornblende is also extensively altered into pale-green Fe-rich chlorite (± epidote ± sphene). The alteration also decreases with depth and fresh grains of hornblende are more common in deep samples (e.g., sample 7; 4205 m).

Secondary quartz, which occurs in fractures, altered feldspars, and biotites, is common in all the samples studied. This quartz is nonluminescent compared to the blue to deep blue colors of the primary quartz. Pinkish luminescence color is observed in some apparently recrystallized primary quartz in the outcrop samples and mainly in sample 4 (1002 m) of the borehole.

Discussion and Conclusions

The results of this study suggest that most of the alteration features observed were produced at low temperature conditions. The exact range of temperature is difficult to estimate by only using the available data, but values less than 300°C are expected, as shown below. Among the most important features of low temperature alterations are:

1. Secondary albite and K-feldspar showing luminescence colors of low-temperature varieties.

2. Occurrence of pumpellyite and prehnite, which have compositions and parageneses comparable to those occurring in the zeolite-pumpellyite facies (cf. Liou 1979; AlDahan 1988).

3. Occurrence of nonluminescent quartz typical of low-temperature variety.

4. Most chlorites replacing biotites show color (pale-green) and composition (Fe-rich) comparable to that found in low temperature environments (e.g., diagenetic chlorites; AlDahan and Morad 1985).

Experimental alteration of granitic rocks (Savaga et al. 1987) indicated that smectite developed on feldspars occurs at a temperature of 150-250°C, which could also indicate a similar range in the rocks investigated. A temperature as low as 150°C was also deduced by AlDahan (1988) from the pumpellyite-prehnite assemblage in sample 7 (4205 m). The occurrence of fracture-filling albite, quartz, calcite, and chlorite might represent precipitations at temperatures even lower than 150°C (cf. Valley et al. this Vol., AlDahan et al. this Vol.).

Apart from the pumpellyite-prehnite assemblage, most of the alteration features observed could be related to a post-impact episode. Activated hydrothermal cells might have been generated due to the impact shock heating and fracturing. The rather expected deep (down to at least 7 km) and extensive fracturing (possibly 50 km beyond the original rim of Siljan; Grieve, this

Vol.) have facilitated the penetration of meteoric waters to deep zones. Such waters are the most expected reacting fluids, which have caused most of the low temperature alterations observed in the investigated rocks (cf. Savaga et al. 1987; Valley et al. this Vol.; Komor et al. this Vol.).

Acknowledgments. The Swedish State Power Board (Vattenfall) is acknowledged for access to borehole samples and financial support to A. A. AlDahan. Miss Kersti Gloersen kindly typed the manuscript.

References

Åberg G, Fredriksson G (1984) Radiometric dating of the postorogenic Järna granite, south central Sweden. Geol Foeren Stockholm Foerh 106:171-174

AlDahan AA (1988) The paragenesis of pumpellyite in granitic rocks from the Siljan area, central Sweden. Mineral Petrol (in press).

AlDahan AA, Morad S (1985) Mineralogy and chemistry of diagenetic clay minerals in Proterozoic sandstones from Sweden. Am J Sci 286:29-80

Bottomley RJ, York D, Grieve RAF (1978) ^{40}Ar-^{39}Ar ages of Scandinavian impact structures: I Mien and Siljan. Contrib Mineral Petrol 68:79-84

Fredriksson K, Wickman FE (1963) Meteoriter. Svensk Natur 16:121-127

Hjelmqvist S (1966) Beskrivning till berggrundskarta över Kopparbergs län. Sver Geol Unders Ca40:217

Hughes W (1982) Igneous petrology. Elsevier, Amsterdam 551 pp

Liou JG (1979) Zeolite facies metamorphism of basaltic rocks from the east Taiwan ophiolite. Am Mineral 64:1-14

Lundqvist Th (1979) The Precambrian of Sweden. Sver Geol Unders C766,:87

Matter A, Ramseyer K (1985) Cathodoluminescence microscopy as a tool for provenance studies of sandstones. In: Zuffa GG (ed) Provenance of Arenites D Reidel Publ Comp MA USA pp 191-211

Savaga D, Cave M, Milodowski A, George I (1987) Hydrothermal alteration of granite by meteoric fluid: an example from the Carnmenellis Granite, United Kingdom. Contrib Mineral Petrol 96:391-405

Svensson NB (1973) Shatter cones from the Siljan Structure, central Sweden. Geol Foeren Stockholm Foerh 95:137-143

Mössbauer Spectroscopic Data of Granitic Rocks from Gravberg-1 Well, Siljan Ring Structure, Central Sweden

A. A. ALDAHAN, T. ERICSSON and B. COLLINI[1]

Introduction

Mössbauer spectroscopy (MS) is a technique based on "recoil-free" emission and absorption of γ-quantas, especially the 14.4 -keV transition in ^{57}Fe is extensively used. As a consequence of the hyperfine interaction combined with an extremely high energy resolution this nuclear method is frequently used for studying problems in atomic and solid state physics. MS can be used to determine the valence state of iron but also to give information about coordination numbers, site distorsions and magnetic fields acting at the iron nucleus. A big advantage is that the amount of sample needed for an investigation is normally quite small, typically less than 100 mg.

Apart from its wide use in mineralogical investigations, MS can also be used for identifying Fe-bearing minerals in a rock sample. Furthermore, the technique will then normally also give Fe^{2+}/Fe^{3+}-ratios in the different minerals constituting the rock. These ratios cannot be given by any competing technique. The Fe^{2+}/Fe^{3+}-ratio for the whole rock sample obtained by wet chemical methods is normally not as accurate as by MS, due to problems with oxidating and reducing processes in the chemical treatment of the sample. In cutting samples, iron contaminations from drilling pipes can easily be detected by the MS and thus be excluded from the analyses.

In this investigation MS has been used to evaluate oxidation conditions in samples from different depths of the Gravberg-1 well. Three cores and three cutting samples were investigated. The mineralogical composition of these rock samples was typically: quartz (20-40 vol.%), feldspare (55-65 vol.%), biotite and chlorite (3-7 vol.%), and minor amounts of sphene, epidote, hornblende, sercite, magnetite and hematite. In the latter structurally closely related Fe-Ti oxides are also included.

The transmission Mössbauer spectrometer was used of conventional type, working in constant acceleration mode. The source, mounted on the vibrator, was ^{57}Co in a Rh matrix kept at room temperature. The powder absorbers contained 75-150 mg/cm^2 and the

---

[1] Department of Mineralogy, Institute of Geology, University of Uppsala, Box 555, 75122 Uppsala, Sweden

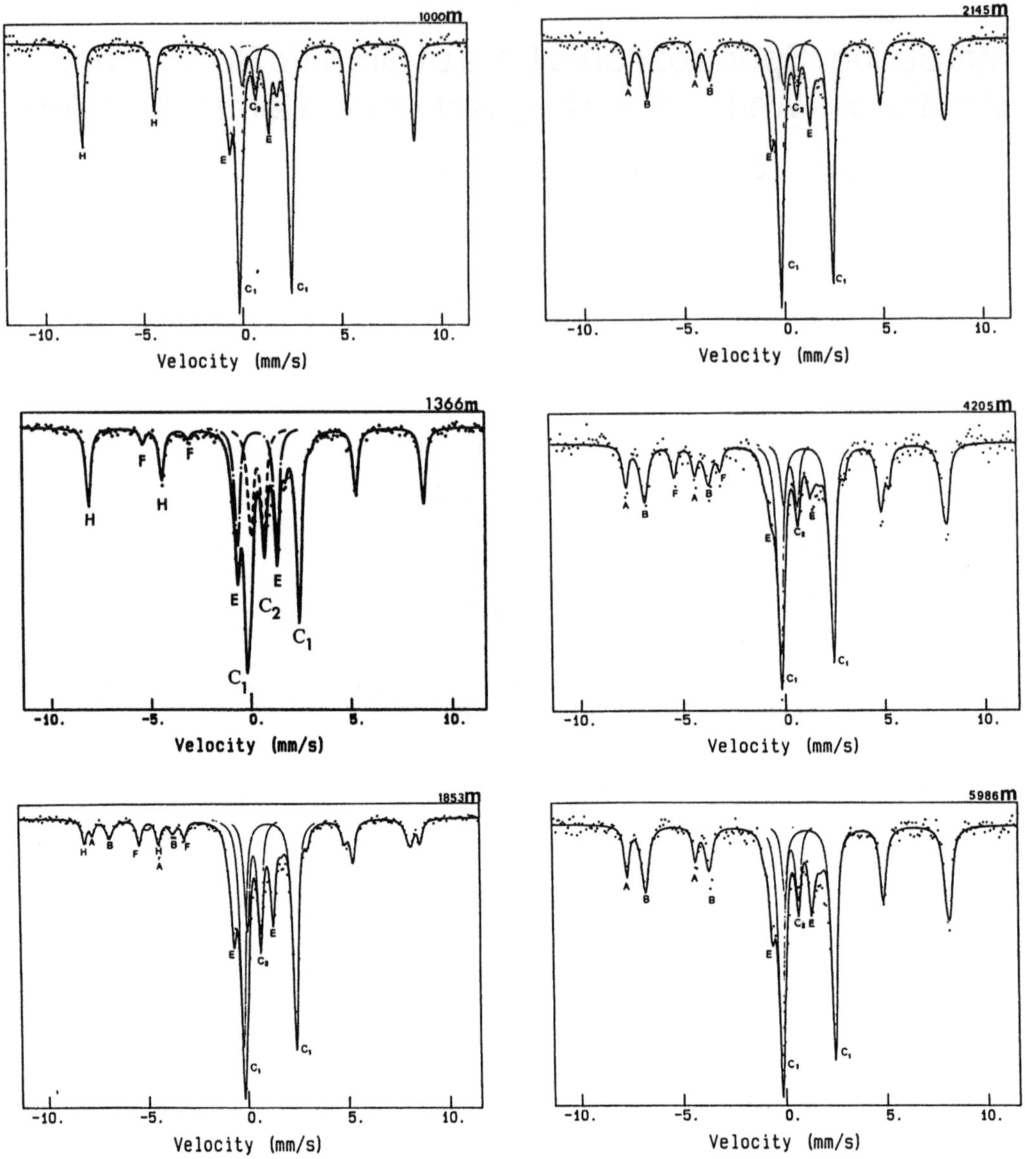

Fig. 1. Mössbauer spectra recorded at room temperature. Numbers above the spectra indicate depth in m. Some diagnostic peaks are assigned: H=hematite; A, B A- and B positions in magnetite; E=Epidote (Fe^{3+}); F=free iron (α-Fe); C_1 and C_2=Fe^{2+} and Fe^{3+} respectively in chlorite, biotite and hornblende

absorption in the spectra was around 1%, thus it was not necessary to make thickness corrections in the fittings of the spectra. The centroid shift, CS, is given relative α-Fe at room temperature. The Mössbauer parameters of synthetic hematite (from Mercks) were used as standard values in fitting hematite found in the investigated rock samples.

Table 1. Distribution of iron (in %) and its oxidation state in the studied samples

Depth (m)	Hematite Fe^{3+}	Magnetite Fe^{3+}	$Fe^{2.5+}$	Iron α-Fe	Bi,Ch,Hb Fe^{2+}	Fe^{3+}	Epidote Fe^{3+}	Fe^{3+}/Fe^{2+}
CR 1000	34	0	0	0	46	7	13	1.17
CT 1366	25	0	0	5	33	18	20	1.96
CT 1853	7	4	9	8	39	19	14	1.11
CR 2145	0	11	22	0	43	7	17	0.85
CR 4205	0	13	26	10	35	10	5	0.84
CT 5986	0	15	30	0	38	7	10	0.87

CR = core sample, CT = cutting sample, Bi = biotite, Ch = chlorite, Hb = hornblende

Fitting Procedure of Mössbauer Spectra

Microscopic examinations indicated that the content of iron-bearing minerals varied from about 4-13 vol.%. In the fittings of the Mössbauer spectra (Fig. 1) the following assignments were done:

1. In some of the spectra Mössbauer parameters, typical for hematite (eventually with a small amount of Ti) were found. (A sextet with CS = 0.38 mms^{-1} and hyperfine magnetic field B = 56.8 tesla).

2. Two magnetic sextets, with CS/B = 0.28/49.3 and CS/B = 0.64/46.0. The intensity of the former is found to be close to half of the latter. These parameters are typical for iron in magnetite. The former sextet emanates from ferric iron at the tetrahedral A-position and the latter from iron at the twice as frequent octahedral B-site. The Mössbauer parameters of the latter indicate that the iron is in 2.5-valent state (fast electron hopping between Fe^{2+} and Fe^{3+}; cf. Häggström et al. 1978). Titano-magnetite having a small concentration of Ti gives the same Mössbauer pattern and is also included in these signals.

3. There was also sometimes a sextet showing parameters characteristic for α-Fe. This sextet was considered to emanate from iron fragments scraped away from the drilling pipes.

4. In all the samples studied, paramagnetic doublets having CS/QS = 1.13 mm/s/2.62 mm/s and CS/QS = 0.38/0.66 were obtained (The quadrupole splitting, QS, is defined as the peak separation in the doublet). These signals are typical for Fe^{2+} and Fe^{3+} in chlorite, biotite and hornblende (cf. Bancroft and Brown 1975; Ericsson et al. 1977).

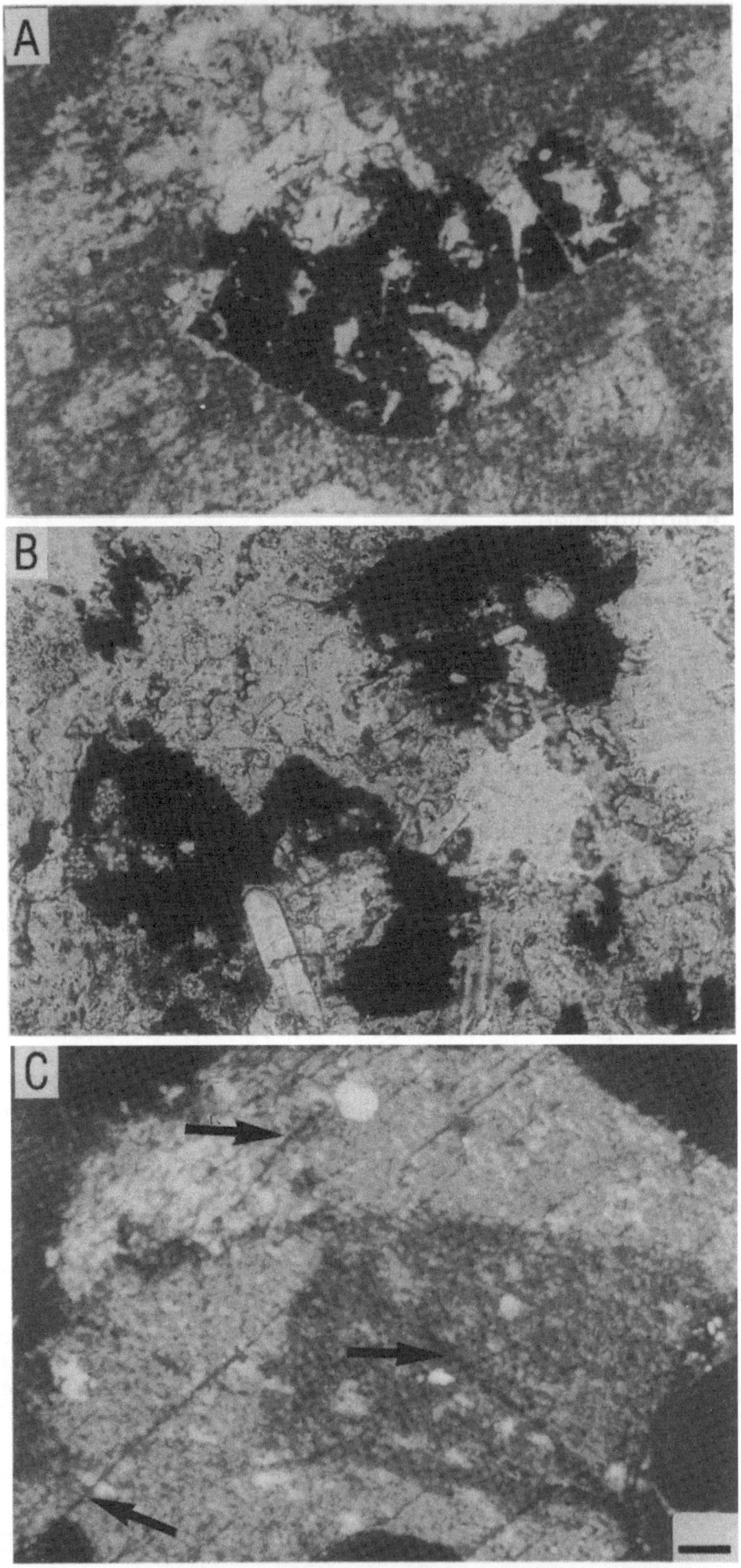
A
B
C

5. There was also another paramagnetic doublet with CS/QS = 0.36/1.97. These parameters are in agreement with published values for ferric iron in epidote and were accordingly assigned to it (e.g., Bancroft et al. 1967).

The fittings above suggest about 40-60% of the iron to be in chlorite, biotite or hornblende. In epidote there is around 5-20% and in magnetite and hematite about 20-45%. The obtained intensities, as well as the assignments to ferric and ferrous iron are given in Table 1. Microprobe analyses done on some of the samples gave compositions for the different mineral phases in agreement with the MS assignments above.

Discussion and Conclusions

The ferric-ferrous ratio, determined from Mössbauer spectra, for the different samples is shown in Table 1. As seen, there a general variation in oxidation degree with depth. This seems to stem from two main sources, namely the ratio of hematite to magnetite (including Fe-Ti oxides) and the amount of epidote in the studied samples.

The change in the hematite-magnetite ratio is related to the degree of oxidation. Fresh crystals of (titano) magnetite were altered into lumps of pseudomorphic (titano) hematite in the hematite dominated samples (Fig. 2 a and b). Furthermore, in such samples hematite occurs as fracture fillings in feldspars and quartz (Fig. 2 c). Therefore, it is clear that highly oxidizing conditions prevailed in the two uppermost samples picked from depths of 1000 and 1366 m; they show no magnetite in the Mössbauer spectra (Table 1). Conversely, in samples picked from depths of 2145, 4205, and 5986 m, no hematite was found in the MS specra. In a sample taken from 1853 m there was a coexistence of hematite and magnetite. Therefore a hematite-magnetite boundary zone at about 2000 meter seems to exist, as shown in Fig. 3. The amount of ferric iron in epidote does not show a clear variation with depth (Table 1). However, the occurrence of the mineral indicates oxidizing and H_2O buffered environments where epidote is common in extensively fractured, thus porous, samples.

A possible explanation for the high oxidation conditions prevailing in the uppermost 2000 m is circulating meteoric water. However, there are exceptions to that role, as some cutting samples picked from a depth of more than 3000 m also show

◀ **Fig. 2 a-c. a** Fresh crystals of magnetite **(dark)** in sample picked at 2145 m. Plane-polarized light. **b** Lumps of hematite and titanium oxides **(dark)** occurring as pseudomorphs after decomposed magnetite and Fe-Ti oxides. Sample picked at 1001 m. Plane-polarized light. **c** Hematite **(arrows)** filling fractures in a microperthite grain. Sample picked at 1000 m. Crossed nicols (bar scale = 50 μm)

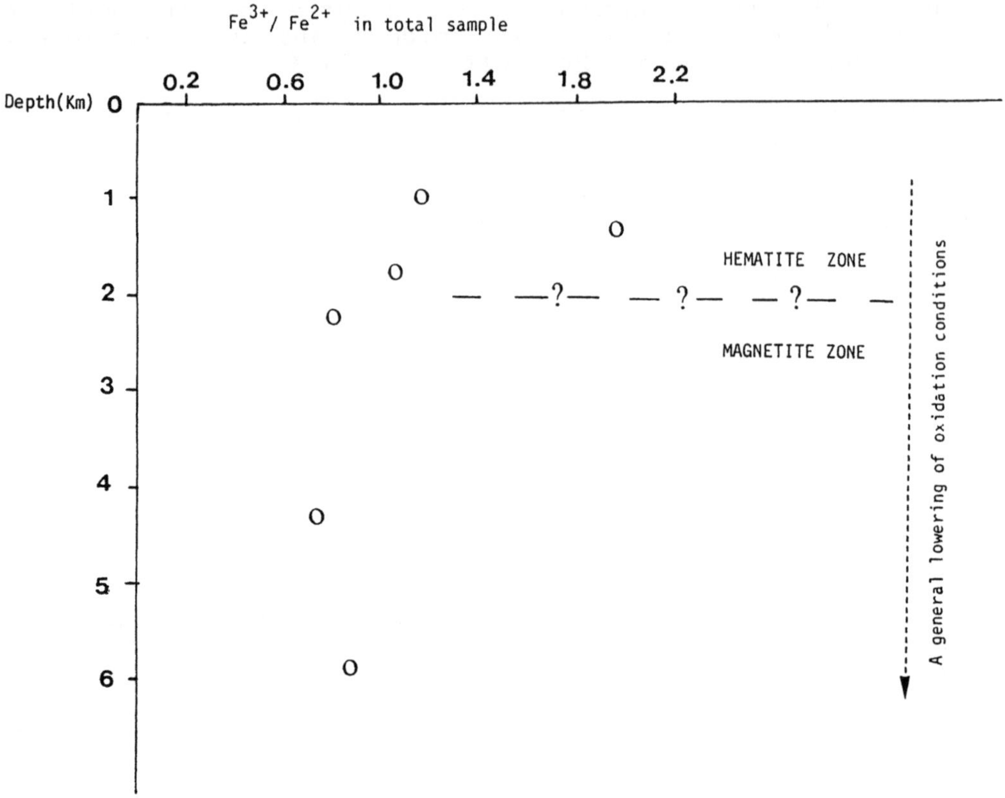

Fig. 3. Distribution of Fe^{3+}/Fe^{2+} values versus depth in the studied samples

occurrence of hematite. These samples were extensively deformed and fractured, thus reasonably more buffered with oxygen from hydrothermal/meteoric water than the surrounding rock. It appears that fracture and deformation zones, which were encountered at depth down to 6000 m in the borehole, act as oxidizing environments and probably also have connections to circulating meteoric water as shown in Fig. 4 (cf. Grieve this Vol.). Oppositely, if methane and related gases had been trapped in these fracture zones, then reducing conditions would have prevailed instead. Thus, MS is a convenient method of studying such situations.

Free iron was detected in some of the samples (Table 1), probably contaminations produced during the drilling operation. Although hydrogen gas was detected at some depth levels of the borehole, it does not seem possible that the free iron was produced as a result of highly reducing conditions. Especially, in the sample taken from a depth of 4205 m the free iron coexists with magnetite and epidote, both containing ferric iron. As it is probably a contamination, it is very important

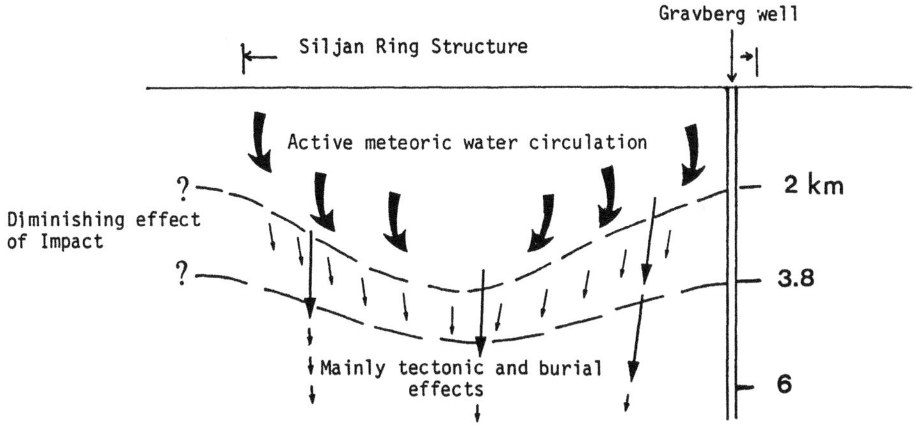

Fig. 4. A model expected extent of meteoric water circulation in relation to depth, fracture systems (long arrows) and impact effects

to consider the amount of free iron in the chemical analyses of the samples. MS, being a nondestructive method, is quite suitable for such a purpose.

Acknowledgment. The authors would like to thank the Swedish State Power Board (Vattenfall) for financial support.

References

Bancroft GM, Brown JR (1975) A Mössbauer study of coexisting hornblendes and biotites: quantitative Fe^{3+}/Fe^{2+} ratios. Am Mineral 60:265-272

Bancroft GM, Maddock AG, Burns RG (1967) Application of the Mössbauer effect to silicate mineralogy-1. Iron silicates of known crystal structure. Geochim Cosmochim Acta 31:2219-2246

Ericsson T, Wäppling R, Punakivi K (1977) Mössbauer spectroscopy applied to clay and related minerals. Geol Foeren Foerh 99:229-244

Häggström L, Annersten H, Ericsson T, Wäppling R, Karner W, Bjarman S (1978) Magnetic dipolar and electric quadrupolar effects on the Mössbauer spectra of magnetite above the Verwey transition. Hyperfine interact 5:201-214

Investigations of Rocks from the Gravberg-1 Well: Petrological and Geochemical Results

D. KARLSSON[1]

Introduction

The surficial bedrock of the central part of Dalarna in middle Sweden is composed of Dala granites (Proterozoic) in the middle surrounded, to the east and south, by old migmatites, granites and leptites (Proterozoic) and to the west and north by old sedimentary rocks and porphyries (Proterozoic).

In the Siljan Ring are there is also a band of younger sedimentary rocks (Paleozoic) surrounding the impact centre.

The Dala granites are (in the Siljan Ring area) dominated by two types, the so-called Järna and Siljan granite. The Siljan granite is a more acid type than the Järna granite.

A general range of some oxides in the two types area:

% by weight	SiO_2	Al_2O_3	CaO
Järna	66	15-16	2-3
Siljan	74	12-13	0.5-1

Gravberg is situated close to the eastern boundary of the distribution area of the Dala granites shown above.

Samples and Analysis Methods

Before spud-in, some forecasts about the types of rocks that could be expected during the drilling of Gravberg-1 were made.

Both Siljan and Järna granites were expected to be found below a thin porphyry layer. The possibility to reach Svecokarelian basement rocks before TD, both migmatites and granites, was suggested. Also, the possibility to hit a nearly vertical dolerite was proposed.

Before spud-in, a preliminary study of some outcrop samples was made, in order to determine suitable cuttings size-grades for the on-site petrographical examination by reflected light

1 Hagconsult AB, Stjärnvägen 2, S-181 83 Lidingö, Sweden

microscope. These tests resulted in the choice of 0.25-0.5-mm-large cuttings as the best fraction for the on-site analyses.

During the drilling, stained cutting samples from every sixth metre were examined by the site geologists and point-counted for their mineral composition.

Further cutting samples at every 100 m and at intervals of special interest were examined with a polarizing microscope. At the same intervals chemical analyses of major and some trace elements were performed.

The analyzed samples cover a depth interval of about 6600 m and represent mainly acid igneous rocks, besides about 100 m of basic intrusions. The analyses of the cuttings samples were supplemented by the analyses of six conventional and eight side wall cores from the well.

Pump pressure, flow rate, mud weight, hole dimension and depth are some parameters that have great impact on settling velocities in the well. It is not possible to completely compensate for this factors, and the problem becomes more difficult with the limited amount of cores available for calibration. Coring has been discouraging in Gravberg and out of 14 attempts only six were successful.

Results

The rocks penetrated in the borehole to the present TD can as of now be divided into the following igneous types, mainly depending on the silica content.

1. Rocks showing granitic to quartz-monzonitic composition and SiO_2 content higher than 72%.

2. Rocks showing granodioritic composition and SiO_2 content between 66-72%.

3. Rocks showing quartz-dioritic composition and SiO_2 content less than 66%.

4. Dolerites.

Porphyry. A fine-grained porphyry which is known from earlier outcrop mapping and from shallow core drilling (in 1984) occurs in the upper part of the well. The porphyry is rather thin and the boundary to the underlying granite (GI) was established through wireline logging to lie between 165 and 250 m. This uncertainty is due to the fact that the mud logging was not fully operating above 400 m.

GI Granite. The GI granite (Fig. 1), which is supposed to be a so-called Siljan granite, is a homogeneous granite with a SiO_2 content generally above 76% and CaO less then 1%. The mineralogical composition shows a high content of K-feldspar (50-60%), a quartz content of 30-35%, a plagioclase content of 15-20% and

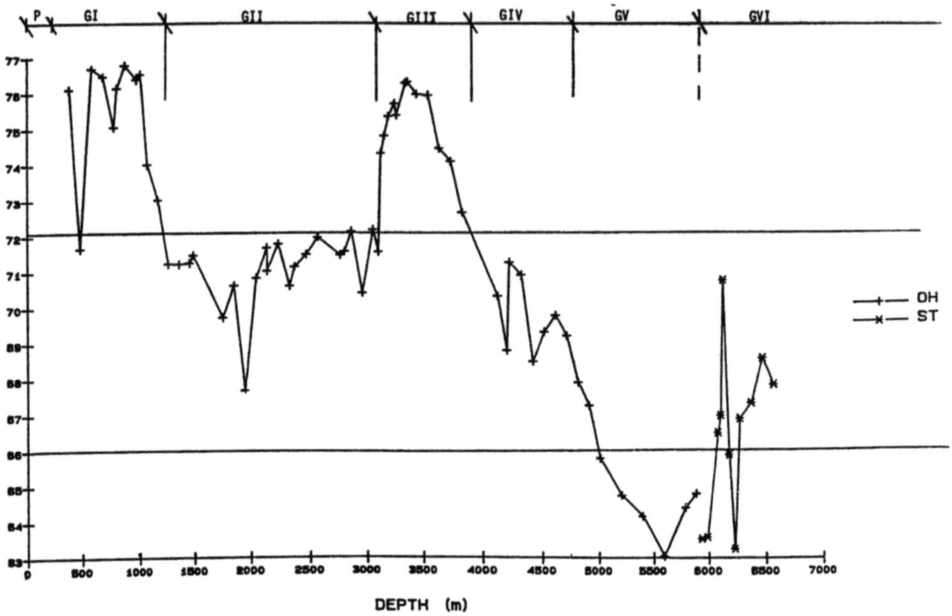

Fig. 1. Interpretation of rock types in Gravberg-1

minor biotite, chlorite and opaques. This GI granite continues down to 1100-1200 m, where a trend break appears in the geochemical properties. This is also supported by the geophysical gamma log. The interpretation is that there is a boundary between the GI and a new GII granite at this depth.

GII Granite. The GII granite, which is close to a Järna granite in composition, has a lower SiO_2 content about 71% and a higher CaO content than the GI granite. This GII-granite ends at about 3100 m where a change in the chemical composition occurs. A significant reversed drilling break and very high uranium and thorium contents also indicate that there is a boundary at about 3100 m depth.

GIII Granite. The GIII granite is very similar in composition to the previous GI granite, with a Silica (SiO_2) content of about 76%, and a lower CaO content than the previous GII granite. Bit-life decreased dramatically when that GIII granite was drilled. The GIII granite continues down to about 3900 m, where again a new relationship appears in the element distribution and the appearance of a new GIV granite. At the 3900-m level a lot of difficulties occurred in the drilling and the hole was much enlarged. This is probably directly related to a boundary between the GIII and GIV granites.

GIV Granite. The GIV granite corresponds in its general composition to the GII granite, with a SiO_2 content of about 71% and higher CaO content than the GIII granites. The GIV granite seems to continue down to about 4800 m. At this depth the geochemical pattern again starts to be somewhat different, also verified by the geophysical logging. The SiO_2 content in the GIV granite decreases with depth and is at 4800 m about 66%. This decrease in the silica content could probably be due to a magmatic differentation process.

GV Granite. It is doubtful whether this possible magmatic differentiation results in a new granitic type or not. But the element distribution is so different that it could be assumed that the granite under 4800 m is a new type, GV granite. This GV granite continues down to about 5800 m where the element distribution is getting more heterogeneous and does not fit in with the pattern of the upper part of the well.

GVI Granite. The granites down to 5800 m, probably belong to the Dala granite sequence. The rock under 5800 m depth shows some indications for being a Svecokarelian granite or migmatite. This pattern is also supported by some crossplots, for example Zr - TiO_2 (Fig. 2), which show the possibility for the occurrence of six different rocktypes beneath the porphyry. The granites found in the well are what generally is known as I-granites, which means they have an igneous origin. Further investigations concerning the plagioclase type will of course give a much more precise terminology for the investigated rocks. Also, datings will give us further ideas about the relationship between the granites.

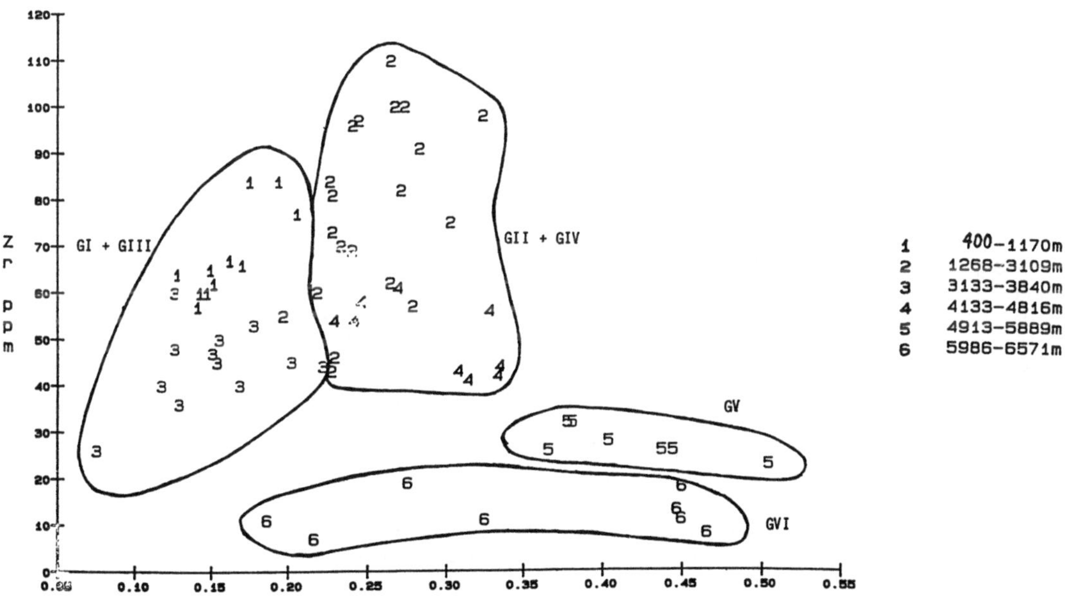

Fig. 2. Crossplot Zr - TiO_2

Dolerites. Apart from the different granites there occur three major doleritic intrusion zones closely related to the seismic reflection horizons, R1, R2 and R3, at about 1500, 2700 and 4700 metres.

When drilling through the first dolerite, in the R1 reflector, many doubts were raised whether the dolerite and the fractured area had any connection to each other or if it was just a coincidence finding a close to vertical dolerite in the reflector. This interpretation was made from what was known from outcrop mapping in the Siljan ring area. Unfortunately, the contact between granite and dolerite was not preserved in core no. 2, which partly was cored in a dolerite. After wireline logging this interval, it was suggested that the dolerite had a rather flat inclination.

When drilling through the R2 reflector it was almost accepted that most of the dolerites must be subhorizontal. After drilling through the R3 reflector there were no longer any doubts about the inclination. In core 6, a contact between dolerite and granite with an angle of about 10° was preserved. The conclusion must be that the occurrence of the dolerites is directly connected to the heavy fractured areas in the rock mass.

One important question to be solved in the future whether the fracturing is due to the intrusion of dolerites or if the dolerites intruded into an already existing fracture system from close to vertical feeding dikes. No evidence has been shown so far that the dolerites and the fracturing should be impact generated. K-Ar age determinations of the dolerites indicate an age just over 900 Ma. This is far older than the supposed impact age of 360 Ma. The petrological analyses of cuttings from the well show that most of the quartz grains are rather clear, inclusion-free, though in some grains spherical and acicular inclusions occur.

Albitization of K-feldspar and plagioclase is a common feature, particularly in the upper 2000 m. The albite may heal fractures in feldspar grains or may replace the feldspar grains almost completely. Albitized K-feldspare grains usually show isolated pathes of albite which either cover a large area of the K-feldspar crystal or compose a small portion. The content of plagioclase in the cuttings generally increases with depth. This relation is partly related to albilization of K-feldspar crystals. Complete albilization of a K-feldspar crystal reduces the volume by about 13%; this is theoretically expected to enhance porosity in the rock. The common clay minerals replacing feldspars are sericite and sericite-smectite.

Chlorite commonly replaces biotite in many parts of the well. Fresh biotite grains are more common in the deeper samples. Within the cuttings, a few grains of accessory minerals appear. The most common ones are epidote, Fe-Ti oxides, sphene, and less commonly calcite and hornblende. Calcite particles are generally very few in the cuttings.

From the wireline logging, besides the three reflection horizons, a number of small but similar fracture zones can be traced. This is supported by the presence of cataclastic fragments in cuttings. Cataclastic fragments (microbreccia) indicate deformation of rocks. The presence of these fragments as well as other deformation features indicate that fault and shear planes were formed in the bedrock. Varying grain size and mineralogical composition of the cataclastic fragments are observed. Also, biotite grains showing deformation and both kink and slip bands are observed. Almost all fractured zones seem to have been healed or filled with secondary minerals.

References

Hjelmqvist S (1966) Description to map of the pre-Quaternary rocks of the Kopparberg County, central Sweden

Vattenfall (1985) The Siljan Ring structure-investigations and evaluations

Vattenfall (1986) Investigations and planning work carried out in 1985.

Geochemical Logging in the Siljan Ring

J. A. GRAU[1], D. V. ELLIS[1], M. M. HERRON[1], J. S. SCHWEITZER[1] and

G. RISSLER-ÅKESSON[2, 3]

Introduction

A valuable set of data was obtained from logging measurements performed in the exploration well Gravberg-1, which was drilled to investigate the possible existence of deep deposits of abiogenic gas originating from the Earth's mantle. These measurements, which were made with a recently developed set of gamma-ray spectroscopy sensors, were used to obtain elemental concentrations for ten elements contained in the rock. The Geochemical Logging Tool (GLT) includes measurements of natural activity for determining Th, U, and K concentrations, delayed activation measurements for determining Al concentrations, and measurements of prompt gamma rays following thermal neutron capture for determining Si, Ca, Fe, S, Gd, and Ti concentrations. The details of the spectrometers used to measure the gamma-ray spectra have been presented previously (Hertzog et al. 1987). The conversion of gamma-ray intensities to elemental concentrations is obtained in one of two ways: through direct calibration, for Th, U, K, and Al, or through a geochemically based closure model, for the elements measured by prompt thermal neutron capture gamma rays. While continuous measurements of elemental concentrations in the rock are of great value in their own right, recent research in geochemically derived properties from elemental concentrations in sedimentary environments (Herron 1986, 1987a, b, c; Herron and Grau 1987) has provided quantitative analyses of rock properties that could be previously achieved only through detailed analyses on core samples. The data set obtained in Gravberg-1 forms the basis for extending the geochemical research into crystalline rock properties.

We first review the details of the closure model to understand the limitations on the accuracy of derived elemental concentrations imposed by the possible mineralogical variation in the rock. Then the results for the elemental concentrations determined from the spectroscopic measurements are presented and

---

1 Schlumberger-Doll Research, Old Quarry Road, Ridgefield, CT 06877, USA

2 MicroLine Consulting, Snäckeskärsgatan 9, 421 57 Västra Frölunda, Sweden

3 On a contract for the Swedish State Power Board in cooperation with the Gas Research Institute

compared with the limited available elemental concentrations measured in cuttings. Finally, the results of the mineralogical analysis are presented.

Geochemical Closure Model

Certain types of spectroscopic measurements in rock formations can be directly calibrated in weight percent of the measured element. The measurement of natural activity, for which there is no neutron source, can be most easily calibrated (Serra et al. 1980). When a neutron source is used, determining the required neutron flux distribution in the rock is very difficult. However, as discussed by Hertzog et al. (1987), if the neutron source is of known intensity, as is the case for measurements of delayed activity from thermal neutron capture reactions with a ^{252}Cf source, certain measurable or calculable parameters (neutron slowing-down length, Ls; formation and borehole thermal neutron macroscopic cross sections, Σ_f and Σ_b; and borehole size) can be used to directly transform the detected gamma-ray intensities into elemental concentrations.

The analysis of prompt gamma rays following neutron capture has so far eluded a direct calibration in terms of elemental concentrations, especially when an accelerator source of 14-MeV neutrons is used. For these measurements a model is needed to convert the observed capture gamma-ray flux attributable to a particular element into a concentration for that element. Previous attempts (Senftle et al. 1978) at such an approach have had success in limited applications, such as the measurement of elemental concentrations in coal seams where all significant elements, except for oxygen, could be measured through thermal neutron capture gamma rays. Attempts to relate elements measured through different types of reactions (Clayton and Coleman 1985) have likewise been successful under only very limited mineralogical conditions. The model we use has met with significant success in clastic sediments (Grau and Herron 1987). The model is a combination of oxide and carbonate representations that relates the concentrations of measured elements to the concentrations of unmeasured elements and includes elements measured by natural radioactivity, thermal neutron capture and thermal neutron-induced delayed activity. One key to the improved results comes from considering only elements that are contained in the rock matrix, excluding those that might be present in the pore fluid or in the borehole fluid.

The model is a simple closure relation which can be expressed mathematically as:

$$F\left\{\sum_i X_i \frac{Y_i}{S_i}\right\} + X_K W_K + X_{Al} W_{Al} = 1, \qquad (1)$$

where F is the calibration factor to be determined at each depth from this equation, Y_i is the fraction of the measured prompt gamma rays that are attributed to element i, S_i is the relative weight-fraction detection sensitivity for element i,

X_i is the ratio of the weight of the associated oxide or carbonate to the weight of element i, W_{Al} is the weight fraction of aluminum determined from the activation measurement, and W_K is the weight fraction of potassium determined from natural activity.

The S_i are measured in the laboratory and contain most of the nuclear physics involved in the relationship between the Y_i and the respective elemental concentrations. The Y_i are determined (Hertzog et al. 1987, Grau and Schweitzer 1987) from a least-squares fit of the measured spectra with elemental standard spectra. Thus, if we are indeed measuring the elements needed to describe 100% of the rock minerals, then the only unknown, F, can be trivially determined from the above equation. The terms FY_i/S_i are then the weight percent of the elements determined through prompt gamma ray, neutron capture measurements.

There are two essential components of the model which must be ascertained in each geological environment: whether or not the elements measured, together with their associated carbon and oxygen, include a high enough fraction the rock elements present, and whether or not the assumed relationships between the measured elements and the unmeasured carbon and oxygen are valid. Typical values of the oxide factors X_i associated with measured elements are shown in Table 1. In certain cases, such as for silicon, the X_i value of 2.139 is just the mass of SiO_2 divided by the mass of Si. In other cases, such as for Fe, the chosen value of 2.075 is significantly different from the 1.430 value which would be appropriate for Fe_2O_3. This is due to the presumed association of bound water, in the form of hydroxyls, with the occurence of Fe in the form of Fe_2O_3. Fortuitously, if Fe occurs as $FeCO_3$, which has no hydroxyls, the value of 2.075 is still appropriate. Thus, if the presence of minerals other

Table 1. Oxide factors for closure normalization

Element	Theoretical	Optimized
Si	2.139 - SiO_2	2.139
Ca	2.497 - $CaCO_3$ 1.399 - CaO	2.497
Al	1.889 - Al_2O_3	2.741
Ti	1.668 - TiO_2	1.668
K	1.205 - K_2O	0.615
Fe	1.287 - FeO 1.430 - Fe_2O_3 2.075 - $FeCO_3$	2.075
S	1.125 - $CaSO_4$ 0.064 - FeS_2	1.125

Table 2. Inherent closure calculation for common minerals

Mineral	Closure sum	Mineral	Closure sum
Pyrite	156.7	Anhydrite	100.0
Ca Feldspar	129.1	Siderite	100.0
Muscovite	104.7	K Feldspar	100.0
Glauconite	103.3	Kaolinite	100.0
Illite	102.8	Biotite	96.4
Montmorillonite	101.8	Na Feldspar	95.2
Chlorite	101.4	Ankerite	79.0
Sandstone	100.0	Dolomite	54.2
Limestone	100.0		

than those of the model used in clastic sediments is known, they can be easily incorporated.

Table 2 shows the accuracy with which the chosen values of X_i can account for the unmeasured elements contained in different minerals. For most minerals agreement is found to within ±5%. Of the minerals which show worse agreement, pyrite can be treated separately by logic imposed on the Fe and S yields. The Mg-rich minerals dolomite and ankerite can be correctly evaluated (Hertzog et al. 1987) by comparing a measurement of the photoelectric cross section with a calculation of the expected photoelectric cross section based on the initially determined elemental concentrations. This comparison provides a determination of the Mg concentration as well as the corrected values for the elements measured by neutron capture gamma rays. Ca-feldspar is not fit well with the current version of the model because Ca is assumed to occur as $CaCO_3$ rather than CaO. Fortunately, however, Ca-feldspar is not common in the subsurface, and thus is not likely to cause large errors in the results. If there were 10% Ca-feldspar in the rock, for example, the elemental concentrations would be in error by only 3% (i.e., 10% of 29%).

This model has been successfully applied in many wells where good agreement has been demonstrated in comparison with elemental analyses performed on core samples. Gravberg-1 is the first attempt at in-situ measurements for elemental concentrations and application of this model to crystalline rocks.

Elemental Analysis for Gravberg-1

The results of the GLT measurements in Gravberg-1, with the application of the closure model to obtain concentrations, are shown for the ten elements in Fig. 1. Also shown on the figure are the results of cuttings analyses at 13 depths over the logged intervals. There is, in general, excellent agreement, including the interval near 1500 m where many of the elemental concentrations vary by over a factor of 2. All regions showing higher Ti concentrations were noted in the cuttings descriptions as having Ti-bearing minerals. Both the variations in elemental concentrations and their absolute values are in good agreement with the measurements on cuttings and the expectations from the minerals present.

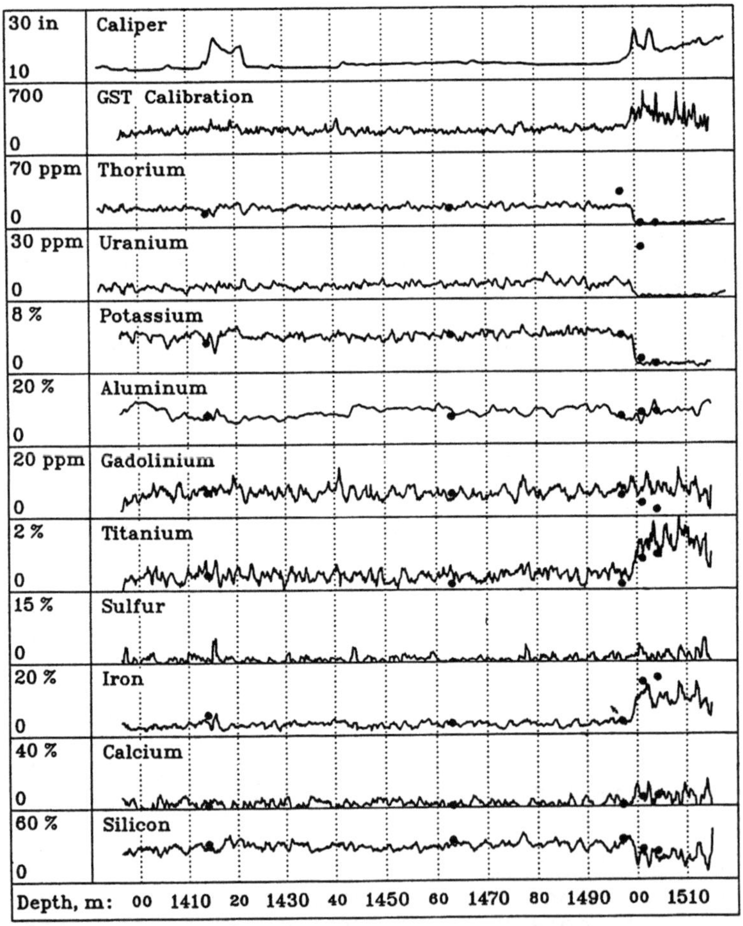

Fig. 1 a. Log-derived concentrations for 10 elements from the Gravberg-1 well, covering the depth interval 1390 to 1520 m. Laboratory analyses of cuttings are shown as **dark circles**. Included also is the calibration factor for the 6 GST elements (Si, Ca, Fe, S, Ti, and Gd) as determined from the closure equation.

Two additional comments on the cuttings analysis should be made. First, the values plotted in Fig. 1 for Gd were actually inferred from measurements of La, another rare-earth element which tends to associate rather uniformly with Gd, in the ratio 5.55 parts La to 1 part Gd. Second, the cuttings analysis of uranium does not appear to be very sensitive (and possibly not very accurate) as many values were quoted as less than 20 ppm. These values were plotted on Fig. 1 as 10 ppm. Some of the uranium cuttings analyses which appear to be missing from Fig. 1 actually would have plotted off-scale, indicating possible thin sections of high uranium deposition. We do not consider the uranium cuttings analysis to be a valid comparison to our logging results.

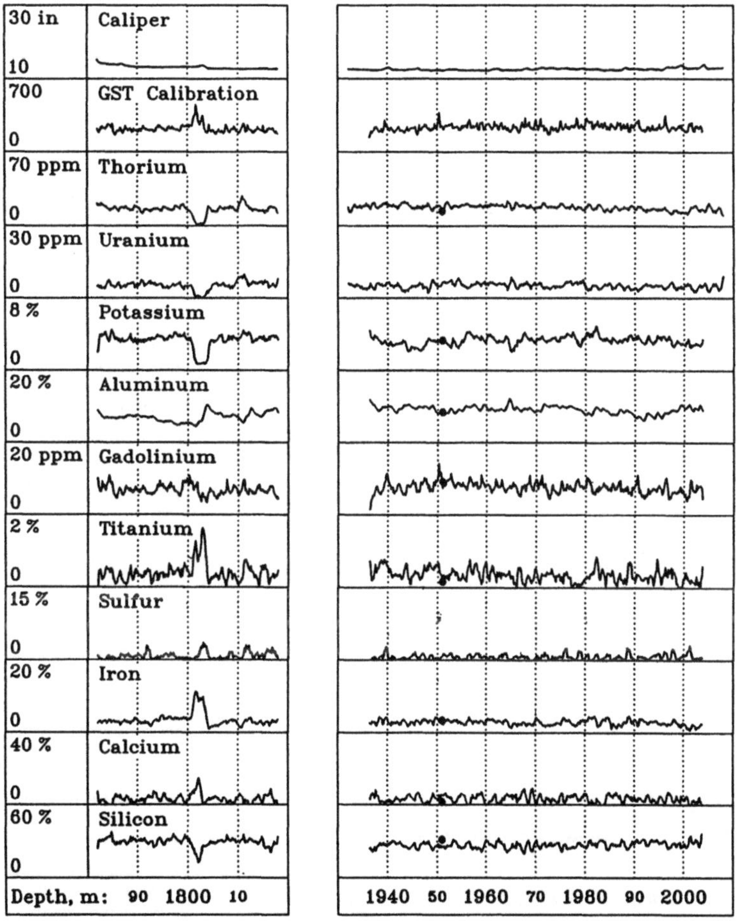

Fig. 1 b. Log-derived concentrations for 10 elements from the Gravberg-1 well, covering the depth intervals 1780 to 1820 m and 1930 to 2010 m. Laboratory analyses of cuttings are shown as **dark circles**. Included also is the calibration factor for the 6 GST elements (Si, Ca, Fe, S, Ti, and Gd) as determined from the closure equation.

The results shown in Fig. 1 have a somewhat larger statistical uncertainty than is usually associated with GLT measurements, because a boron-carbide-loaded external sleeve, which is normally used for GLT measurements, could not be acquired when this well was logged. This omission had two detrimental effects on the measured neutron capture spectra. First, a substantial portion of the total spectrum was due to gamma rays originating in the tool and in the surrounding fluid, reducing the total signal intensity from the elements in the rock. Second, the specific Fe signal from the tool pressure housing had to be removed to obtain the net Fe contribution from the rock itself. While the agreement of the derived Fe with the cutting analysis is quite good, the approximate nature of the correction for the

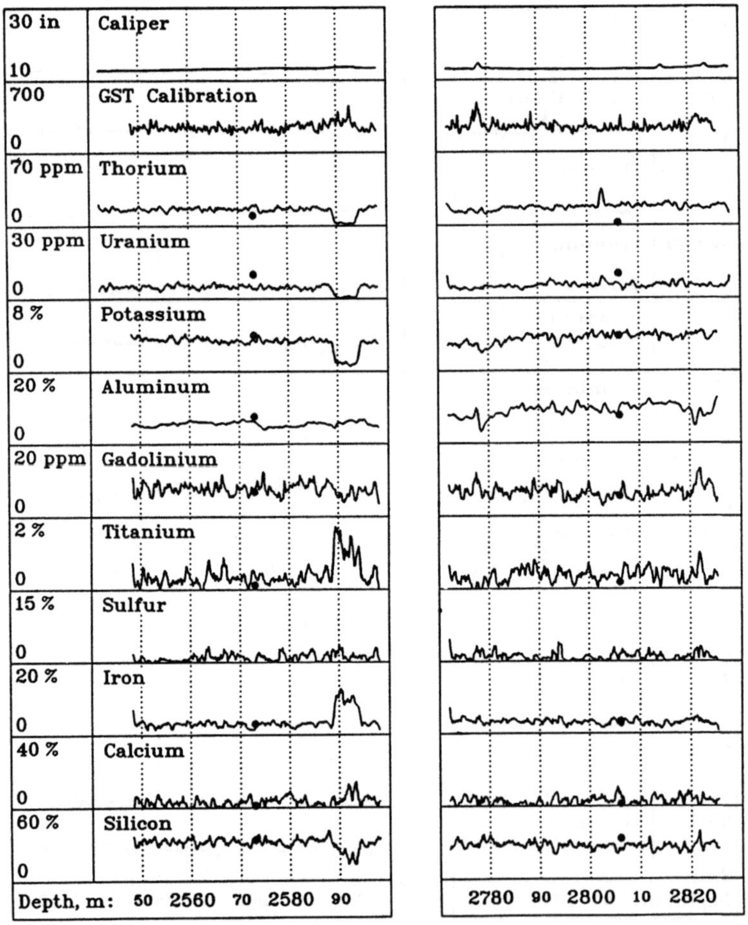

Fig. 1 c. Log-derived concentrations for 10 elements from the Gravberg-1 well, covering the depth intervals 2540 to 2600 m and 2760 to 2830 m. Laboratory analyses of cuttings are shown as dark circles. Included also is the calibration factor for the 6 GST elements (Si, Ca, Fe, S, Ti, and Gd) as determined from the closure equation

Fe contribution from the pressure housing causes a larger than normal uncertainty in the derived Fe concentrations. Both of these problems disappear when the external sleeve is used, improving the statistical precision on all capture elements and improving the accuracy of the derived Fe concentrations.

The calibration factor, F, in Eq. (1) depends upon many borehole parameters. Near 1500 m, for example, the borehole size varies dramatically, resulting in changes of F of over a factor of 2. However as noted earlier, the variation in elemental concentrations through this region is well described by the final results. Thus it appears that the calibration factor in our model successfully accounts for many of the large intensity variations caused by changes in the borehole environment.

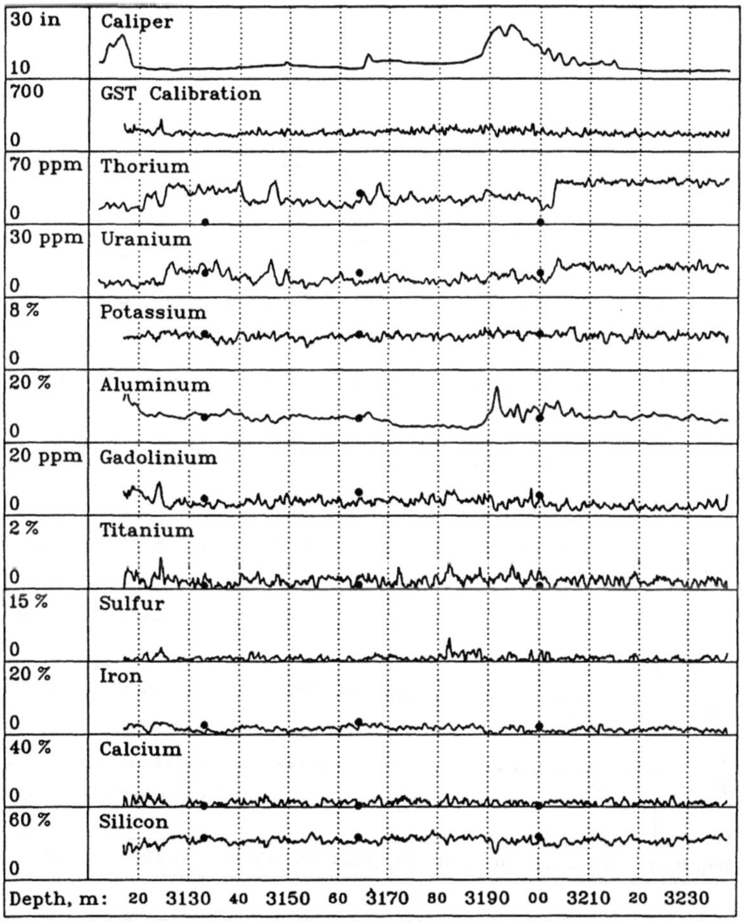

Fig. 1 d. Log-derived concentrations for 10 elements from the Gravberg-1 well, covering the depth interval 3110 to 3240 m. Laboratory analyses of cuttings are shown as dark circles. Included also is the calibration factor for the 6 GST elements (Si, Ca, Fe, S, Ti, and Gd) as determined from the closure equation

Some special comments should be made concerning the derived aluminum concentrations. The normal algorithm for deriving the aluminum concentration from the measured count rates (Hertzog et al. 1987) could not be reliably used here because of the adverse hole conditions. The algorithm requires a slowing down length and a caliper (assuming a circular hole). Because of the extremely large borehole size (beyond the range of data from which the algorithm was derived), the slowing down lengths were unreliable. For the same reason and also because of the elliptical cross section of much of the borehole, the caliper correction could not be reliably applied. We fabricated an expected slowing down length by assuming a rock with only a few percent porosity, with the grain density variation tied to the

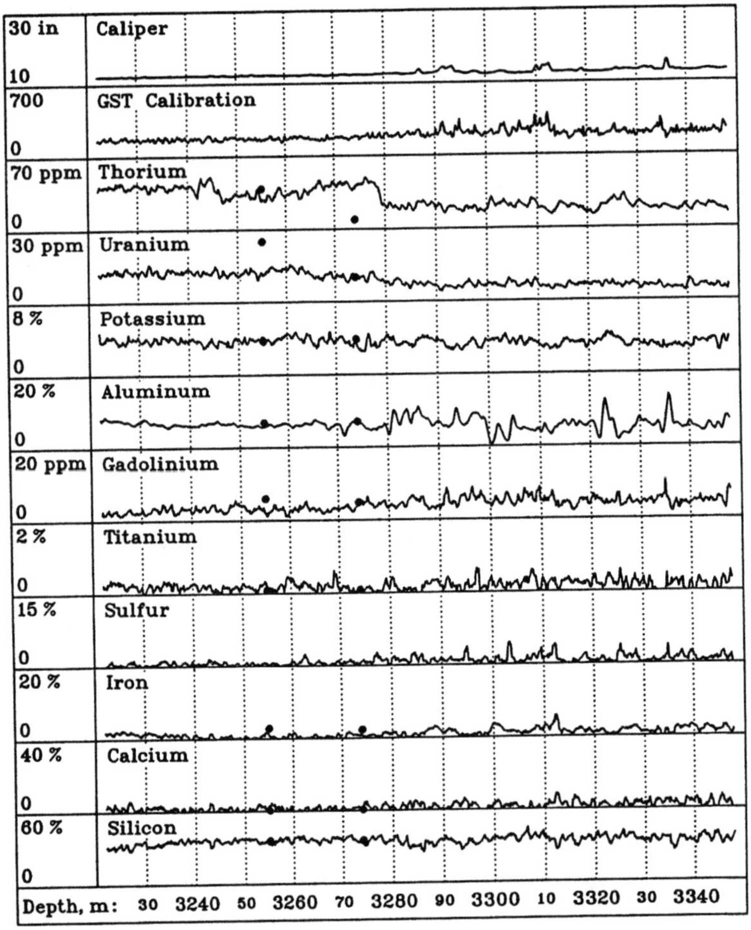

Fig. 1 e. Log-derived concentrations for 10 elements from the Gravberg-1 well, covering the depth interval 3220 to 3350 m. Laboratory analyses of cuttings are shown as **dark circles.** Included also is the calibration factor for the 6 GST elements (Si, Ca, Fe, S, Ti, and Gd) as determined from the closure equation

variation in bulk density (where reliable). After making an approximate borehole size correction, we then compared the derived Al concentrations to those obtained from the cuttings. It appeared that a normalization factor of 1.5 was required to obtain good agreement between the logging results and the cuttings analysis, and this value was included in the data shown in Fig. 1. While this uncertainty in the absolute value of the derived aluminum concentrations is larger than would normally occur, it has no measurable affect on the other elemental concentrations as derived from Eq. (1).

Mineralogical Analysis for Gravberg-1

It is common in igneous petrology to express whole rock chemical data in terms of normative minerals. In this procedure, oxide concentrations are sequentially apportioned as abundances of minerals of defined chemical composition. For example, all the P_2O_5, as well as a stoichiometric amount of CaO, would first be calculated as the mineral apatite $Ca_5(PO_4)_3F$. This procedure is followed sequentially through all of the measured oxides with the result being a suite of mineral abundances that would produce a chemical analysis identical to that of the sample. Depending on the thermodynamic history of the sample, the normative minerals may or may not come close to the true mineral assemblage. Regardless, the normative mineral analysis is a convenient means of condensing chemical data into a form that can be readily used in comparing samples.

The suite of elements measured by the GLT is substantially different from the assemblage measured in a normal whole rock analysis. As a result, it is necessary to define a new procedure for expressing the log data in terms of chemical-mineral equivalents. A general procedure has been developed for geochemical log data in clastic, sedimentary strata (Herron and Grau 1987). The Gravberg-1 well offers an opportunity to extend this procedure to igneous rocks. The procedure we will follow, similar to that of Herron (1986), begins with a factor analysis of variance using the log concentrations and porosity terms. The factor analysis results will be used as a guide for establishing the mineral suite to be used in determining the chemical-mineral equivalents.

Factor Analysis Results. The elemental concentrations, bulk density, caliper, thermal neutron porosity, and epithermal neutron porosity were analyzed using an r-mode factor analysis with varimax rotation (Herron 1986). Six factors were retained prior to rotation using the criterion that the unrotated eigenvalues be greater than unity. The elements of the rotated factor matrix, squared so that the values are proportional to the fractional variance explained for each variable and with sign retained, are plotted in Fig. 2. An element plotted on the diamond perimeter would have all of its variance explained by some combination of the two factors shown. An element plotting at the center of the figure has zero variance associated with the two factors shown.

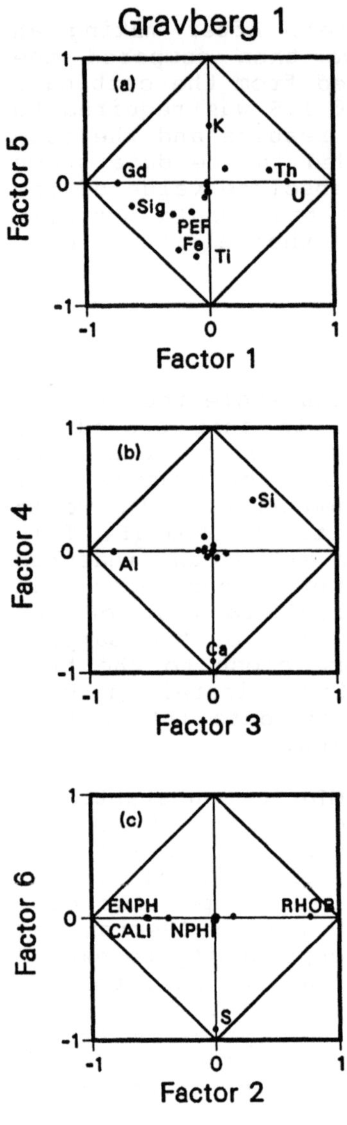

Fig. 2. Significant components found in a factor analysis of elemental concentrations, bulk density, caliper, and thermal and epithermal neutron porosity. Six factors were needed to adequately describe all the variance of the measured data

Figure 2 a shows the factor matrix elements for factors 1 and 5. Factor 1 is loaded positively for Th and U, and negatively for the rare earth element Gd and sigma, the thermal neutron capture cross section. From this we can conclude that: a single factor accounts for most of the variance of these variables; Th and U covary; Gd and Sigma covary; and Th and U anticorrelate with Gd and Sigma. Thorium and uranium are expected to covary under nonoxidizing conditions due to their chemical similarity (Adams et al. 1959). The source of these elements in igneous rocks is believed to be trace heavy minerals, commonly zircon, but the core chemistry shows no correlation between Zr and Th or U. Gadolinium, a rare earth element (REE), has the largest thermal neutron capture cross-section of any element. In the absence of clay minerals and large volumes of saline water, Gd should dominate the total sigma. Like Th and U, rare earths are incompatible elements which are excluded from most major rock-forming minerals and they tend to be concentrated in pegmatites or in accessory minerals.

Factor 5 shows a similar pattern with K loaded positively and Fe and Ti loaded negatively. The photoelectric absorption cross-section (PEF), associated with heavy elements like iron, is also negatively loaded on this factor. The Fe and Ti correlation presumably represents Fe-Ti oxide minerals such as ilmenite and titanomagnetite. A mafic mineral could also be responsible for some of the iron content. Common igneous minerals which contain abundant potassium are K-feldspars (orthoclase, microcline) and muscovite mica. Of course, the anti-correlation of K with Fe and Ti makes sense, since one does not expect orthoclase to coexist with mafic minerals.

Figure 2 b shows the factor matrix elements for factors 3 and 4. Factor 3 is strongly loaded negatively for aluminum, with some positive loading for Si. Factor 4 is similarly strongly loaded negatively for Ca, with some positive loading for Si. Likely sources of the Al variance, excluding that associated with K-feldspar, are plagioclase or any alteration products such as epidote or clay minerals. Likely sources of the Ca variance include Ca-rich plagioclase and carbonate cements such as calcite or dolomite. All aluminosilicates and quartz are sources for Si, but the factor analysis indicates that the Ca source and the Al source together control the Si variance.

Figure 2 c shows the factor matrix elements for factors 2 and 6. Factor 6 is strongly loaded only with sulfur. A likely source for the sulfur among the major sulfide minerals is pyrite. Factor 2 is strongly loaded positively on bulk density (RHOB) and negatively on both neutron porosity measurements (ENPH and NPHI) and the caliper. This factor seems to represent borehole conditions adversely affecting borehole response.

The factor scores for each depth interval are shown in Figs. 3 and 4. Factor scores are linear combinations of the normalized variables at each depth interval and the factor matrix elements. As such, they have a mean of zero and a standard deviation of one, and are presented in terms of deviation from the mean at each depth. Factor 1 is generally negative above 2800 m, indicating larger than average Gd and sigma, and smaller than average Th and U values. At about 2800 m, the factor scores begin to move above zero and stay above zero for almost the entire 3100-3400 m depth interval, due to the lower Gd and sigma values and higher Th and U concentrations. This zonation suggests a large-scale change in conditions such as a slight change in the accessory mineral content of the rock, or a zonation reflecting alteration.

Factor 2 shows large borehole sizes, along with lower bulk densities and higher neutron porosities, when the scores move more negatively. The major variations are due to poor hole conditions, while some of the lesser variations may reflect true changes in porosity.

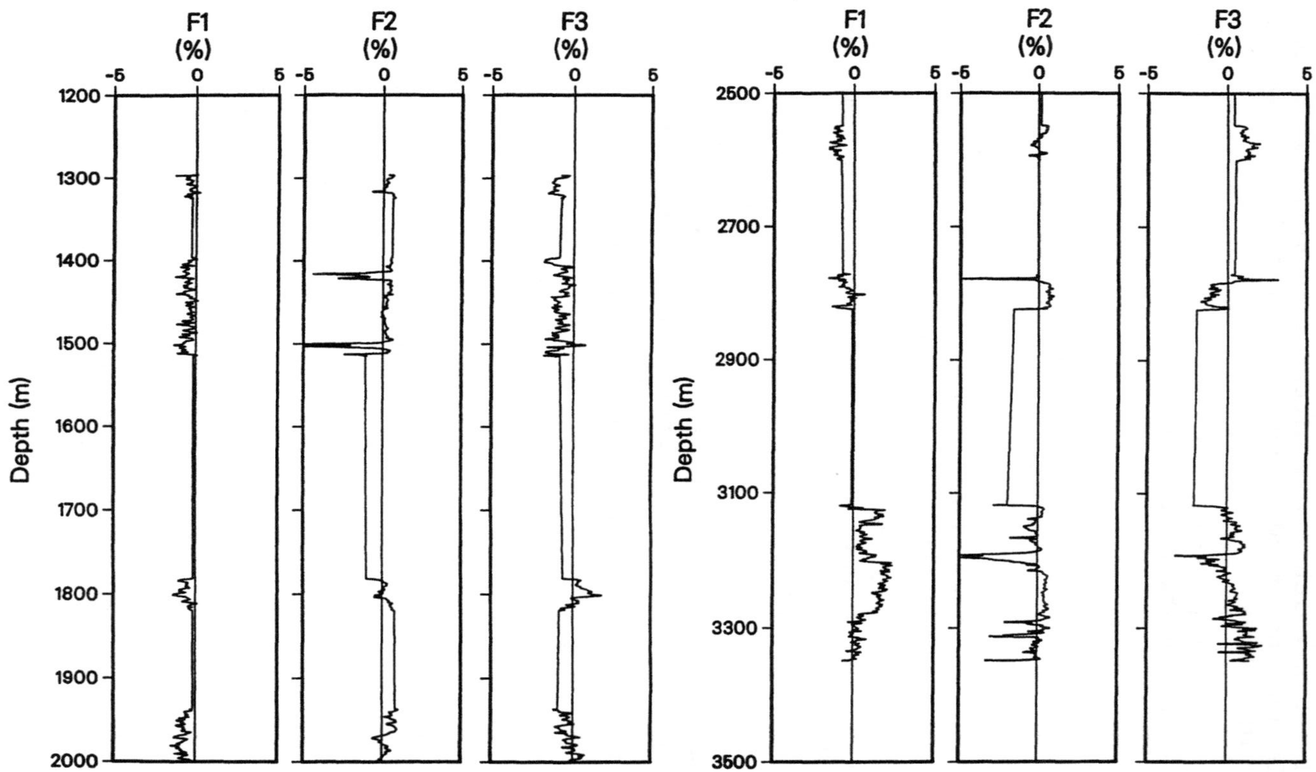

Fig. 3 a, b. a Values for factor-analysis factors 1-3 as a function of depth, covering the depth interval 1200 to 2000 m. **b** Values for factor-analysis factors 1-3 as a function of depth, covering the depth interval 2500 to 3500 m

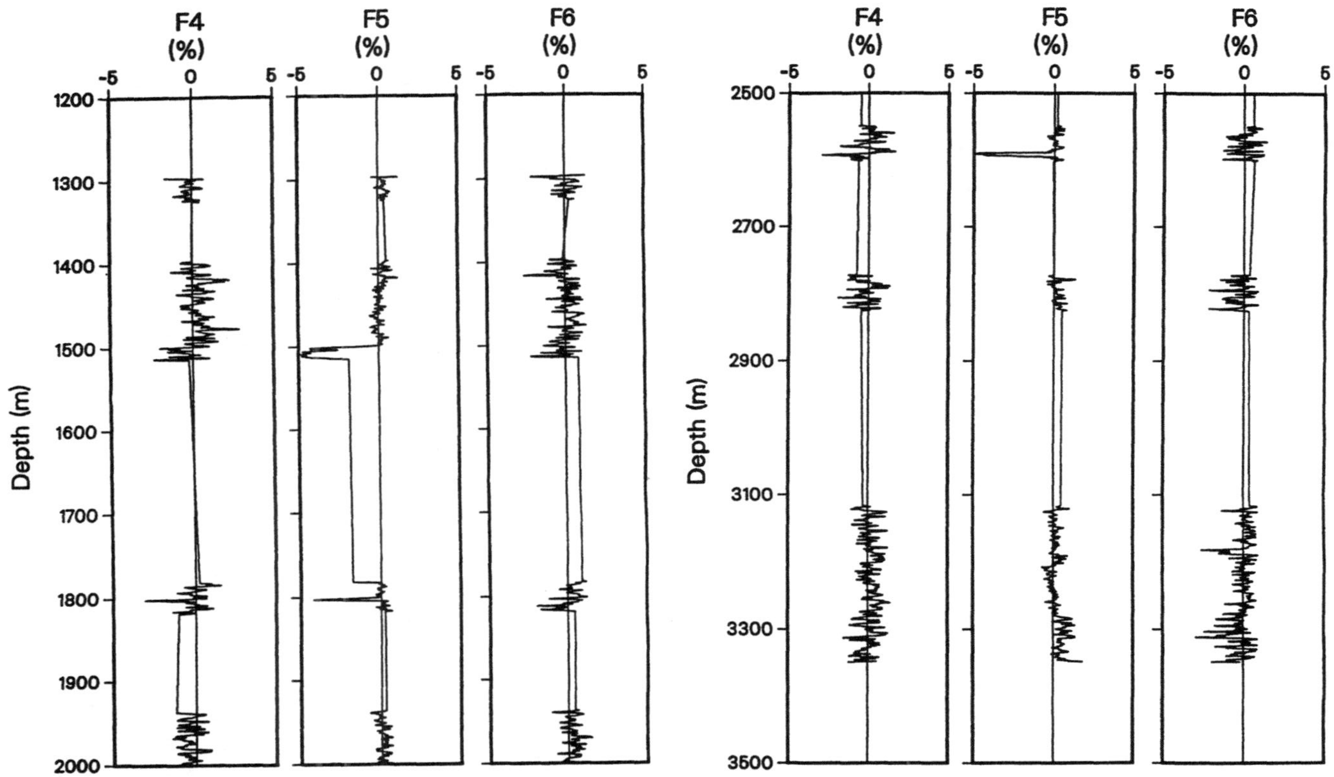

Fig. 4 a, b. a Values for factor-analysis factors 4-6 as a function of depth, covering the depth interval 1200 to 2000 m. **b** Values for factor-analysis factors 4-6 as a function of depth, covering the depth interval 2500 to 3500 m

Factor 3 is more positive for lower Al concentrations and has a slight dependence on higher Si concentrations. The depth interval near 1800 m shows the impact of the lower than average Al concentrations at 1785-1815 m and again over the 2550-2600 m interval, as well as a broad trend of decreasing Al concentrations from 3110 to 3230 m, interrupted by apparent caliper-related problems (Fig. 1 d) between 3190 and 3210 m. It is likely that plagioclase and alteration products such as clay minerals govern the Al variance that is not related to potassium.

Factor 4 scores are more negative for higher Ca concentrations, with a slight dependence on Si. The calcium concentrations in this well are fairly low, averaging only 2.6% according to the core data, and so some of the Ca variance probably reflects noise. There are a few significant streaks of high Ca (negative factor 4 scores) at about 1500 m, 1800 m, and 2600 m. These may represent fracture fillings of carbonate cement.

Factor 5 scores are more positive for higher orthoclase abundances and lower Fe-Ti oxide abundances. Here many of the features occur over a rather short depth interval, such as the 1500-1513 m, 1802-1804 m, and 2589-2593 m depth intervals, and are marked by high Fe and Ti concentrations and low K, Th, and U concentrations (Fig. 1 a-c).

Factor 6 scores are more negative for higher S concentrations and show a number of thin intervals of more negative values throughout the well, a generally decreasing trend at the 1800 m zone, and significant negative peaks at 3182 m and several between 3260 m and 3340 m. These peaks in S content probably represent thin intervals of mineralization of sulfides, including pyrite.

Chemical Mineral Equivalents. The factor analysis results can serve as a useful guide for setting up a suite of minerals for a preliminary recasting of the chemical data in terms of chemical-mineral equivalent (CME) abundances. From the six factors, we identify nine possible mineral contributors and the caliper-porosity signal. The minerals could be: quartz, orthoclase, plagioclase, a heavy mineral we will by default call zircon, the REE mineral, ilmenite, a mafic mineral such as an iron-rich pyroxene, pyrite, and calcite. Rather than a sequential normative mineral analysis where each element is absolutely assigned to a given mineral, we will utilize the matrix inversion procedure outlined by Herron (1986). The column matrix of mineral abundances, M, is solved from the inverse of the composition matrix, C, and the column matrix of elemental concentrations from the logs, E.

$$[M] = [C]^{-1}[E]. \tag{2}$$

Table 3 gives reasonable end-member compositions for the major rock-forming minerals that can be used in the element-to-mineral conversion. The orthoclase, quartz, ilmenite, pyrite, and calcite compositions all follow from standard chemical formulae. Plagioclase is assumed to be a mixture of equal parts

Table 3. Elemental composition of the chemical-mineral equivalents

Mineral	Al (%)	Si (%)	Fe (%)	K (%)	Ti (%)	S (%)	Th (ppm)	Gd (ppm)	Ca (%)
Orthoclase	10	30	0	14	0	0	0	0	0
Quartz	0	47	0	0	0	0	0	0	0
Plagioclase	14	26	0	0	0	0	0	0	7
Ilmenite	0	0	36.8	0	31.6	0	0	0	0
Mafic	0	19	42.3	0	0	0	0	0	0
Pyrite	0	0	47	0	0	53	0	0	0
HM (Zircon)	0	0	0	0	0	0	4000	0	0
REE Mineral	0	0	0	0	0	0	0	100	0
Calcite	0	0	0	0	0	0	0	0	40

of anorthite and albite; cuttings analysis could refine this estimate. For a potential mafic mineral, we have chosen an iron-rich pyroxene. For the Th-U component, we have chosen a heavy mineral with a Th content of 4000 ppm, typical for the ubiquitous heavy mineral zircon. The REE mineral is somewhat unclear. A common source might be monazite, although this phosphate mineral generally contains abundant (ca. 10%) thorium (Adams et al. 1959), and the factor analysis results suggest that the Th and Gd factors anti-correlate. Therefore, we have named the Gd-containing constituent the REE mineral and arbitrarily assigned a Gd content of 100 ppm to the pure end-member. It is to be remembered that these chemical-mineral equivalents are for the purpose of rapidly distilling the chemical information provided by the logs, and may or may not correspond to the true mineral assemblage. The occurrence of other minerals noted on the cuttings report, epidote, biotite, hornblende, sphene, and magnetite, will manifest itself in the other chemical-mineral equivalents.

The Gravberg data were initially processed through a seven-element model, using Eq. (2) and the values listed in Table 3, but excluding the element Ca and the mineral calcite. At every depth interval, the derived abundances were summed and, if the total was less than 100%, a residual abundance was determined. At each depth, the derived plagioclase abundance was multiplied by the 7% Ca content shown in Table 3 and compared to the measured Ca. The excess or deficiency of Ca was divided by the Ca content of calcite (40%) and expressed in terms of apparent calcite. The chemical mineral equivalent (CME) abundances are shown in Figs. 5-8.

Figure 5 shows the CME abundances of orthoclase, quartz, and pyroxene, the major constituents of the Gravberg-1 well, as a function of depth. Also shown in Fig. 5 are the mineral abundances measured on the cuttings. In general, the CME orthoclase and cuttings orthoclase show fairly good agreement over the few comparable depth intervals, although the average cuttings orthoclase is slightly greater than the CME orthoclase. In view of the good agreement between core and log potassium and the stable K content of orthoclase, this suggests that the cuttings orthoclase may be slightly overestimated. The CME quartz and plagioclase show much more variability over the logged intervals than do the cuttings mineral abundances, even though average abundances are similar. This probably reflects the homogenization of cuttings before they are sampled at the surface, and demonstrates a significant advantage of continuous logging information as a function of depth. The quartz content of the lowest logged interval is slightly greater than in the rest of the well, but it also shows large variability. The log-derived plagioclase averages slightly greater than the cuttings plagioclase, and this may be due to the inappropriate choice of the albite-anorthite content of the CME plagioclase, or it may reflect the overestimation of the cuttings orthoclase and consequent underestimation or the plagioclase. In any event, the cuttings analyses and the CME abundances clearly show that the dominant constituents of the well are orthoclase, quartz, and plagioclase.

Figure 6 shows abundances of the CME mafic mineral, ilmenite, and pyrite. No corresponding quantitative data are available from the cuttings analysis. These minerals are generally present in lower abundances than the orthoclase, quartz, and plagioclase, but show some dramatic occurrences that were previewed in the factor analysis results. Significant mafic and ilmenite peaks show up in the 1500-1513 m, 1802-1804 m, and 2589-2593 m intervals with significant reductions in the quartz and orthoclase contents at the same depths. In the remainder of the well that was logged, mafic concentrations average about 4% and ilmenite abundances about 1%. Pyrite abundances show pronounced peaks an 3182 m and frequently between 3260 m and 3340 m. This interval apparently contains several sulfide mineralization zones, which are reflected in the CME pyrite abundance. Pyrite abundances average about 2% over the logged intervals.

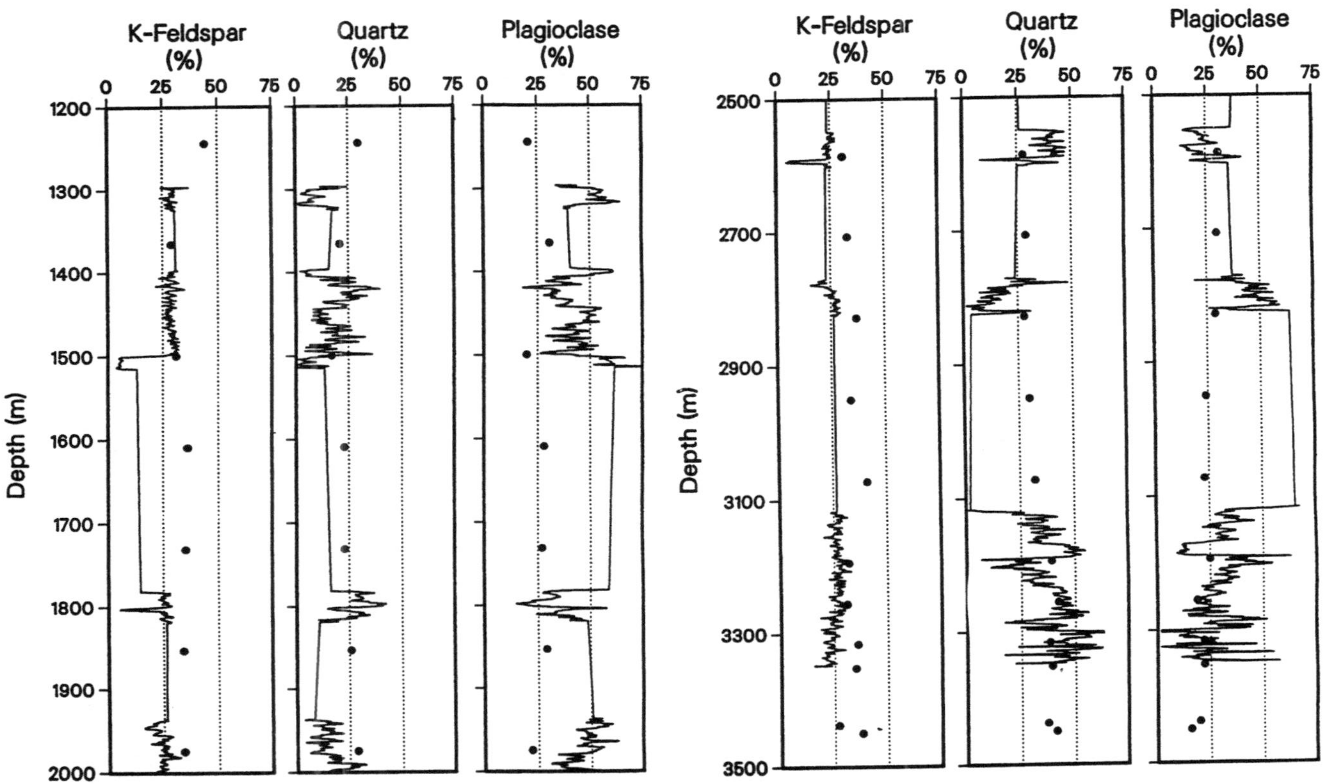

Fig. 5 a, b. a Weight-percent concentrations for K-feldspar, quartz, and plagioclase as derived from the elements-to-minerals transformation, covering the depth interval 1200 to 2000 m. The **solid circles** show the results of cuttings analyses. **b** Weigt-percent concentrations for K-feldspar, quartz, and plagioclase as derived from the elements-to-minerals transformation, covering the depth interval 2500 to 3500 m. The **solid circle** show the results of cuttings analyses

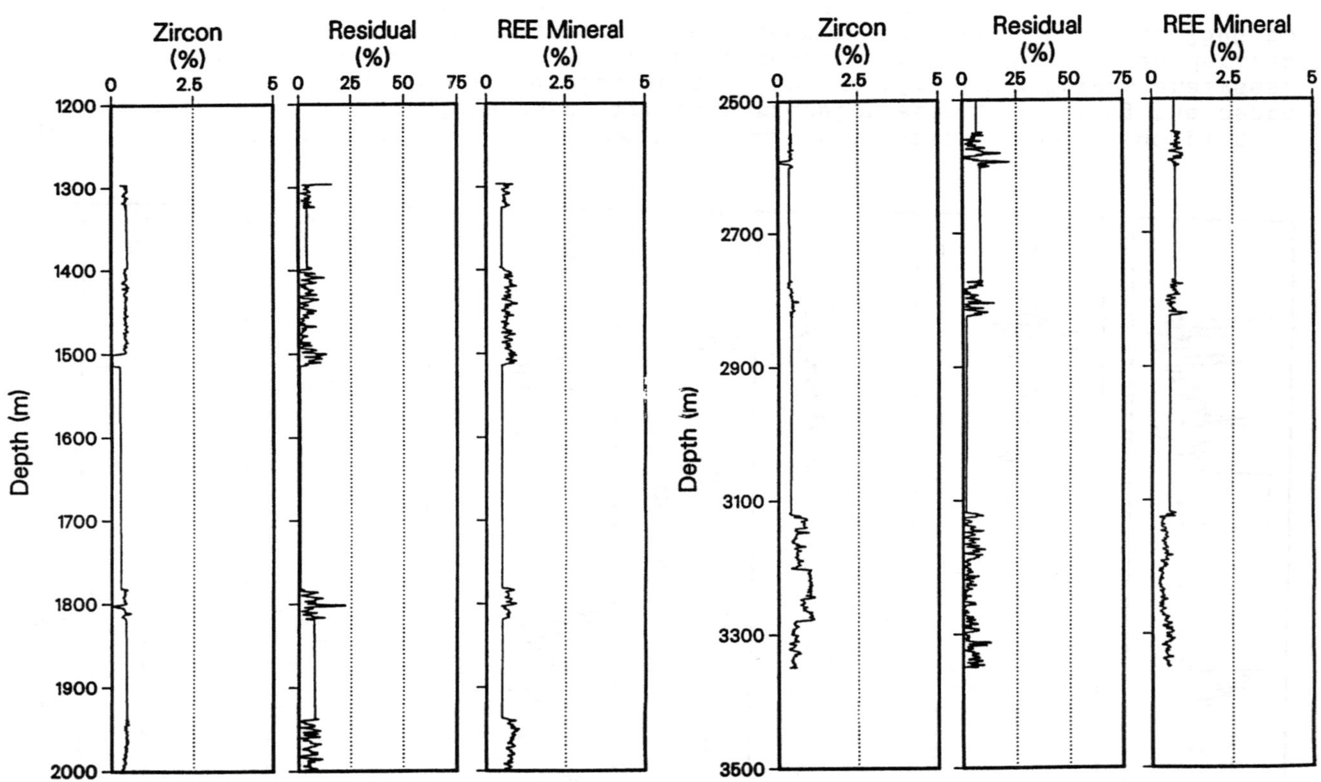

Fig. 6 a, b. a Weight-percent concentrations for zircon, the residual, and the rare-earth mineral as derived from the elements-to-minerals transformation, covering the depth interval 1200 to 2000 m. b Weight-percent concentrations for zircon, the residual, and the rare-earth mineral as derived from the elements-to-minerals transformation, covering the depth interval 2500 to 3500 m

Figure 7 shows the abundances of the CME zircon, the REE mineral, and the residual contribution. Above 3100 m, the zircon abundances average about 0.5%, and the REE mineral abundances average about 1%. In the 3100-3300 m zone, however, the zircon abundance increases to about 1%, while the REE mineral abundance decreases to about 0.5%. This interval, also marked by apparent sulfide mineralization, shows several trace element differences from the rest of the well in addition to the highest quartz abundances of the logged intervals.

Figure 7 shows that the residual mineral abundance averages about 5% over the entire well. Some of the deviation in the residual mineral abundance may come from the existence of additional minerals which were not part of the model or from deviation between the chemical concentrations listed in Table 3 and the true concentrations, most notable for plagioclase. Figure 8 shows the excess Ca log, expressed as apparent calcite (Ca/40%) as a function of depth. There are few intervals where the apparent calcite is negative, which most likely indicates an overestimation of the plagioclase content or the Ca concentration in the plagioclase. Positive deviations may indicate the opposite, or may indicate the existence of calcite. It may be noted that the apparent calcite curve and the residual mineral abundance curve show many parallels, with peaks at 1800 m, in the 2550 m interval, and below 3300 m. These indicate that a Ca-bearing mineral, possibly calcite or sphene, is responsible for both the residual mineral abundance and the apparent calcite abundance.

In summary, the chemical mineral abundance curves show many features over the sparsely logged intervals that are not easily seen or interpreted from the cuttings analyses. The dominant minerals in the well are orthoclase, quartz and plagioclase, but the log data clearly show low abundances of orthoclase and quartz in several intervals. These same intervals are usually composed of significant mafic mineral and ilmenite concentrations, such as might be caused by intrusive events. The major element composition is significantly different in these zones from the rest of the well. The Th and U source, here called zircon, and the Gd source (REE mineral) show fairly stable abundances in the upper part of the well, but reverse abundances below 3100 m. This same deep interval shows the highest quartz abundances and indications of several zones of significant sulfide content.

Conclusion

A geochemically-based closure model has been described that allows relative measurements of thermal neutron capture gamma-ray spectral contributions to be converted into elemental concentrations of Si, Ca, Fe, S, Ti, and Gd in conjunction with directly calibrated measurements of Th, U, K, and Al. The application of this procedure to logging data from Gravberg-1 has resulted in elemental concentration logs that are in good agreement with the available cuttings analyses. A factor analysis performed on these elemental concentrations and on the porosity

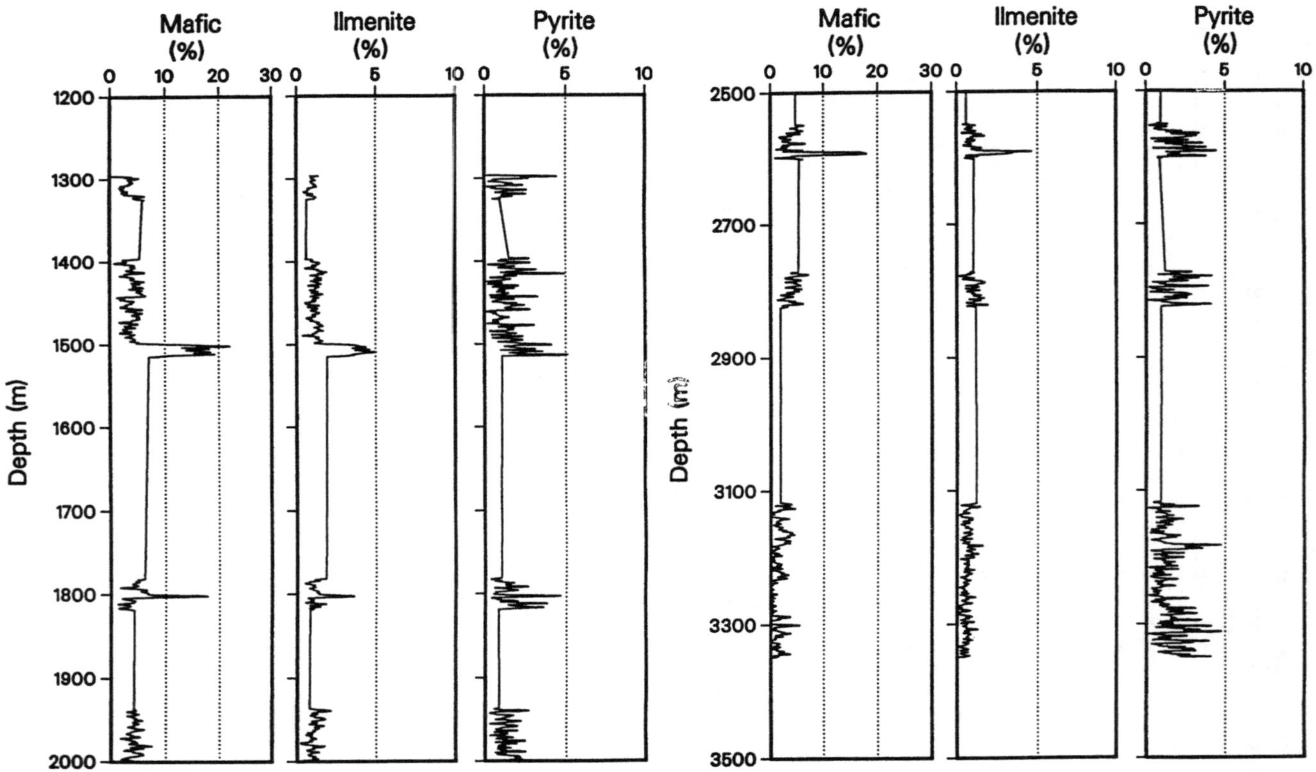

Fig. 7 a, b. a Weight-percent concentrations for mafic, ilmenite, and pyrite as derived from the elements-to-minerals transformation, covering the depth interval 1200 to 2000 m. b Weight-percent concentrations for mafic, ilmenite, and pyrite as derived from the elements-to-minerals transformation, covering the depth interval 2500 to 3500 m

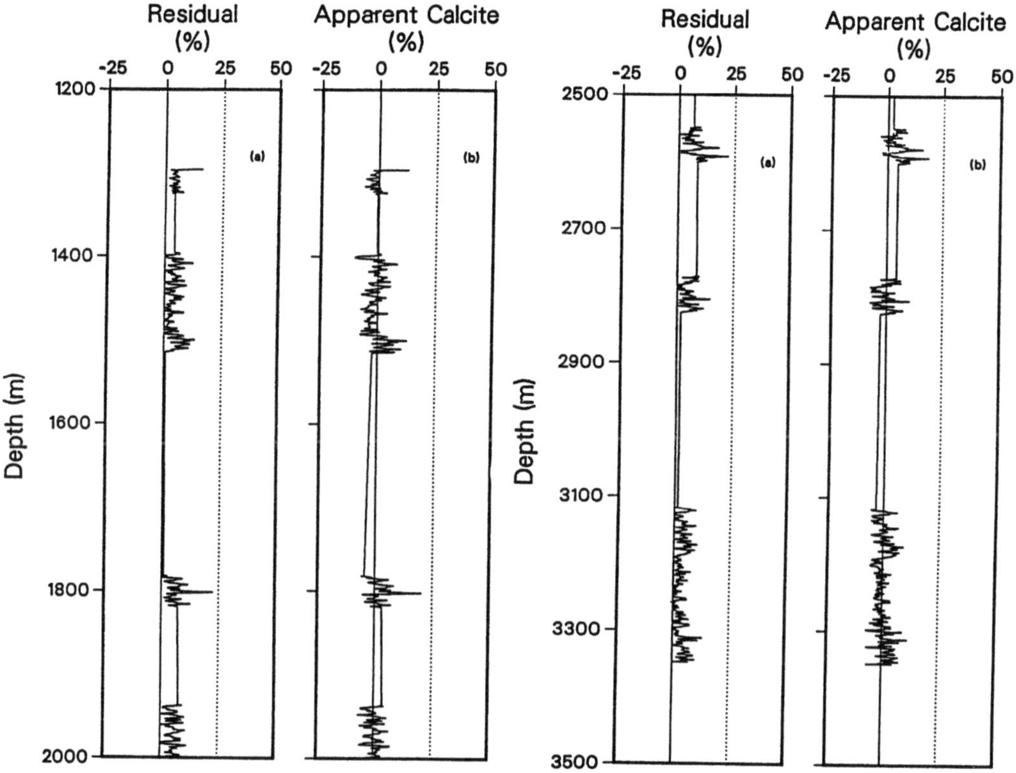

Fig. 8 a, b. a Weight-percent concentrations for apparent calcite as derived from the elements-to-minerals transformation, covering the depth interval 1200 to 2000 m. All nonplagioclase calcium was assumed to be present as calcite. b Weight-percent concentrations for apparent calcite as derived from the elements-to-minerals transformation, covering the depth interval 2500 to 3500 m. All nonplagioclase calcium was assumed to be present as calcite

terms identified nine possible mineral contributors, allowing for a transformation from elemental concentrations to mineral concentrations. The comparison of these derived mineral concentrations to those measured from the cuttings, where available, was generally favorable. This first extension of these spectroscopic measurements and analysis techniques to crystalline rocks has resulted in a successful determination of elemental concentrations from logging measurements, and the initial estimates of mineralogy indicate that further research has significant potential for improving the description of crystalline rock.

Acknowledgement. The authors wish to thank the Swedish State Power Board and the Gas Research Institute for providing the opportunity to publish this data.

References

Adams JAS, Osmond JK, Rogers JJW (1959) The geochemistry of thorium and uranium. Phys Chem Earth 3:298

Clayton CG, Coleman CF (1985) Analysis of neutron flux distributions generated by sources of fast neutrons in a range of bituminous coals. Int J Appl Radiat Isot 36:757

Grau JA, Herron MM (1987) Absolute elemental concentrations estimated from geochemical well logging using neutron-induced gamma-ray spectroscopy and a geochemical model. AAPG Annu Conv, Los Angeles

Grau JA, Schweitzer JS (1987) Prompt gamma-ray spectral analysis of well data obtained with NaI (Tl) and 14-Mev neutrons. Nucl Geophys 1:157

Herron MM (1986) Mineralogy from geochemical well logging. Clay clay Minera 34:204

Herron MM (1987a) Future applications of elemental concentrations from geophysical logging. Nucl Geophys 1:197

Herron MM (1987b) Estimating the intrinsic permeability of clastic sediments from geochemical data. SPWLA 28th Annu Logging Symp, London, Pap HH

Herron MM (1987c) Geochemical classification of terrigenous sands and shales from core or log data. J Sediment Petrol (submitted)

Herron MM, Grau JA (1987) Clay and framework mineralogy, cation exchange capacity, matrix density, and porosity from geochemical well logging in Kern County,California. AAPG Annu Conv, Los Angeles

Hertzog R, Colson L, Seeman B, O'Brien M, Scott H, McKeon D, Wraight P, Grau J, Ellis D, Schweitzer J, Herron M (1987) Geochemical logging with spectrometry tools, Tech Symp Soc Pet Eng, Dallas, SPE 16792

Senftle FE, Tanner AB, Philbin PW, Boynton GR, Schram CW (1978) Insitu analysis of coal using a ^{252}Cf-Ge(Li) borehole sonde. Min Eng (AIME) 30:666

Serra O, Baldwin J, Quirein J (1980) Theory, interpretation and practical applications of natural gamma-ray-spectroscopy. SPWLA 21st Annu Logging Symp, Lafayette, Pap Q

Capture of Deep Earth Gas in Water-Filled Fractures

G. GUSTAFSON[1]

Introduction

It is known that large accumulations of hydrocarbon gases can be derived from deeply burried material. Thus, gold (1979) advanced the hypothesis that hydrocarbon gases of abiogenic origin are abundant, and could be exploited as an energy source. In 1980, and ad-hoc committee of the National Academy of Sciences (USA) (Geotimes 1980) studied the subject of abiogenic gas, but it did not select a specific locality for possible deeply buried hydrocarbon accumulations.

Evidence for an ascending microflow of gas - containing methane from deeply buried sources has been demonstrated in boreholes at three sites in the Precambrian bedrocks of Sweden (Swedish State Power Board 1985). Despite a large number of geological, geophysical and geochemical studies to support the Deep Earth Gas Project (DEPG) at Siljan, the key issue has been the lack of a conventional caprock which would prevent the escape of earth gas from the formation.

In this chapter, a model is presented which differs from the conventional caprock model of a reservoir. In our model, deeply buried gas accumulations can under certain circumstances be trapped in the hydrocarbon reservoir fracture system by capillary forces. The theory for pressure buildup is based on our existing knowledge on the state of stress, strength and the hydraulic conductivity of fractured hard rocks in the upper part of the Earth's crust.

Pressure Buildup in Fractured Rock

The theory for pressure buildup in fractured rock provides that both the water phase and the gas phase of the fluid are discontinuous in the rock mass. Hence, gas bubbles and porewater can pass each other in a fracture. If a capillary meniscus can then develop in each of the fractures in a fracture net, so that a water meniscus is formed in each fracture, a hydraulic cup will be formed whose boundary is defined by the boundary between the water and gas phases. A qualitative model for a system of gas

---

1 VIAK AB, Mölndalsvägen 85, 41285 Göteborg, Sweden

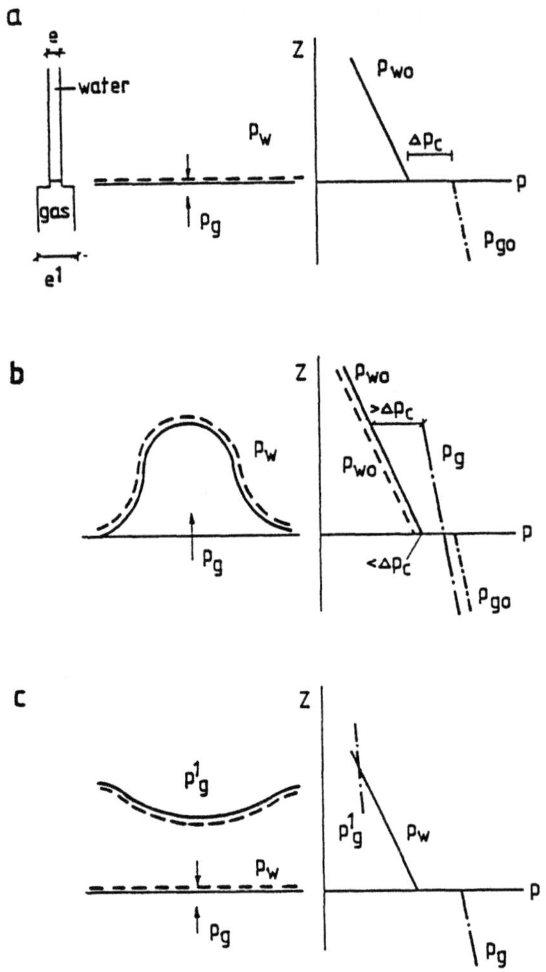

Fig. 1 a-c. Model of pressure gradients for a system of gas and water in fractures where gas is supplied from below. a Initial condition with pressure difference from capillary forces. b Gas migration into fracture and rise of gas bubble giving lower gas pressure (volume expansion) and higher water pressure. c Closing of gas bubble with a water miniscus and a gas pressure acting from below.

and water in fractures, where gas is supplied from below is illustrated in Fig. 1 a-c.

The theory which describes the model is as follows. Natural rock fractures have a variable but finite length, λ, and an aperture that varies both for the same fracture and between different fractures. The variation in fracture aperture, e, can be expressed in terms of a statistical distribution, F(e), such that $P(\epsilon < e) = F(e)$ and the frequency distribution $dF(e)/de = f(e)$.

If two fractures join, one of which has aperture e, and is water-filled, and the other has aperture e_1, and is gas-filled, a pressure difference Δp_c can occur if $e>e_1$. We then have

$$\Delta p_c = \frac{2F \cos \Theta}{e} = \frac{\alpha}{e}, \tag{1}$$

where F is the surface tension, Θ is the contact angle, and α is the capillary coefficient. The probability for this to occur is $p(e<e_1) = 1-F(e)$.

Suppose there are N fractures over a vertical distance L. The number of fractures having an aperture between e and e + de is then given by

$$dN = N \cdot \frac{dF(e)}{de} \cdot de = N \cdot f(e)\, de. \tag{2}$$

Considering the probability for a pressure difference to occur, the total pressure difference of N fractures will be:

$$\Delta p_{tot} = \Sigma dN \cdot \Delta p_c \cdot p(e<e_1) = \int_0^\infty N \cdot f(e) \cdot de \cdot \frac{\alpha}{e} [1-F(e)]. \tag{3}$$

Let the number of interesting fractures be described as the product of the distance, L, and some fracture junction frequency, γ, to give the equation

$$N = L \cdot \gamma. \tag{4}$$

If the pressure difference in Eq. (3) is measured over a distance dz, then the vertical pressure gradient due to capillary forces is given by the following equation:

$$\frac{dp_c}{dz} = \gamma \cdot a \cdot \int_0^\infty f(e) \cdot [1-F(e)] \cdot \frac{de}{e}. \tag{5}$$

To evaluate the integral in Eq. (5), the distribution function, f(e), which describes the distribution of fracture aperture widths e must be known. We suppose that f(e) is defined by the (log-normal distribution) equation, Dagan (1979), where y = lne:

$$f(e) = \frac{1}{e\, \sigma_y \sqrt{2\pi}} \cdot e^{-(\ln e - \mu_y)^2/2\sigma_y}. \tag{6}$$

Equations (5) and (6) then give, for the vertical pressure gradient due to capillary forces,

$$\frac{dp_c}{dz} = a\,\gamma \cdot \frac{1 - R(\sigma_y)}{e^{(\mu - \sigma^2/2)}},$$

$$\text{where, } R(\sigma) = \frac{1}{2\pi} \int_{-\infty}^{\infty} e^{-(t+\sigma)^2/2} \int_{-\infty}^{t} e^{-x^2/2}\, dx\, dt. \tag{7}$$

Let S denote the volume fraction of water in the fracture volume. Then the equation of equilibrium for the mixture of water and gas in the fracture volume contained in a slab of vertical thickness dz can be written in the form

$$\frac{dp_h}{dz} = (S\rho_w + (1-S)\rho_g)g, \tag{8}$$

where ρ_w and ρ_g is the density of water and gas respectively.

To determine the pressure gradient due to capillary forces, Eq. (7), we need to know the fracture aperture e and the frequency of joining fractures. These quantities are extremely difficult to measure directly. They must therefore be determined by an alternative method. Thus, the rock mass permeability K for granitic rocks is directly related to the fracture aperture width e. The fracture aperture width e can then be estimated indirectly from pump tests, Braester (1977), and by application of the standard equation

$$e = \sqrt[3]{\frac{12\ \mu\ K}{f \cdot \rho_w \cdot g}}, \tag{9}$$

where μ is the viscosity of water and f is the fracture frequency.

The hydraulic conductivity K is known to be log-normally distributed, as are the fracture apertures. Therefore the following relation will hold

$$\mu_{lne} = \frac{\mu_{lnK} + \ln \frac{12\ \mu}{\rho\ g\ \gamma}}{3}, \tag{10}$$

where μ_{lne} is the mean of logfracture aperture and the standard deviations are related as follows:

$$\sigma_{lne} = \frac{\sigma_{lnK}}{3}. \tag{11}$$

Fracture mapping of large exposures of rocks in open pit mining has revealed that the fracture length shows a negative exponential distribution. The frequency of joining fractures can then be regarded as being of the same order of magnitude as the inverse of the characteristic fracture length.

If the parameters of Eq. (7) and (8) can be determined a hypothetical pressure gradient due to microseepage from below can be calculated. The situation will now be tested on the deep borehole at Siljan.

Hydraulic Conductivity in Gravberg-1 and its Influence on the Pressure Buildup

During the drilling of the deep borehole at Siljan several hydraulic tests, drillstem tests and leak off tests were performed in the interval 1250.5 m - 3165 m. In Fig. 2, the evaluated hydraulic conductivity (Gustafson 1986a) is plotted against the centre of each tested interval.

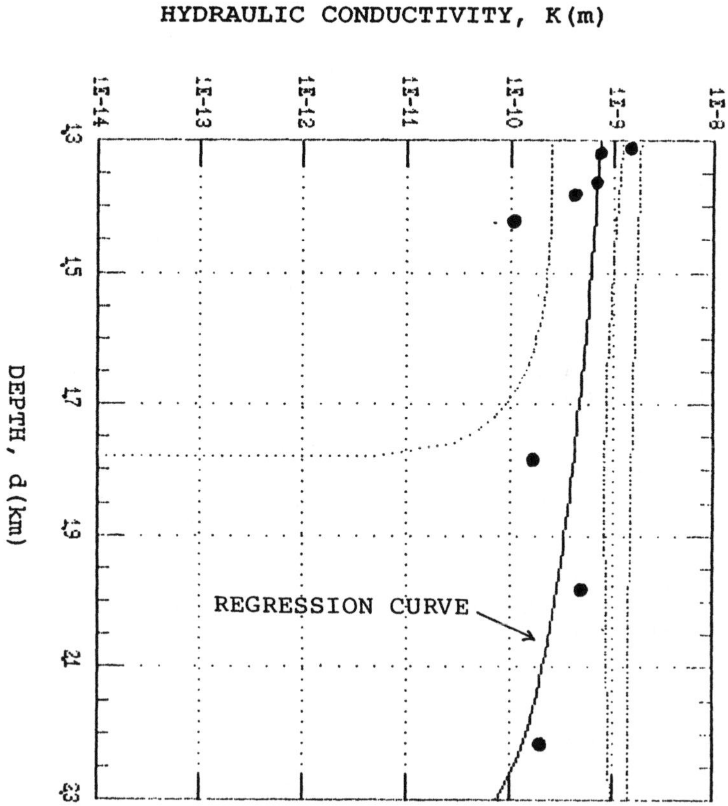

Fig. 2. Hydraulic conductivity in Gravberg-1

As can be seen from Fig. 2, most of the data lies within the interval $K = 10^{-10} - 10^{-9}$ m/s and there is a decreasing trend of about one order of magnitude per km depth.

In order to determine the possible pressure gradient due to gas capture, the parameters of Eq. (7) must be determined. The surface tension of water against natural gas is $F = 0.04$ N/m and the contact angle is assumed to be $\Theta = 0°$, which means active gas seepage from below. With a water viscosity of $\mu = 10^{-3}$ Pas, a water density of 990 kg/m^3 and a frequency of open fractures of $f = 0.1$ 1/m as found in the shallow boreholes, Swedish State Power Board (1985) the effective fracture opening can be calculated to be $e = 1.1 \times 10^{-5}$ m for a hydraulic conductivity of $K = 10^{-10}$ m/s. Other investigations (Gustafson 1986b) have shown that the standard deviation of the natural logarithm of the hydraulic conductivity normally is about $\sigma_{lnk} = 3$, which gives a factor $R(\sigma) = 0.25$. The frequency of joining fractures can be assumed to be of the same magnitude as the fracture frequency, $\gamma = 0.1$ m^{-1}. With this assumption the possible pressure gradient due to capillary forces can be calculated to be $dp_c/dz = 0.9$ kPa/m, which can be compared to the hydrostatic gradient of $dp_w/dz = 9.7$ kPa/m. It is thus not likely that a gas reservoir can exist within the uppermost 3 km of the rock in the area.

Possible Depth to a Gas Reservoir

As was found in Fig. 2, there is a decreasing trend with depth of the hydraulic conductivity of about one order of magnitude per km. It can also easily be seen from Eq. (9) and (7) that the possible pressure gradient increases with decreasing hydraulic conductivity.

It is also well established that the hydraulic conductivity decreases with increased rock stress, and the trend shown in Fig. 2 is probably due to this.

The total stress in the rock can be subdivided into a pore pressure and an effective stress. One interesting question is what happens if the pore pressure gradient due to gas trapping in a tight rock is the same as the vertical total stress gradient. In that case it can be assumed that the growth of the effective stresses with depth will cease and the mean effective conductivity will be constant. With a total vertical stress gradient of 27 kPa/m, the corresponding hydraulic conductivity can be calculated to be $K = 4 \times 10^{-12}$ m/s, a figure that can be compared to the results of the hydraulic tests in the borehole. They show a conductivity in the range of $K = 10^{-10}$ m/s at 2300 m depth and a decreasing trend. A sufficiently tight rock can thus appear within the target depth of 7500 m.

References

Braester C (1977) STF Symp Groundwater in Hard Rocks, Stockholm

Dagan G (1979) Water Resourc Res 15

Geotimes (1980) Nov

Gold TJ (1979) Petrol Geol 1:3-19

Gustafson G (1986a) Hydraulic tests, Gravberg-1, Rep Oct

Gustrafson G (1986b) BeFo 84:1/86, Stockholm

Astroblemes

Physical Aspects on Cratering and Slumping

K. B. Broberg[1]

Introduction

Three hundred and sixty million years ago a large meteorite hit the Earth, causing a giant ring depression, the vestige of which now contains some lakes, among them Siljan, which gives its name to the structure (Fredriksson and Wickman 1963, Rondot 1975, Vattenfall 1985). Many questions are connected with the event, particularly the collapse (the "slumping"), which will be discussed in the present paper.

Gravity action is generally assumed to explain rapid crater modification and its scale dependence (Cintala et al. 1976, Hartmann 1972, Melosh 1982, Settle and Head 1979, Croft 1981b). Several mechanisms have been suggested, such as rim slumping, ejecta fallback and crater floor rebound.

Settle and Head (1979) studied the morphology of modified lunar craters and reconstructed transient craters by assuming volume conservation. The reconstruction was made under the assumption that the maximum wall inclination of the transient crater was larger than the angle of repose (about 45°). They assumed 60° inclination, but presented cavity reconstruction results also for 45°.

Melosh (1982) similarly assumed that slumping from a static state would require a wall inclination larger than the angle of repose, but noticed that the average wall inclination appeared to be appreciably smaller for transient craters. To overcome this apparent contradiction, he argued that the collapse of a transient crater could not be treated as static, but that acoustic fluidization exists for a short time at the transient crater, leading to a dramatic decrease of the internal friction.

The assumption of a wall inclination larger than the angle of repose as a necessary condition for slumping from a static state, however, does not seem to be correct. In fact, shear collapse can, in principle, occur even below a horizontal surface, such as a crater floor. The necessary condition, of course, is the appearance of stresses that can cause sliding. This will later be discussed to some detail.

---

[1] Division of Solid Mechanics Lund Institute of Technology Box 118, S-221 00 Lund, Sweden

A very interesting discussion about the different events during the collapse of the transient crater is given by Croft (1981b). He assumes that the central uplift has been initiated and even grown significantly before the radial growth of the crater along the ground surface is completed. This would imply that the concept of a transient crater is very imprecise. However, the cratering action is certainly a faster process than the uplift, which, most probably, is still in the initial stage at the time when the radial growth is completed. Therefore the useful idea of a transient crater, quiescent for a moment, and constituting the interface between cratering and collapse, appears to be very reasonable.

The Transient Crater at Siljan

The dimensions of the transient crater at Siljan can be estimated from a reconstruction, using the known data of the present dimensions of the ring structure. This involves estimates both of the modification due to slumping and of the long-time modification due to erosion, etc.

The present dimensions of the ring structure at Siljan are shown in Fig. 1. There is a gravity anomaly at the central parts, indicating lower than normal density. This conclusion is supported by evidence from drillcore data (Vattenfall 1985).

The rock is granitic except in the Ring, where sedimentary rock is found (Hjelmqvist 1966), probably swept down during several ice-ages from a previous sedimentary overburden. The height of this overburden was probably more than 1 km. (Collini personal communication).

Several investigations (Cooper 1977, Dence 1973, Pike 1977, Settle and Head 1979) indicate that the depth-diameter ratio of transient craters decreases with increasing scale. For a terres-

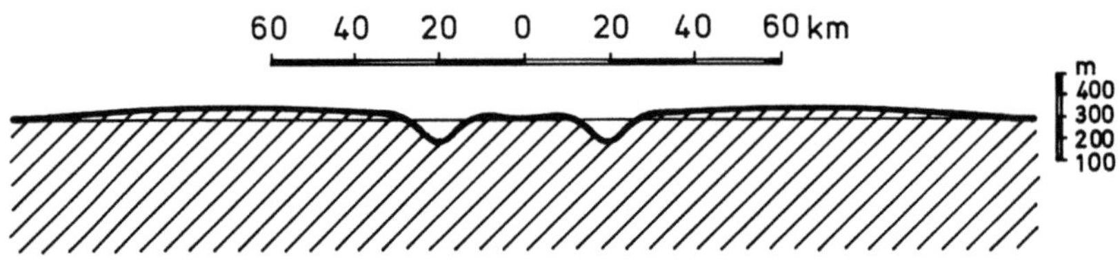

Fig. 1. Present profile of the ring structure at Siljan. The figure is based upon averaged altitudes for different radii. Numbers on vertical scale give altitudes above sea level. For each radius the average was obtained after exclusion of the five highest and five lowest values in 20 different directions (spaced 18°). (Data provided by Mr. Tord Lindbo, Swedish State Power Board)

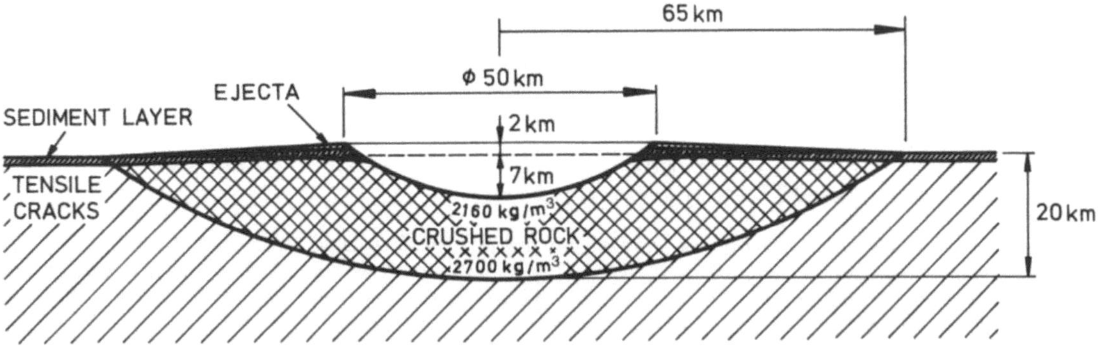

Fig. 2. Assumed shape and size of the transient crater at Siljan

trial crater of the Siljan size, the depth below the ground level can be estimated at about 0.14 and the rim-crest height at about 0.04 rim-crest diameters.

The shape of transient craters appears to be close enough to a spherical bowl that this easy-to-handle shape can be used in the model (see Fig. 2). In addition, the rim shape is assumed to be simply conical. As explained by several authors (e.g., Croft 1981a) the transient cavity below the original ground level is due both to ejecta that has been thrown out beyond the rim crest and to interior mass flow, causing uplift of the ground surface outside the rim crest. In other words, the rim consists both of material uplifted from below and of ejecta thrown from the excavated crater over the rim crest (see Fig. 2).

Fig. 3. Crater produced by a small explosive charge in plasticine (block built by cubes). One observes by counting the number of cubes along vertical lines that virtually no material has been ejected

For convenience, the ejecta fall back is here included in the concept of the transient crater, even though some of it does not occur until the slumping process is well started.

An illustration of the mass movements which cause uplift of the material outside the rim crest is given in Fig. 3, which shows the result of an experiment with plasticine, a material which gives a very small amount of ejecta.

On a very small scale, virtually all ejecta are thrown far away from the crater, because the length of throw depends essentially on the initial velocity, which is dependent on relative position, but fairly independent of the scale. Thus, with increasing scale, less and less ejecta is thrown out beyond the rim region, and at a sufficiently large scale some ejecta should also be expected to fall back inside the rim crest. Housen et al. (1983) arrive at a different result; however, in their dimensional analysis the fact that the initial ejecta velocity is not primarily gravity-scaled seems to be overlooked. For a crater of the Siljan size the amount of ejecta that is thrown out beyond the rim region is certainly small. It is estimated at 1% of the mass that occupied the transient crater plus the meteorite mass (which is even smaller, of the order of a tenth % of the mass that occupied the transient crater).

Quite obviously, mass flow, of the magnitude required for formation of the transient crater and the rim, cannot occur in hard rock without previous crushing. The volume in which the rock is crushed must be much larger than the transient cavity (cf. Croft 1981a). An estimate can be obtained by investigating how mass is conserved. The balance equation simply tells that the mass of the region in which material is crushed has to be the same before as after the formation of the transient crater, apart from the small and insignificant amount of ejecta that is thrown out beyond this region.

Mass conservation cannot be exchanged by volume conservation because crushed rock has lower density than solid rock. The rock is assumed to be crushed above a certain spherical surface as indicated in Fig. 2. In more peripheral parts of this region, large blocks are expected, whereas crushing to smaller blocks (or grains) must be assumed for the neighbourhood of the walls and bottom of the transient crater.

The peripheral blocks have been exposed to an essentially radial translation, with little rotation, implying only small decrease of the overall density. The blocks near the crater walls and bottom, on the other hand, have certainly taken part in a rather turbulent flow. Therefore the overall density in this region is probably significantly lower than for solid rock. However, a very wide distribution of block (and grain) size is expected in this region, a condition that is known to imply higher density of "granular" material than a narrow distribution.

For the model to be studied, the density in the crushed region is specified to 2160 kg/m^3 for parts near the walls and bottom of the transient crater. This corresponds to a porosity of 20%. (For solid rock the density is about 2700 kg/m^3).

There is also impact melt in the transient crater. The amount can be estimated to around 250 km^3 for a crater of the Siljan size (Broberg 1985).

Regardless of the scale, the crushing front lags behind the shock front at a distance that increases with time. This is due to the delay time for fracture, which increases with decreasing stress. It is thus a material characteristic, independent of the scale. At an event of the Siljan size, the duration of the compressive part of the shock wave is so long compared to the delay time that the crushing front, on the scale considered, almost coincides with the shock front. In other words, crushing occurs while the shear stresses are still approximately equal to their peak values.

It is sometimes argued that most of the crushing in hard rock, like granite, is caused by tensile stresses and thus not by shear stresses. The argument appears to be that critical tensile stresses are of the order of 10 MPa and critical shear stresses of the order of 200 MPa. A little reflection, however, shows that the ambient pressure has to be added to the critical tensile stress. At 20 km depth it is about 530 MPa, and thus the tensile stress carried by the shock wave must be at least 540 MPa, not 10 MPa, to cause fracture. It is true that the critical shear stress should also be adjusted, but not more than to about 700 MPa at this depth, and a shock wave carrying

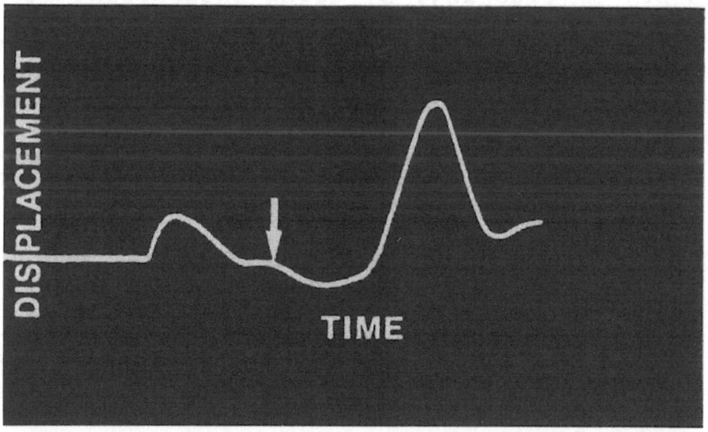

Fig. 4. Displacement at the lower surface of a thick granite plate, the upper surface of which is subjected to an explosive load. The displacement was measured straight below the charge by a capacitive method after gold-plating of the lower surface. Arrow shows arrival of waves reflected from the plate boundary. The plate was chosen to be much thicker than the depth of crushing in order to avoid spalling cracks

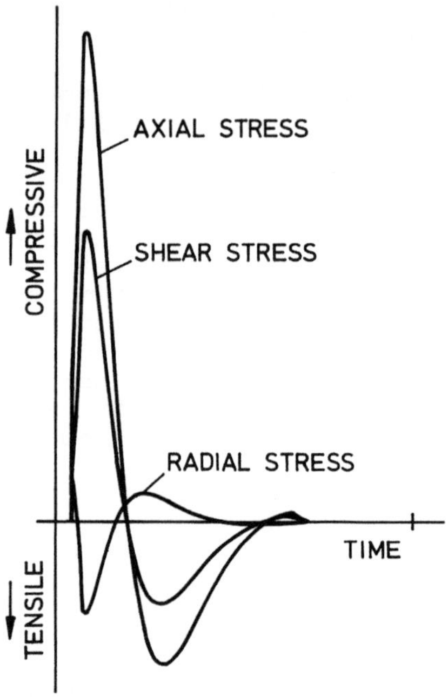

Fig. 5. Stresses calculated for a point just below the region of crushed granite in the case described in Fig. 4

this shear stress at its front probably does not carry as much as 540 MPa tensile stress at its tail. This opinion is supported by experimental results from measurements of displacements at the lower surface of granitic plates, the upper surface of which was subjected to detonation of small explosive charges. Figure 4 shows one result for a plate much thicker than the depth of crushing. By using theoretical results (Broberg 1959) it is possible to calculate the force-time relation of the point-load on the surface that would cause the same deflection under assumptions of linear elasticity throughout, and then, based on this relation, to calculate the stresses at different depths below crushed material. A result is shown in Fig. 5, and it is obvious that the maximum shear stresses are considerably higher than the maximum tensile stresses. Scale effects could hardly change this conclusion.

On the other hand, near the surface, where the ambient stress is low, the most far-reaching fractures are certainly those caused by tensile stresses. These fractures, however, probably do not form a three-dimensional network such that the rock can be regarded as crushed (cf. Croft 1981a). Movements within this outer region of tensile fractures are therefore much restricted.

The dependence of the critical stresses for crushing of the rock on the ambient pressure explains to one part the fact that crater depth diameter ratios decrease with increasing scale. In addition, the work against gravity during mass movements at the formation of the transient crater consumes a relatively larger portion of the impact energy the larger the scale is.

Considering the scale effects, it is reasonable to assume that the region of crushed rock in a crater of the Siljan size extends much further out horizontally than vertically.

Based on the considerations mentioned, the following assumptions about the dimensions of the transient crater are made (see Fig. 2). The rim-crest diameter is 50 km, the crater depth below the original ground level is 7 km, the rim-crest height is 2 km, the rim extends to 65 km from the point of impact, and so does also the region of crushed rock in the horizontal direction, whereas it extends to 20 km in the vertical direction. (The latter figure is not crucial in the present study). A simple calculation shows that the volume of the rim is about 10,000 km^3 and the volume of the cavity below the original ground level is about 5700 km^3. The difference is about 3% of the original volume of crushed rock (which can be matched to the previous assumption of a porosity varying from zero at the peripheral parts of this volume and 20% close to the walls and bottom of the transient crater). Consideration of the disappearance of some ejecta does not change this figure noteworthy.

Stresses below the Transient Crater

In a rock mass (for instance crushed granite) under static equilibrium the stress state is not necessarily hydrostatic, but the stress difference, i.e., the largest difference between the principal stresses, is limited by friction laws. Since friction forces exist in a rock mass the stresses at static equilibrium are not uniquely determined by boundary conditions.

The fact (or rather the assumption) that the rock mass around the transient crater is at rest during a moment of time does not necessarily imply that static equilibrium prevails. In regions where the stress difference exceeds the highest permissible static equilibrium value internal motion is initiated.

The problem of finding the initial conditions for mass flow after the transient crater has been formed, consists of calculating the stresses in the surrounding rock mass, assuming this mass to be motionless. However, the stresses are not uniquely determinable, but depend upon the previous history. Since this cannot be obtained with any useful degree of precision, an estimate will be made, based on investigations of different admissible stress fields. Which kind of admissible stress field comes closest?

Melosh (1977) and McKinnon (1978) choose a plastic field at static equilibrium under gravity load. Two basic objections can be raised to this choice. The first - and least important - is that the usual elastic-plastic relations are not appropriate unless one allows the yield condition to be pressure-dependent. The second one is that, if one accepts the fact that collapse actually occurs, then it is not a question of static equilibrium, even if the transient crater is at rest for a moment. Some kind of elastic-viscoplastic relations would be more appropriate. At the very first moment, such relations would imply no plastic flow at all. Thus the stress field would come close to an elastic stress field at this moment.

A distinction should be made between the stresses during the outward mass flow when the transient crater is created and the stresses when this mass flow has been arrested. It might be helpful to think of the analogous situation when a stamp is pressed downwards into an elastic-plastic half-space, causing outward plastic flow. When it is removed, the unloading implies return to elastic stresses and, perhaps, to reversed plastic flow. If the stamp is removed quickly, elastic shear stresses exceeding those capable of causing plastic flow can exist for a short moment, whereupon reversed (visco-)plastic flow is initiated and proceeds until static equilibrium is obtained.

Not shown by the analogy is the fact that the load is not only removed but actually reversed: after formation of the transient crater, gravity load prevails. An element at the bottom of the transient crater, for instance, has experienced vertical compression and horizontal elongation during the formation of the transient crater (cf. Fig. 3), whereas immediately afterwards, gravity forces cause a horizontal compressive load on the element. This implies opposite direction of the shear stresses before and after the formation of the transient crater.

The conclusion is that the stress field at the moment when the transient crater is completed ought to be rather close to an elastic stress field at static equilibrium under gravity load. Shear stresses, considerably above the critical ones (for static equilibrium) can exist for a short moment, but any shear stress above the critical value causes almost immediately shear flow in the rock mass.

Linear elasticity is assumed. There is, of course, some dependence on Poisson's ratio. The value of 0.25 is chosen. The stress field is found by means of finite element calculations.

Initiation of Sliding

After having calculated the stresses below the transient crater, the question of interest concerns whether or not these stresses are able to cause sliding. On the basis of several investigations Hoek and Brown (1980) suggest a formula connecting the stress difference ($\sigma_1-\sigma_3$) required for failure with the smallest principal stress σ_3 (here, as usual in rock and soil mechanics, compressive stresses are defined as positive):

$$\sigma_1-\sigma_3 = \sqrt{(m\sigma_3+s\sigma_c)\sigma_c}, \qquad (1)$$

where m and s are dimensionless constants, and σ_c is the uniaxial compressive strength of the intact rock material. This criterion appears to hold reasonably well both for intact rock material and for rock masses, although, of course, with different values of the constants m and s. Hoek and Brown (1980) define several qualities of rock masses, and it is not quite obvious which one corresponds most closely to the rock mass

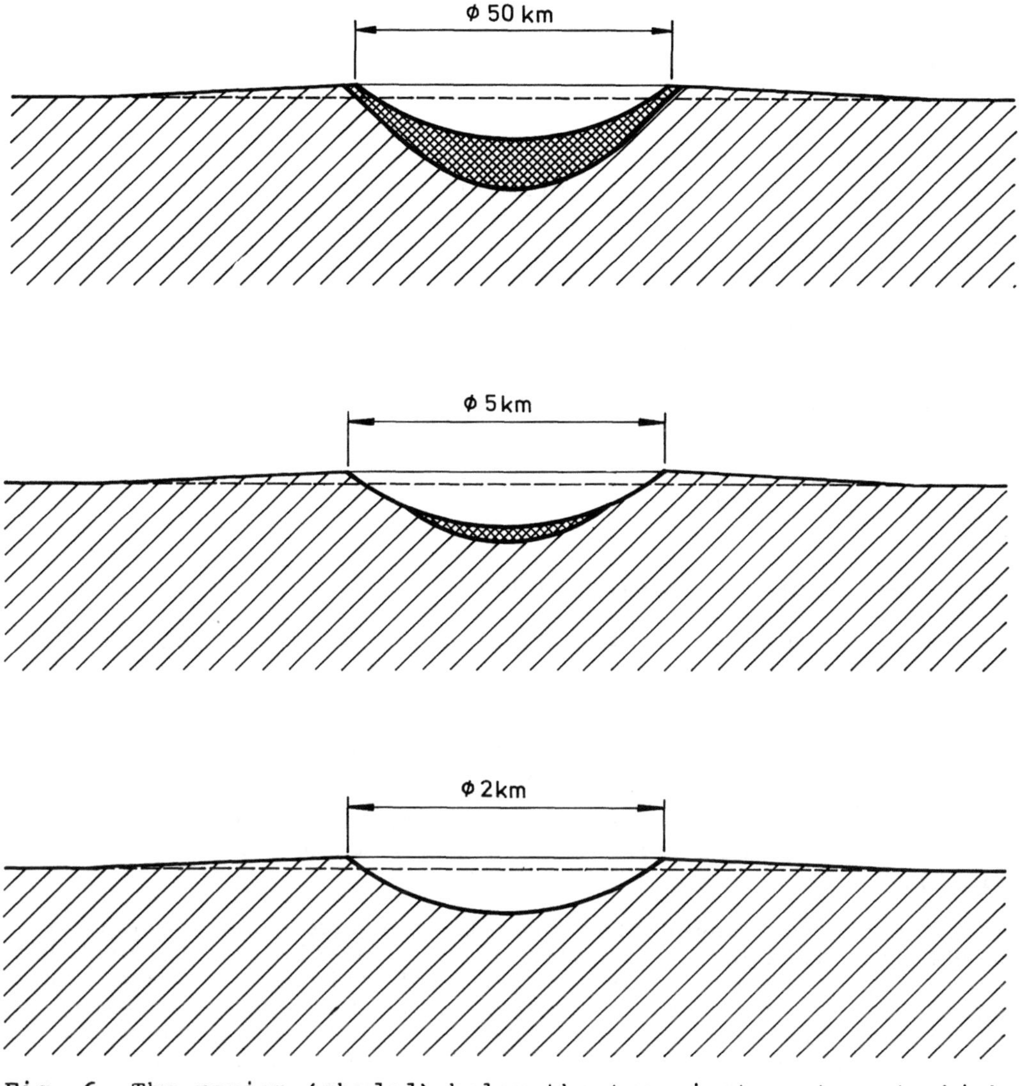

Fig. 6. The region (shaded) below the transient crater at which the stress difference is high enough the initiate sliding. Upper figure crater diameter 50 km (as assumed for the Siljan crater); **middle figure** crater diameter 5 km; **lower figure** crater diameter 2 km. The scale influence is obvious

near the walls ond bottom of the transient crater. The choice made here is a granitic rock mass of "fair quality" ("several sets of moderately weakened joints spaced at 0.3 to 1 m"), for which Hoek and Brown (1980) suggest the values m = 0.5 and s = 0.0001. As regards σ_C compressive strengths around 200 MPa have been reported for granitic material of the kind found at Siljan, and even though there is some uncertainty as to the role of the fairly high temperatures prevailing, this value is chosen here.

The results of the finite element calculations show that criterion (1) is fulfilled in the shaded region below the transient crater shown in Fig. 6. Thus sliding will be initiated in this region.

As a comparison, results for transient craters with 2-km and 5-km rim-crest diameter, geometrically similar tc the one (50-km rim-crest diameter) primarily studied are also shown. The scale influence is obvious. No sliding will be initiated in the 2-km crater and the region of initial sliding is comparatively small in the 5-km crater.

The scale influence is thus simply a consequence of the nonlinearity of criterion (1). The simple linear idealization, the Mohr-Coulomb law with constant (i.e., pressure-independent) friction coefficient, cannot give rise to a scale influence. The nonlinearity of the failure relation is, however, a well-established fact in rock and soil mechanics (cf. e.g., Brace and Martin 1968, Cherry et al. 1968, Heuze 1983, Hoek and Brown 1980, Paterson 1978).

Localization of the Sliding Event

As soon as intense sliding occurs somewhere, localization of the deformation is expected to take place. This implies that large block aggregates move with respect to each other with very little internal deformation. Localization is a phenomenon that appears when the critical stress (here $\sigma_1-\sigma_3$) required to increase the deformation (or the deformation velocity) decreases with increasing deformation (or deformation velocity). It is generally expected to occur whenever friction is involved (since the "dynamic friction" is lower than the "static friction").

For a rock mass under high overall pressure, the tendency towards localization to certain planes of sliding is probably very strong. Whereas the mechanism of sliding in a rock mass under low overall pressure consists of rolling and sliding of individual blocks over each other, the additional mechanism of block fracturing enters at high overall pressure. This obviously contributes to a lowering of the resistance to sliding, further accentuated at high sliding velocities, because of the fact that high velocity impacts facilitate fracturing of rock material.

In a simple landslide, localization can occur along one single (usually curved) slip layer. However, due to the obvious constraints inherent in the geometry, localized flow during a crater collapse must take place along several different planes. Thus one should expect localization along both essentially vertical, radial planes and essentially horizontal planes. The first system tends to deform the geometry from a circular to a polygonal shape, such as has actually been observed in many slumped extra-terrestrial craters. The second system gives rise to terracing, also frequently observed in slumped extraterrestrial craters. A nice example of both polygonial shape and terraced walls is given by the lunar crater Timocharis (see e.g., Settle and Head 1979).

It is obvious that the network of intersecting planes of localization must be much more dense near the crater bottom, where the largest overall distortions are expected, than near the rim. It is even questionable whether any localization at all to individual planes takes place near the bottom, but this does not exclude the likelihood of heterogeneous deformation inside the rock mass.

The stress calculations give only the initial conditions for the collapse event. The motion that follows is very complicated, involving localized flow, friction loss and time-dependent material characteristics, difficult to estimate. Therefore far-reaching simplifications have to be made.

The motion is assumed to proceed inwards, i.e., from larger to smaller horizontal radii. This implies a height increase of the moving mass (the shaded region in Fig. 6). This height increase ought to be more pronounced at a small than at a large horizontal radius. It is simply assumed that the relative height increase during each short time interval decreases linearly towards the periphery, and that it is zero at the (currently) most peripheral part of the moving mass. The latter part is thus simply sliding in the direction towards the centre.

In this way, the morphology of the collapse event is completely specified, but not the time sequence. The morphology change has to be consistent with a potential energy decrease of the moving mass during the early phases of the motion, and it is. This potential energy decrease is accompanied by a kinetic energy increase (initially) and an energy loss due to friction.

The potential energy of the moving mass is known (i.e., it is possible to calculate) at any stage of the motion as long as this proceeds in the specified way described. It decreases towards a minimum, whereupon it increases. If the friction loss is large, the motion comes to a stop before this minimum is reached. Here it is assumed - rather arbitrarily - that the motion stops just as the minimum is reached.

The result is shown in Fig. 7. The motion has resulted in a central uplift. The potential energy of the mass that was moving is now smaller than its initial value, but larger than its lowest possible value. The latter would be obtained if the mass occupied the position in which its free surface is horizontal. The way in which the motion was specified simply does not allow that this position is reached. (If it did, another criterion for deciding when the motion stops would have to be used in order to arrive at a reasonable morphology, i.e., a central peak, reaching to the neighbourhood of the original ground level.)

Now it must be checked whether the stresses in the new configuration actually imply arrested motion. A new finite element calculation shows that this is not the case, but condition (1)

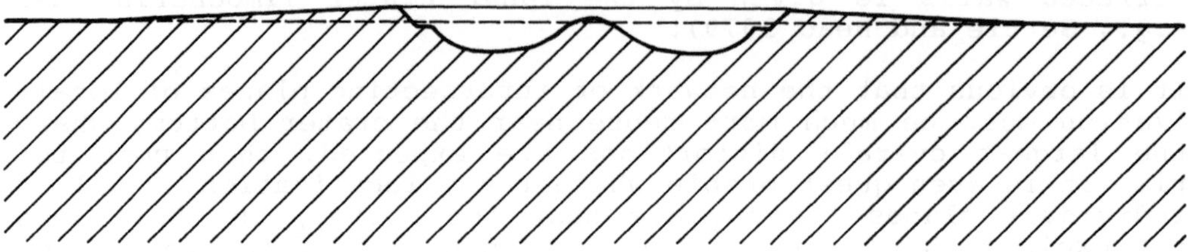

Fig. 7. The first stage of the collapse process, according to the model described in the text

predicts motion of the rock mass shown by the shaded region in Fig. 8. The previous procedure is then repeated, resulting in a new configuration at the potential minimum of the new moving rock mass (see Fig. 9). The procedure has to be repeated once more (see Figs. 10 and 11), until no further motion is predicted by condition (1). Figure 11 thus shows the final morphology after the collapse is completed.

One can, of course, argue that the whole process should take place in essentially one step, and not in three distinct steps as implied by the sequence shown in Figs. 7-11. This is probably correct, but it is not likely that the final result would differ very much from the one obtained by the step-by-step approach. Of course, it cannot be claimed that the final configuration obtained here agrees in a detailed way with the real one, but it can be regarded as one reasonably good exemple of permissible final states.

The irregular morphology shown in Fig. 11 might be a result of the step-by-step procedure to some extent, but not completely. Several "corners" (rings) can be traced to "corners" of the original moving mass. In reality, of course, a more irregular structure is to be expected, due, for instance to irregularities in the flow localization.

The impact melt was probably initially much hotter than the melting point and, consequently, highly fluent. To where did it

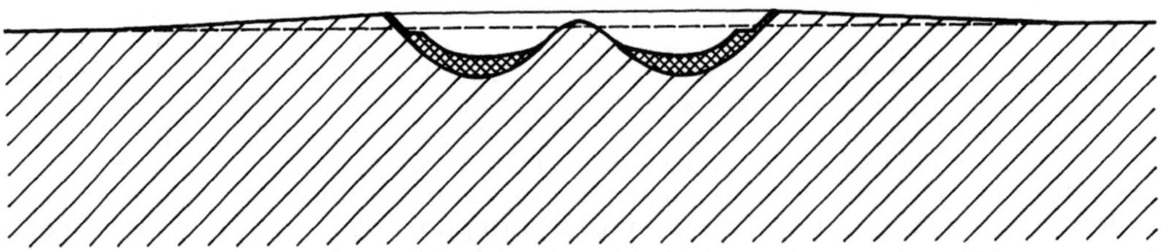

Fig. 8. The region (shaded) below the crater after the first stage of the collapse process at which the stress difference is high enough to initiate sliding

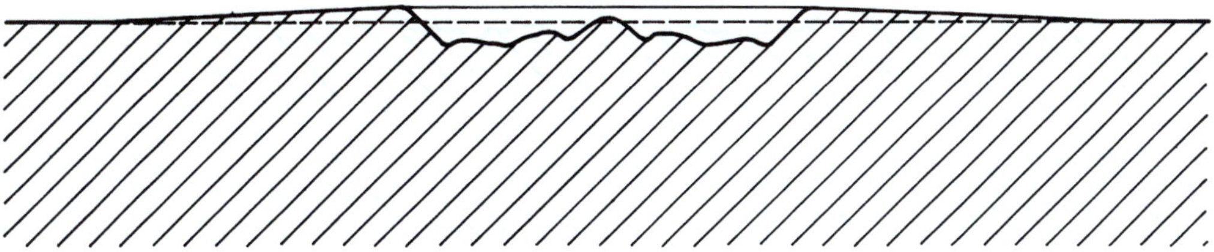

Fig. 9. The second stage of the collapse process, according to the model described in the text

go? Since it originated from the neighbourhood of the impact site, it was probably forced towards the walls and bottom of the transient crater. (A substantial part, of course, disappeared as ejecta that did not fall back inside the crater.) During the collapse, most of it probably maintained its position on the surface, but a good deal of mixing with underlying rock certainly also occured.

The amount of impact melt was very large. Its thickness would have been several 10 m, even if it had been evenly distributed inside the crater. This highly fluent lava layer must have rushed down the slopes towards the lowest parts.

It is possible to estimate the collapse time to a few minutes. It should be only a few times larger than the time of free fall, which is less than 1 min for 9 km height.

Long-Time Modification

A great number of slow processes have led to almost complete levelling of the ground - the altitude differences are now of the order of only 100 m, see Fig. 1. The rim and the central part of the slumped crater consisted of blocks. Near the surface the material was probably thoroughly crushed and therefore comparatively mobile. It could have easily been carried by glaciers, for example.

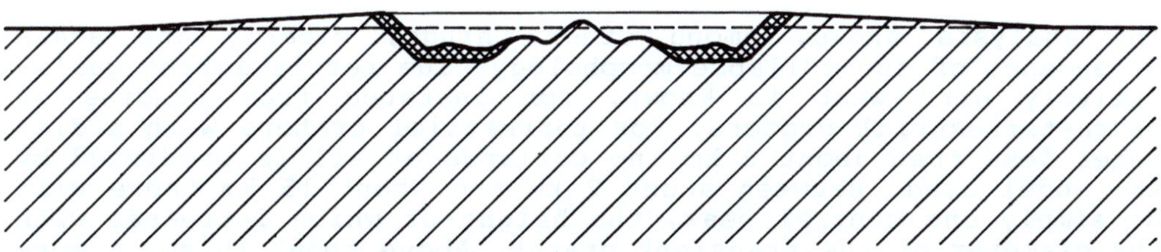

Fig. 10. The region (shaded) below the crater after the second stage of the collapse process at which the stress difference is high enough to initiate sliding

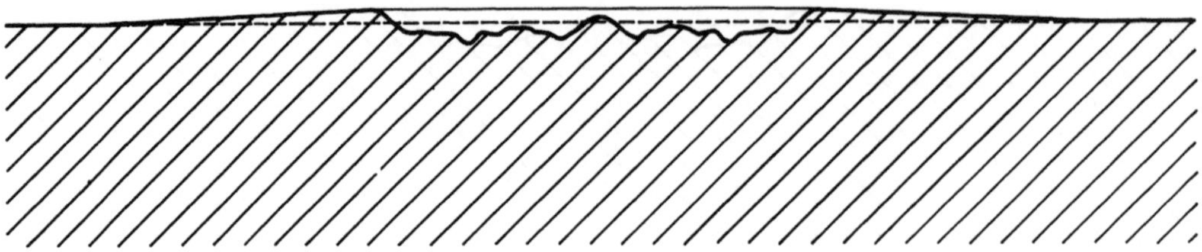

Fig. 11. The third stage of the collapse process, according to the model described in the text. Calculation of the stresses show only insignificant regions in which continued sliding could be expected. The profile shown is therefore expected to give a reasonably good picture of how the Siljan crater could have looked immediately after the collapse

Now, if one compares the result obtained from the study of the collapse (i.e., the result shown in Fig. 11) with the present crater profile, Fig. 1, is it possible to see any agreement? As far as the dimensions are concerned, there seems to be some agreement. The very distinct ring depression shown in Fig. 1, on the other hand is not so clearly seen in Fig. 11, but might have resulted from erosion at sensitive regions. The agreement would have been better if the primary central uplift had been allowed to be higher. It might have been, but, again, a detailed agreement with the real case can hardly be obtained with the methods used and with the information at hand about material behaviour, etc.

The Scale Influence

The scale influence has already been mentioned. If one studies the motion of the moving mass shown in Fig. 6 for the crater with 5 km diameter, one obtains the result shown in Fig. 12. One step is sufficient, since a finite element analysis shows that condition (1) does not predict further motion. Thus the result is simply a central peak.

Discussion

The analysis shows, among other things, clearly the scale dependence of the collapse morphology. The result agrees qualitatively with what is known, for instance from studies of extra-terrestrial craters. For terrestrial craters with a diameter less than about 2 km, no collapse at all is predicted by the model used. Between about 2 km and about 15 km there will be simply one central peak, the height of which increases with the diameter of the transient crater. Above about 15 km, a more complex collapse process occurs, according to the model. Roughly speaking, no very large depths are tolerated, since the gravity then causes stress differences overshooting the critical ones. This implies that the depression after a crater collapse will show only a weak dependence on the depth of the transient crater for very large craters. On the other hand the

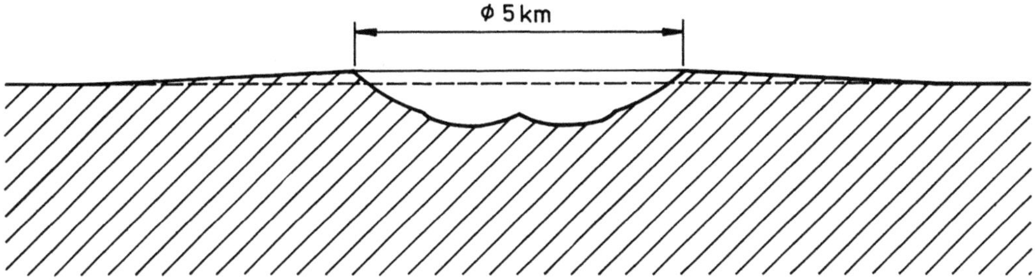

Fig. 12. The morphology after the collapse of a terrestrial crater with 5 km diameter. Only one step is needed, according to the model used. The result is thus only a central peak, instead of the more complex configuraton shown in Fig. 11 for the 50-km crater

depressions obtained here are still considerable - about 3 km - and the rim height is almost 2 km.

Which meteorite causes a terrestrial transient crater with a diameter of 50 km and a depth of 7 km below the original ground level? Estimates have led to the conclusion that a stony meteorite with a diameter of 2 km and normal incidence at 25 km/s could cause a crater of about this size.

References

Brace WF, Martin RJ (1968) A test of the law effective stress for crystalline rocks of low permeability. Int J Rock Mech Min Sci 5

Broberg KB (1959) A problem on stress waves in an infinite elastic plate. Trans R Inst Technol Stockholm Sweden 139

Broberg KB (1985) On the formation of the Siljan crater. Rep Div Solid Mech, Lund, Sweden

Cherry JT, Larson DB, Rapp EG (1968) A unique description of the failure af a brittle material. Int J Rock Mech Min Sci 5:455-463

Cintala M, Head JW, Mutch TA (1976) Characteristics of fresh Martian craters as a function of diameter: Comparison with the Moon and Mercury. Geophys Res Lett 3:117-120

Cooper HF (1977) A summary of explosion cratering phenomena relevant to meteor impact events.

Croft SK, (1981a) In:Schultz PH, Merrill RB (eds) The excavation stage of basin formation: a quantitative model in multi-ring basins. Proc Lunar Planet Sci, Vol 12A. (Pergamon Press, New York, pp 11-14) 207-225

Croft SK (1981b) In:Schultz PH, Merril RB (eds) The modification stage of basin formation: condition of ring formation in multi-ring basins. Proc Lunar Planet Sci 12A Pergamon Press, New York, pp 227-257

Dence MR (1973) Dimensional analysis of impact structures. Meteoritics 8:343-344

Fredriksson K, Wickmann FE (1963) Meteoriter. Svensk Naturvet 16:121-157

Hartmann WK (1972) Interplanet variations in scale of crater morphology - Earth, Mars, Moon. Icarus 17:707-713

Heuze FE (1983) High temperature mechanical, physical and thermal properties of granitic rocks - a review. Int J Rock Mech Min Sci Geomech Abstr 20:No 1

Hjelmqvist S (1966) Beskrivning till berggrundskarta över Kopparbergs län. Svergies Geol Unders Ser Ca No 40 (English Summary 199:211)

Hoek E, Brown ET (1980) Underground excavations in rock. Inst Min Metall, London, p 176

Housen KR, Schmidt RM, Holsapple KA (1983) Crater ejecta scaling laws: fundamental forms based on dimensional analysis. J Geophys Res 88:2485-2499

McKinnon WB (1978) An investigation into the role of plastic failure in crater modification. Proc 9th Conf Lunar Planet Sci, pp 3965-3973

Melosh HJ (1977) In:Roddy DJ Pepin RO and Merril RB (eds) Crater modification by gravity: a mechanical analysis of slumping impact and explosion cratering. Pergamon Press, New York, pp 1245-1260

Melosh HJ (1982) A schematic model of crater modification by gravity. J Geophys Res 87:371-380

Paterson MS (1978) Experimental rock deformation. The Brittle Field. Springer, Berlin Heidelberg New York, pp 39-42

Pike RJ (1977) Apparent depth/apparent diameter relation for lunar craters. Proc 8th Lunar Sci Conf, pp 3427-3436

Rondot J (1975) Comparaison entre les astroblèmes de Siljan, Suéde. et, de Charlevoix. Québec Bull Geol Inst Univ Uppsala 6:85-92

Settle M, Head JW (1979) The role fo rim slumping in the modification of lunar impact craters. J Geophys Res 84:3081-3096

Vattenfall (1985) Swedish State Power Board Siljan Ring Structure. Investigations and evaluations. Gas Proj G2 Rep No 4

Structural Deformation, Breccia Formation, and Shock Metamorphism in the Basement of Complex Terrestrial Impact Craters: Implications for the Cratering Process

D. STÖFFLER[1], L. BISCHOFF[2], W. OSKIERSKI[2,3] and B. WIEST[3,4]

Introduction

The present comprehensive study of the Siljan impact structure by geophysical, geological and drilling exploration can be supported by evaluating previous research on similar terrestrial impact structures. It is the aim of this contribution to summarize observations and interpretations which result from a long-term research program on terrestrial impact craters at our institutes and which in our view are relevant to the analysis of the Siljan structure. The following review will be mainly based on a suite of impact craters with variable target stratigraphy (variable thickness of sedimentary rock strata on top of a crystalline basement), variable crater diameter ranging from about 20 to 40 km, and variable depth of erosion. The basic parameters of these craters (Rochechouart, Lappajärvi, Ries, and Carswell) are given in Table 1 and cross-sections are presented in Fig. 1.

Basic Geology of Selected Complex Impact Structures

The craters which have been selected for this review belong to the group of "complex craters" (Dence et al. 1977; Grieve 1987) which are typified by a central uplift, a flat crater floor, and a complex, tectonically deformed crater rim. The largest of these craters, Carswell, is most similar to the Siljan structure (Fig. 1).

Rochechouart, France (Fig. 2). The impact occurred 186 ±8 Ma ago (Reimold et al. 1983) in the Hercynian Massif Central composed of granites, gneisses, schists, diorite, amphibolite and acid dike rocks (Lambert 1977; Bischoff and Oskierski 1987). The structure is deeply eroded and only remnants of allochthonous polymitic impact breccias (impact melt rocks, suevite).

1 Institut für Planetologie, Westfälische Wilhelms-Universität, Wilhelm-Klemm-Str. 10, D-4400 Münster, FRG
2 Geologisch-Paläontologisches Institut und Museum, Westfälische Wilhelms-Universität, Corrensstraße 24, D-4400 Münster, FRG
3 now at Hewlett Packard, Schickardstraße 2, D-7030 Böblingen
4 now at Alexander v. Humboldt-Stiftung, Jean-Paul-Straße 12, D-5300 Bonn 2

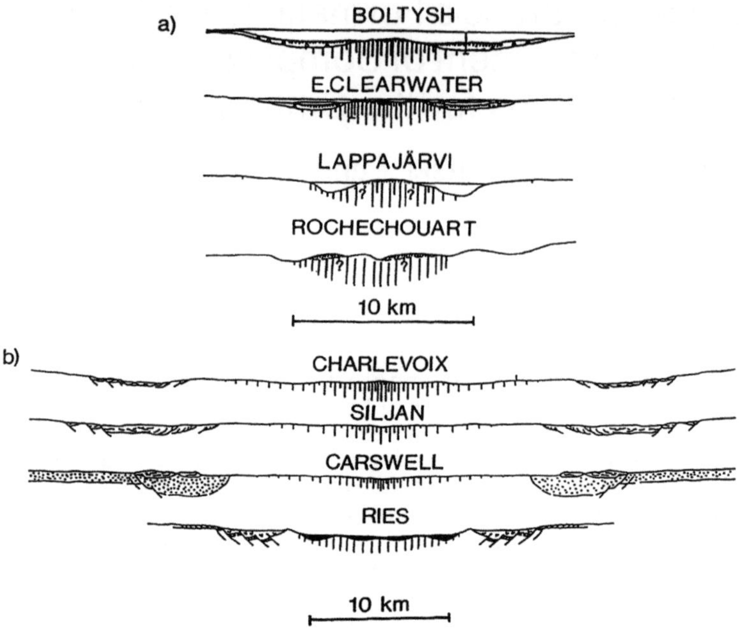

Fig. 1 a, b. Complex impact structures with central uplift in crystalline bedrock (no vertical exaggeration); vertical lines: shocked basement; short diagonal lines in part **a** melt rocks and suevite breccias triangles: fragmental breccias; white top section: post-impact sediments or lakes; areas with dots, wavy lines or diagonal lines in part **b** sedimentary rocks; black: suevite and fragmental breccias; v's: polymict Bunte Breccia; modified and supplemented from Dence et al. (1977); note different for **a** and **b**

fragmental breccias) can be found with a maximum thickness of about 60 m. These breccias overlay discontinuously the monomictly brecciated and fractured crystalline basement (Fig. 3) which contains shatter cones and is dissected by numerous breccia dikes (pseudotachylite and clastic matrix breccias). In the central part of the structure the basement rocks display shock effects up to a peak pressure of about 30 GPa (diaplectic glass; Lambert 1977). A circular negative gravity anomaly of up to 9 milligals was found by Pohl et al. (1978).

Lappajärvi, Finnland (Fig. 4). Lappajärvi crater was formed in crystalline bedrock (granite, micaschist, amphibolite, granodiorite, pegmatite) of western central Finnland 77 ±0.3 Ma ago (Lehtinen 1976; Elo 1976; Jessberger and Reimold 1980; Reimold 1982). Only part of the breccia lens of the crater is preserved. A central coherent layer of melt rocks (karnäite) is exposed on islands in the central region of Lake Lappajärvi; suevite, now covered by the lake, has been found as boulders in the glacial deposits south of the lake. The structure is characterized by a strong circular negative anomaly of up to 10 milligals (Fig. 4).

Table 1. Parameters and characteristics of impact craters discussed in this review; age data from Reimold (1983), Jessberger and Reimold (1980), Staudacher et al. (1982), and Grieve and Robertson (1979); other data from Lambert (1977), Geologica Bavarica (1977); calculated values after Grieve et al. (1981), * based on the formula: uplift (km) = 0.045 - 0.085 times rim diameter; value in parantheses after Grieve et al. (1981): uplift (km) = 0.065 $D^{1.1}$

Crater	Rochechouart	Lappajärvi	Ries	Carswell
Age (Ma)	186 ± 8	77 ± 0.3	14.8 ± 0.3	485
Target stratigraphy	Hercynian crystalline basement	Precambrian crystalline basement	550m of Phanerozoic limestones, shales and sandstones on top of Hercynian crystalline basement	2.800m of Proterozoic limestones, shales and sandstones on top of Precambrian crystalline basement
Diameter of structural rim (km)	~ 25	~ 21	25	~ 38
Calculated depth of excavation (km)	2.1	1.8	2.0	3.5
* Calculated structural uplift (km)	1.1 - 2.1 (2.1)	1.0 - 1.9 (1.8)	1.1 - 2.0 (2.0)	1.8 - 3.4 (3.5)
Peak shock in basement (GPa)	~ 30	?	20 - 28	20 - 28
Allochthonous layers of polymict breccias	impact melt suevite fragmental breccias	impact melt suevite	suevite fragmental breccias	none
Monomict breccias	X	X	X	X
Breccia dikes in the basement	X	?	X	X
Shatter cones	X	X	X	X
Negative gravity anomaly (milligals)	-9	-10	-18	?

Ries, S-Germany (Fig. 5). The Ries impact crater (age: 14.7 ±0.3 Ma; Staudacher et al. 1982) was excavated in a 550 m thick sequence of limestones, shales, and sandstones of Permian, Triassic, Jurassic and Tertiary age underlain by gneisses, granites, amphibolites and granodiorites of Hercynian age. The Ries is the least eroded complex terrestrial impact structure (Pohl et al. 1977 and Geol. Bavarica 1977 and references therein). An uplifted inner ring of crystalline basement blocks (12 km diameter) defines the central crater cavity which is filled by ~300-350 m of post-Ries lake sediments overlaying some 270 m of "fallback" suevite in the center of the structure. The inner ring is surrounded by a zone of allochthonous megablocks followed by the structural rim at a radius of ca. 12.5 km beyond which the subhorizontal layering of the sedimentary target rocks remains undisturbed (Fig. 1). The central crater basement has been explored in great detail by geophysical measurements and by a 1200 m deep drilling, 3.5 km off the center (Geologica

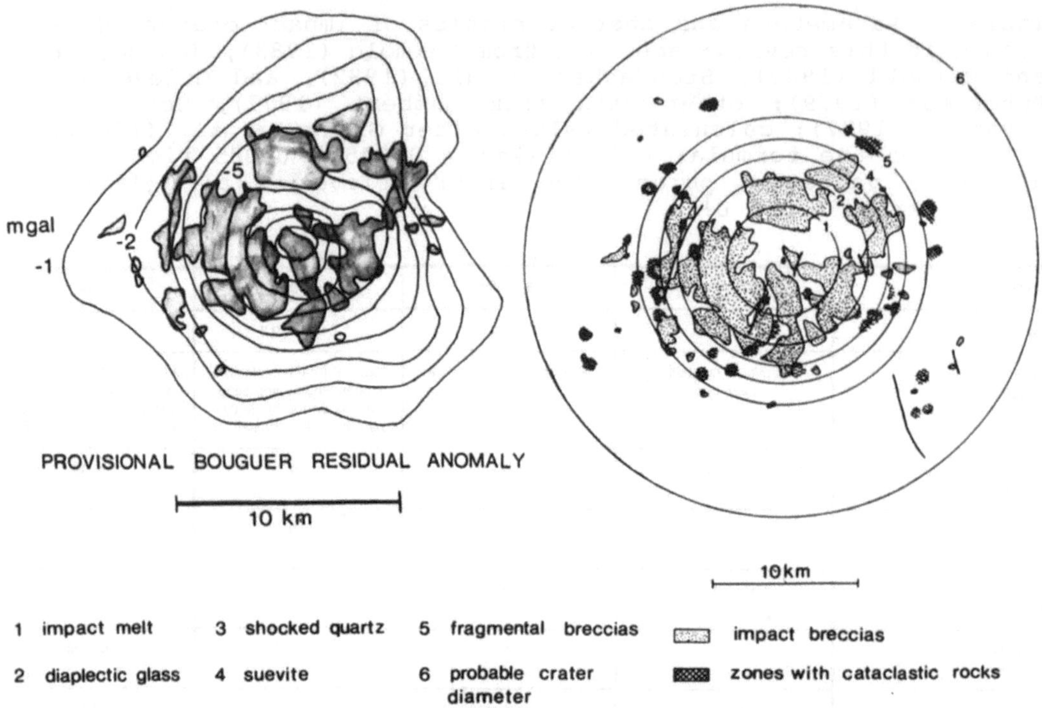

Fig. 2. Gravity a nomaly and distribution of breccias and certain shock effects at Rochechouart crater, France; gravity data from Pohl et al. (1978); breccia distribution and circles 1-5 from Lambert (1977); the circles 1-5 indicate the maximum radial extent of the formations in question

Bavarica 1977). The Ries crater displays a strong negative anomaly of the gravity and magnetic field (Fig. 5).

Carswell, NW-Saskatchewan, Canada (Fig. 6). The Carswell impact structure cuts into a sequence of at least 2200 m, more likely 2800 m of Proterozoic dolomites (Carswell Formation), shales (Douglas Formation), and sandstones (Athasbasca Formation) covering a 1750 to 2320-Ma-old crystalline basement complex of feldspathic to mafic gneisses and amphibolites (Earl River Complex) and metapelitic gneisses of the Peter River Complex (Lainé at al. 1975; Wiést 1987). The central part of the structure consists of a dome-like uplift of the crystalline basement (19 km in diameter) which is surrounded by an annulus of displaced sandstones of the Athabasca Formation followed by two additional annuli of the Douglas and Carswell Formations Fig. 6). The latter forms a circular discordant contact to the undisturbed Athabasca Formation outside of the annulus at a radial distance of about 19 km from the center of the structure. The younger Douglas and Carswell Formations were obviously downfaulted by the crater formation and therefore preserved from erosion. The crater is so deeply eroded that the allochthonous breccias of the crater fill were completely removed. The basement rocks are shocked up to about 28 GPa in the center. Monomict breccias, shatter cones, and several types of breccia dikes occur frequently throughout the crystalline basement and

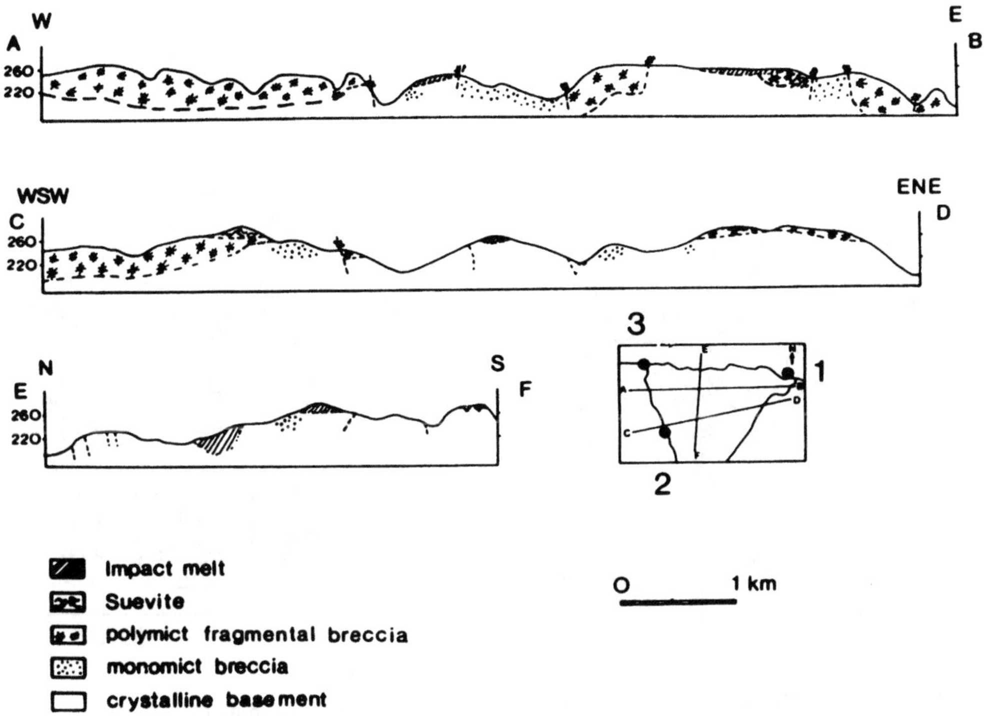

Fig. 3. Geological profiles at Rochechouart crater, France; location of profiles is given in the rectangle (black dots: 1 Rochechouart; 2 Mandat; 3 Pressignac). (After Oskierski 1983)

Table 2. Classification of breccia dikes at terrestrial impact craters. (After Lambert 1981 and Oskierski 1983)

	A) MELT MATRIX		B) CLASTIC MATRIX		
			polymict		monomict
			with melt	without melt	
	pseudotachylite	impact melt breccia	suevitic breccia	fragmental breccia	monomict breccia
Clasts	mineral and rare lithic clasts	angular mineral and lithic clasts	angular to subrounded mineral and lithic clasts		
Thickness	0.1 - cm's	dm - m	dm - m	(a) cm - dm (b) dm - m	(a) cm (b) dm - m's
Time of formation	1st generation	3rd generation	3rd generation	(a) 2nd generation (b) 2nd - 3rd generation	(a) 1st - 2nd generation (b) 2nd - 3rd generation

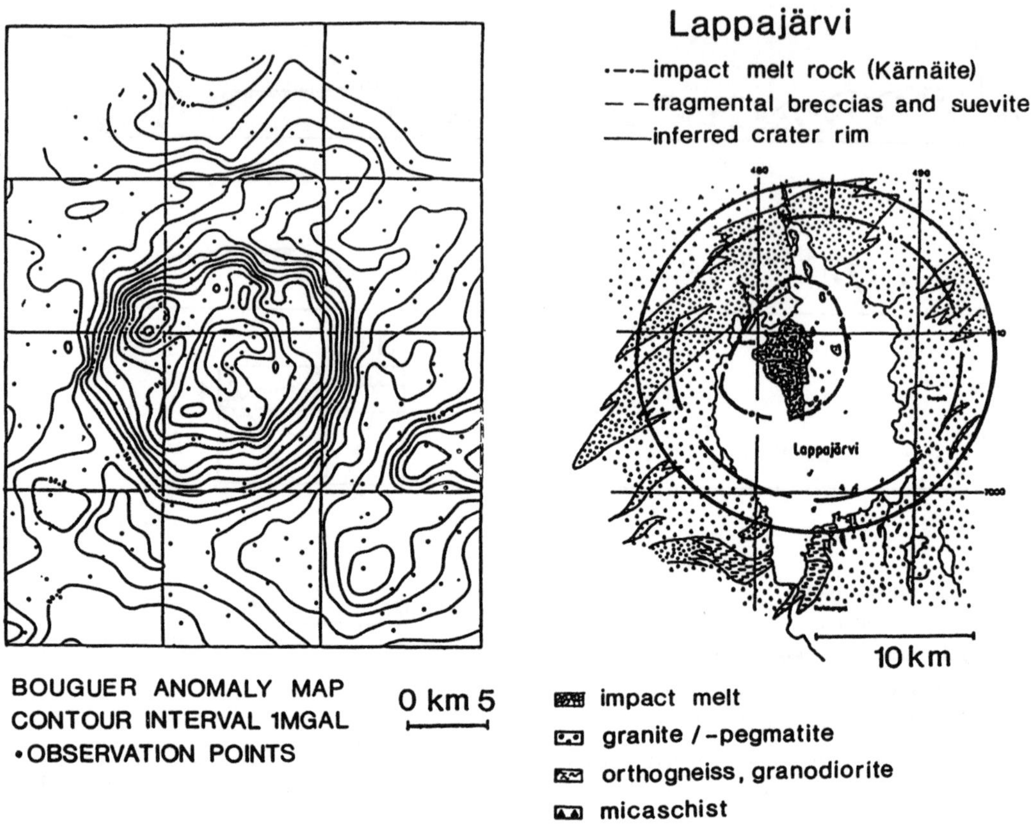

Fig. 4. Gravity anomaly and geological map of the Lappajärvi crater, Finland; gravity map and proposed extent of melt rocks and breccias by Elo (1976)

Examples for the Subsurface Structure and Stratigraphy in Vertical Sections

The crater cavity of complex impact structures is filled by allochthonous polymict breccias: impact melt rocks and fragmental breccias (Fig. 1). They cover the "crater floor" which is represented by a complex transition zone between the allochthonous breccias and the underlaying parautochthonous basement. The reader is referred to other reviews dealing with the allochthonous breccias (e.g. Grieve et al. 1977; Stöffler et al. 1979). This article focuses on the crater floor zone and the basement. Both are well exposed at the eroded Rochechouart structure and in the research drill core of the Ries crater.

From the profiles of Fig. 3 (Oskierski 1983; Bischoff and Oskierski 1987), it is obvious that the Rochechouart impact melt rocks rest directly on the brecciated basement in the central area but at further radial distance they overlay allochthonous polymict fragmental breccias and suevite breccias which appear

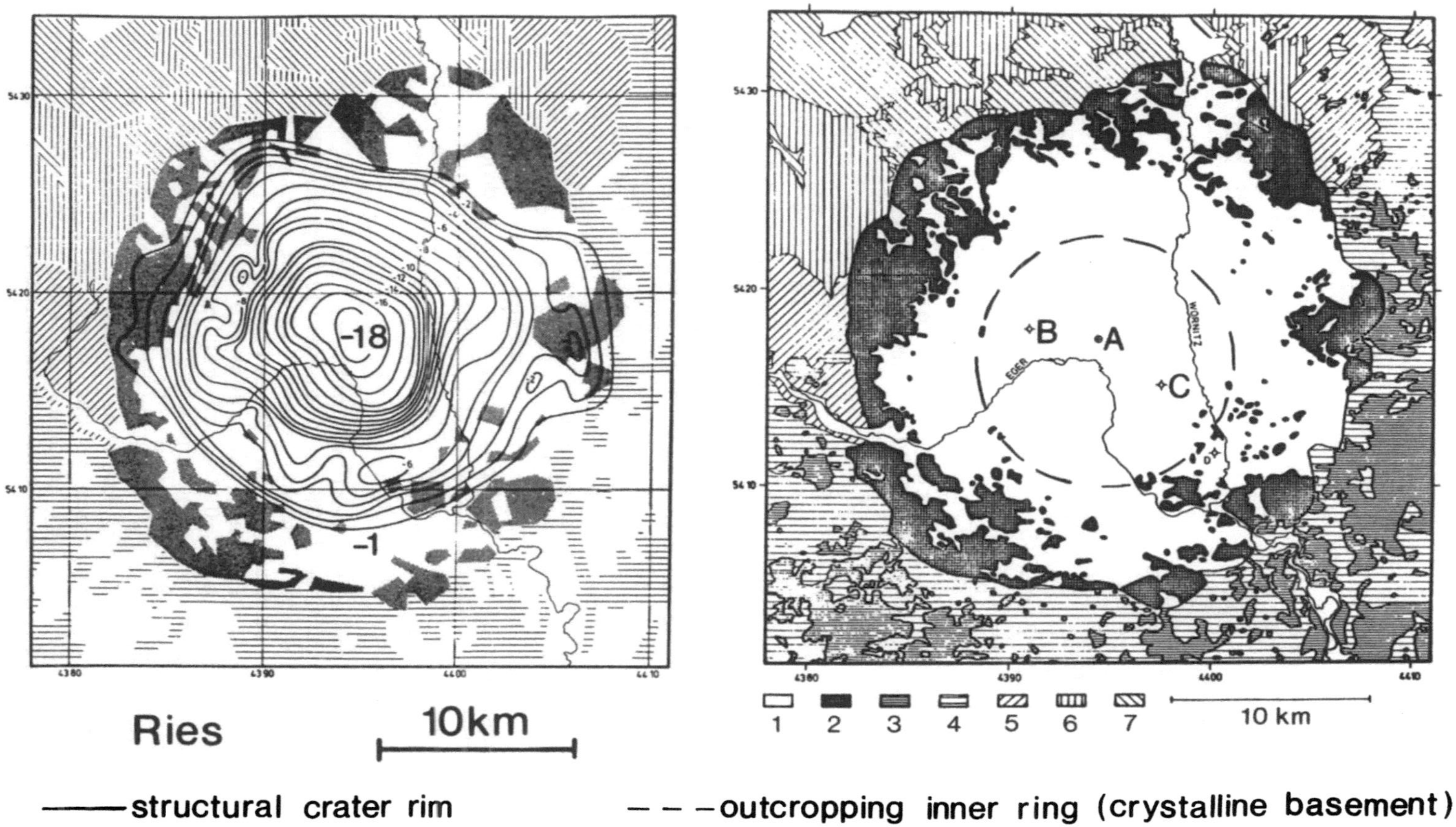

Fig. 5. Gravity anomaly and geological map of the Ries crater, Germany; from Pohl at al. (1977); the geological units (1 post-impact lake sediments; 2 megablock zone (outcropping); 3 "Bunte Breccie" ejecta blanket; 4 Malmian; 5 Dogger; 6 Liassic; 7 Keuper; 2 and 3 are displaced rocks; schematic geology in the gravity map taken from the right figure

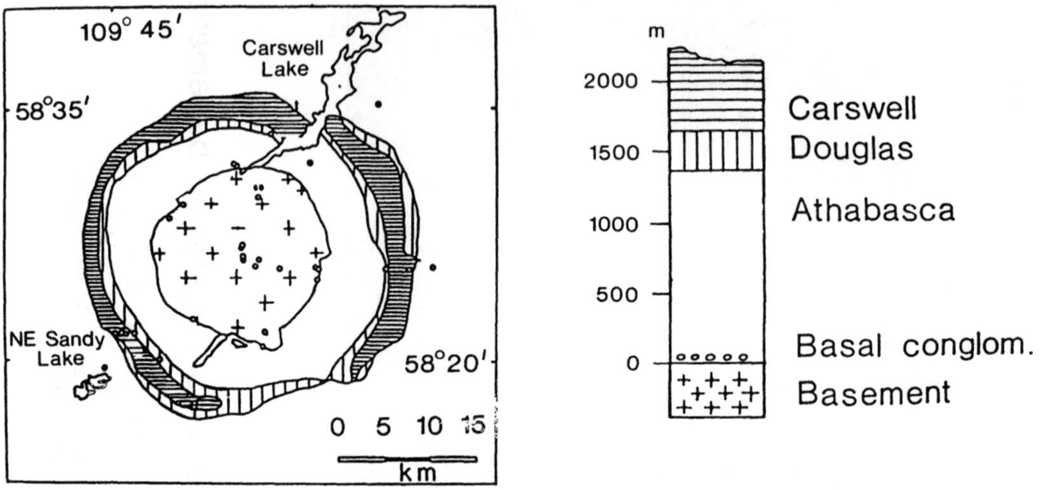

Fig. 6. Geological map and target stratigraphy of Carswell impact structure, Canada (Wiest 1987); **numbered circles:** location of drill holes

to be always on top of fragmental breccias. The mode of deformation of the crystalline basement below these breccia layers is exemplified by Fig. 7. Zones of monomict brecciation (cataclasis and relative displacement of clasts) alternate with zones of intense irregular fracturing and various types of breccia dikes. Shatter cones and pseudotachylite dikes or veins are also common in this zone but not shown in Fig. 7. Very often the pseudotachylite dikes form a complicated network of subparallel veins interconnected by feather joints (Fig. 8). Fragments of pseudotachylite frequently occur in dikes filled with polymict fragmental breccia (Fig. 7). The breccia dikes will be further discussed below (see also Bischoff and Oskierski 1987). It is not clear to what depth breccia dikes occur in the central crater basement. At the Ries the abundance of dikes does not decrease to at least 600 m below the crater floor zone at a distance of 3.5 km from the crater center (Fig. 9).

Another typical case of a well-known crater floor zone can be made by the results of the 1200 m deep drilling in the Ries crater. Although a coherent melt sheet is not observed the transition between the melt-bearing suevite breccia layer and the crystalline basement is very instructive (Fig. 9). In the depth range between 520 m and 602 m, suevitic breccia is intercalated in a dike-like configuration with brecciated and fractured crystalline rocks which appear to belong to one single megablock unit. Similar to the situation in Rochechouart, the lower section of the breccias belongs rather to the type of polymict fragmental breccias than to the suevite type. The Ries drill core provides additional unique information about the attenuation of shock effects and the distribution of dike brec-

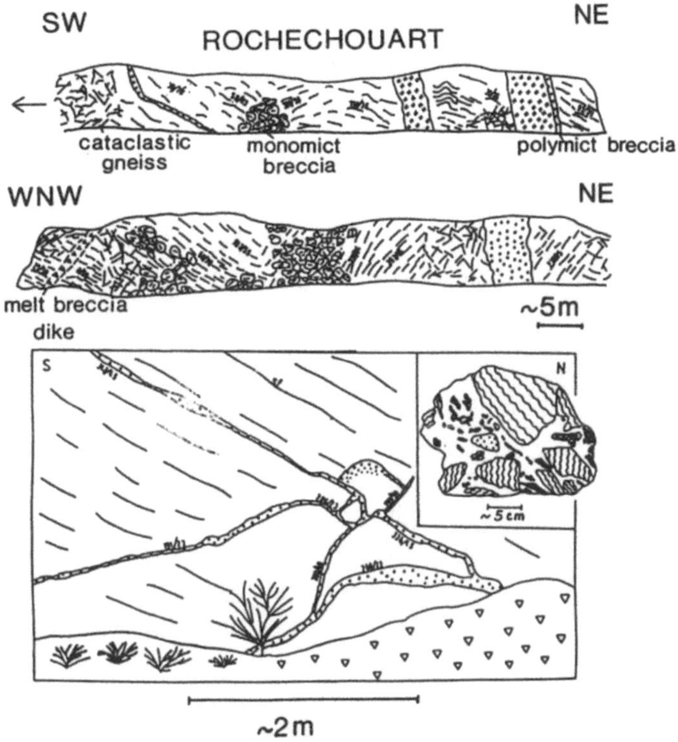

Fig. 7. Geology of three outcrops at Rochechouart crater, France, showing the geometry of breccia dikes; **dotted:** polymict fragmental breccia dikes; **black triangles:** melt breccia dike; **inset** in lower section: hand specimen from a polymict fragmental breccia dike (**dotted:** granite; **waves:** biotite gneiss; **black:** pseudotachylite fragments; **white:** clastic matrix); upper two sections: 1 km S of Babaudus, lower section: quarry of Laurière. (Oskierski 1983)

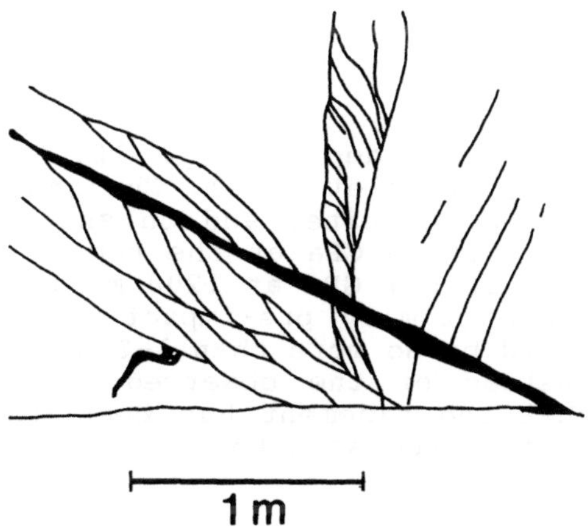

Fig. 8. Geometry of pseudotachylites dikes and veins in the Champagnac quarry of Rochechouart crater, France. (Bischoff and Oskierski 1987)

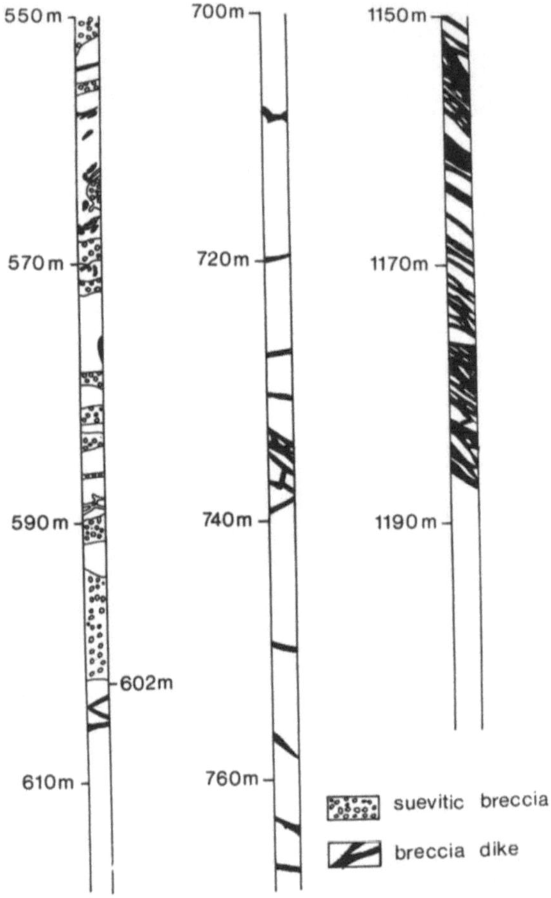

Fig. 9. Research drill core Nördlingen 1973 located 3.5 km west of the center of the Ries crater, Germany; intercalations of polymict breccias in crystalline basement rocks; note transition from suevite breccia dikes to melt-free breccia dikes at 602 m depth; reproduced from Geological Bavarica 1974

cias with depth (Fig. 10; see papers in Geologica Bavarica 1977; Rehfeldt 1983). The shock effects based on quartz, hornblende, biotite, and on shatter cones strongly attenuate from about 20-28 GPa at the top of the crater floor zone (520 m) to less than 1 GPa at 1200 m. If the basement rocks were still at their primary pre-impact location, the calculated attenuation could be no more than 1-2 GPa over the length of the drill core instead of the observed ~25 GPa (Stöffler 1977). This means that the basement has been displaced considerably by the cratering process after the shock wave passage.

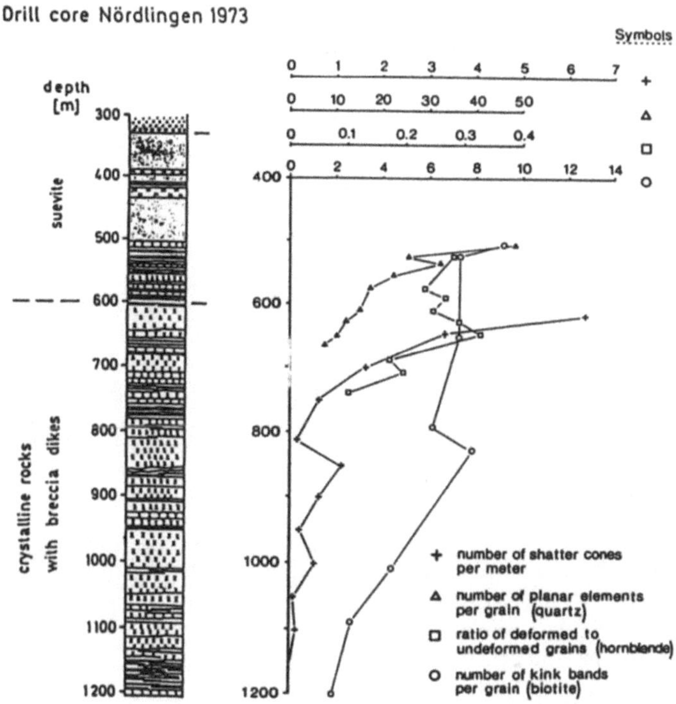

Fig. 10. Variation of shock effects in the crystalline basement rocks of the Ries crater drill core; drill core profile from Stöffler et al. (1977); data on shock effects from Engelhardt and Graup (1977); shatter cone data (averaged) from Hüttner (1977)

Examples for the Variation of the Subsurface Structure and Breccia Distribution in Radial Sections

The radial variation of shock-induced phenomena will be considered at two stratigraphic levels, the crater floor zone and the subcrater basement. As in the foregoing chapter details of the composition and origin of the allochthonous breccia layers in the crater cavity are not of concern.

The Crater Floor Zone. Information about this zone can be mainly drawn from the Rochechouart, Lappajärvi and Ries craters. It appears that there is a systematic zoning with respect to the maximum extent of the main breccia types and to particular shock effects in a horizontal section (Figs. 2, 3 and 4). The situation at Rochechouart and Lappajärvi indicates that at craters in crystalline bedrock the radial extent of allochthonous layers of breccia increases in the sequence melt breccia (melt sheet), suevitic breccia, fragmental breccia. The regions of monomict brecciation which are part of the parautochthonous basement extend even beyond the outer margin of the breccia layers (Fig. 2).

Subcrater Basement. The relevant features which vary with radial range are the shock intensity, which is best calibrated by quartz, the abundance of shatter cones, and the type and abundance of dike breccias, fault systems, shear zones, and zones of monomict brecciation. Shock effects in quartz and feldspar (planar deformation structures, rarely diaplectic glass) indicating peak pressures in the 30 to 7.5 GPa range, are restricted to the innermost zone extending to a distance which is less than the maximum extent of suevite and fragmental breccias but more than the maximum range of the melt sheet (e.g., Fig. 2). Shatter cones reach to at least a radial distance similar to that of the weakest shock effect in quartz, probably somewhat beyond this range (Robertson and Grieve 1977).

The maximum radial extent and distribution of dike breccias is not completely clear. The present data, especially from Carswell, suggest that dikes of polymict fragmental breccia reach far into the peripheral fault system of the structural rim zone (Fig. 11; Wiest 1987). Wiest finds that the intensity of deformation as well as the abundance of dike breccias increase towards the rim fault system, that means near to the contact of the crystalline basement and the displaced Athabasca sandstones (Fig. 6). Figure 11 also provides information about the structural geology of the annular fault system near the crater rim. The innermost part of the annulus (Athabasca) consists of steeply dipping, overturned megablocks (Fig. 11 b); the outer part (Carswell dolomites) exhibits a somewhat less disturbed pattern of blocks dipping at variable angles of 16° to 80° inward and outward with respect to the crater. Occasionally, blocks are overturned. Also fold structures and overthrusting are common in the dolomites of the outer annulus of sedimentary rocks (Fig. 11 c; Wiest 1987).

Systematics of the Main Deformation Features and Breccia Formations of the Crater Floor and Crater Basement

A generalized model section of a complex impact crater in the 20 - 50-km size range can be reconstructed from observations discussed above in combination with additional literature data (e.g., Grieve et al. 1977; Phinney and Simonds 1977; Grieve and Floran 1978; Stöffler et al. 1979). Using the schematic cross-section of Fig. 12, the general characteristics of the central crater region can be explained by proceeding along a vertical section. Four major units may be distinguished from top to bottom:

1. A layer of **impact melt rock** rests upon (a) the brecciated autochtonous to parautochthonous basement, or (b) on a layer of **suevite breccia,** or (c) on a layer of polymict **fragmental breccia** going from the center radially outward.

2. In a transition zone between the allochthonous melt and clastic breccia lens and the parautochthonous basement **"injection" dikes** of impact melt and of suevitic or fragmental brec-

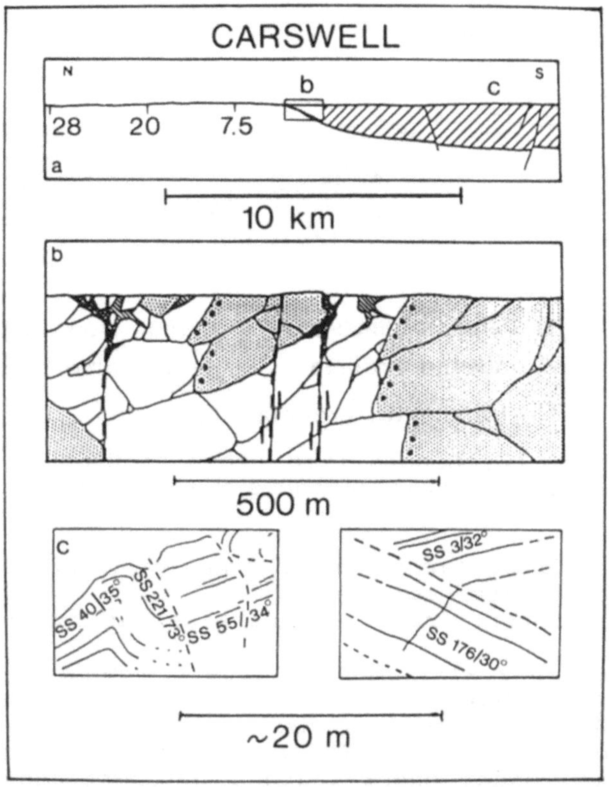

Fig. 11 a-c. N-S profile of the Carswell impact structure, Canada; **a** white: crystalline basement, hatched: sandstones and dolomites; numbers: shock pressure in GPa; **b** and **c** indicate location of schematic profile **b** and outcrops **c**; **b white:** crystalline rocks; **dotted:** Athabasca sandstone (overturned megablocks with basal conglomerate on top; see Fig. 6); **cross-hatched** and **hatched:** fragmental breccia and melt breccia dikes; note also vertical faults; **c** deformation geometry of dolomite strata of the Carswell Formation near the Carswell Lake (see Fig. 6). (Wiest 1987)

cia, mostly with thicknesses in the dm to m range, occur as indicators of the complex crater floor which is not a sharp discontinuity.

3. The underlying upper zone of **parautochthonous basement** is characterized by **shock effects in quartz and feldspar** (planar elements and incipient formation of diaplectic glass at shock pressures of about 30 GPa). The shock intensity decreases with depth. In this zone we observe:

a) pseudotachylite dikes
b) dikes of polymict fragmental breccias

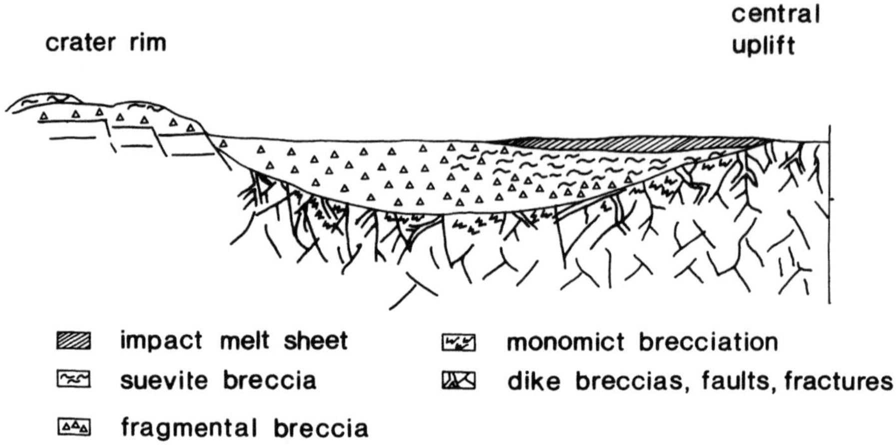

Fig. 12. Generalized cross-section of a complex impact crater in crystalline bedrocks with central uplift (see text)

c) irregular zones or dikes of monomict breccias
d) shatter cones
e) strong fracturing of basement rocks.

4. In a still deeper zone the more or less **autochthonous basement rocks** lack shock effects in quartz (peak pressure <5-10 GPa) but the same deformation features listed under (3) remain characteristic. It can be assumed that the occurrence of dikes and other shock-induced deformations gradually fade out with increasing depth although no direct observation is available because of lack of exposure or insufficient depth of drilling.

The various types of breccia dikes described above are classified and defined in Table 2. They are formed at different times during the cratering process in at least three generations. The term "autochthonous" regarding the crater basement has to be used with caution because the basement consists most probably of large displaced megablocks of tens to hundreds of meters in size (Geologica Bavarica 1977; Wiest 1987).

The above-mentioned shock-induced features also vary with radial range in a subhorizontal section from the center to the rim area of the crater. This leads to a zoning which can be observed in various horizontal levels of erosion.

In an upper stratigraphic section (crater fill and crater floor) the zoning is related to the maximum lateral extent of the following breccia types (Fig. 13):

a) impact melt sheet
b) suevite breccia layer
c) fragmental breccia layer
d) "intrusive" dikes of impact melt or fragmental breccia

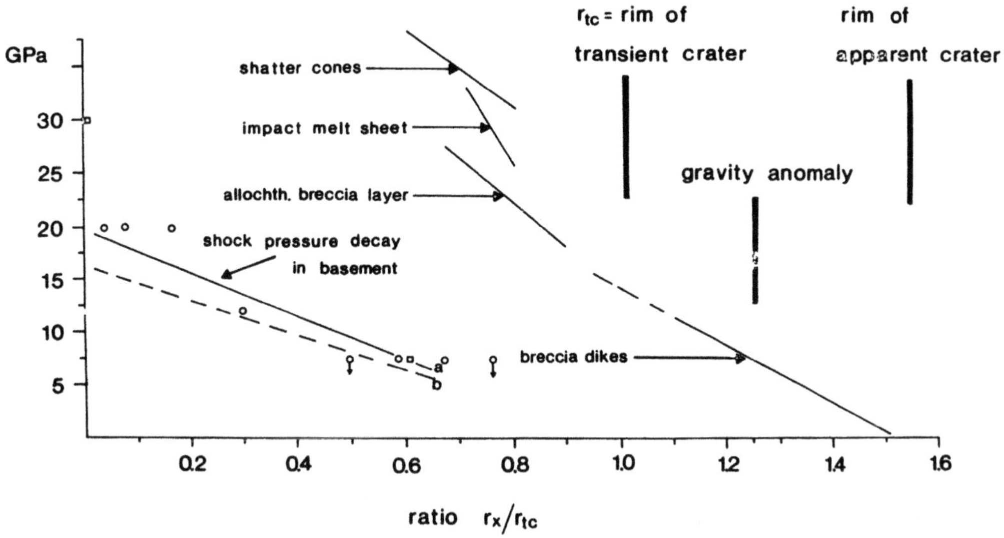

Fig. 13. Shock pressure attenuation and radial extent of impact-induced phenomena in complex craters with central uplift as a function of radial range normalized to the transient crater radius (for data sources see Table 3); pressure scale relates only to the shock pressure decay data (**circles:** Carswell, Wiest 1987; **squares:** Rochechouart, Lambert 1977; **solid and dashed regression lines** for Slate Islands and Charlevoix craters, respectively, Robertson and Grieve 1977); r_{tc} = radius of transient crater

It appears from the presently available information on craters in crystalline bedrocks that the radial range increases from (a) to (d) (cf. Fig. 2.).

In a deeper stratigraphic level (crater basement) the zoning refers to

a) the radii of shock pressure isobars
b) radius of shatter coning
c) radius of the maximum extent of different types of dike breccias, monomict brecciation, and intense fracturing
d) radius of the peripheral fault system near the structural rim.

In the top center of the uplifted basement a maximum shock pressure of about 30 GPa has been observed as indicated by the presence of diaplectic feldspar glass (Grieve and Floran 1978; Lambert 1977). The radial decrease of the peak shock pressure is very steep because the downward displacement and uplifting leads to a shortening of the spacing between the shock isobars as compared to their original position at the end of the excavation stage (Robertson and Grieve 1977). The apparent shock pressure as observed in somewhat eroded complex craters attenuates from ~20 GPa near the center of the basement uplift to less than ~7.5 GPa at a radial distance of about 50-60% of

Table 3. Geometrical parameters of complex craters (km); d_r of Rochechouart and Lappajärvi calculated from the ratio d_{gr}/d_r at the crater floor level of the Ries crater (d_{gr}= 20 km); data sources: Pohl et al. (1977); Lambert (1977); Pohl et al. 1978; Elo (1976); Bischoff and Oskierski (1987); Wiest (1987)

	d_r structural rim	d_{gr} gravity anomaly	d_r calculated	d_{tc} transient crater	d_m/d_{tc}	d_{br}/d_{tc}	d_{sq}/d_{tc}	d_{sc}/d_{tc}	d_{db}/d_{tc}	d_{gr}/d_{tc}
Ries	25	21.5 (20)	–	16	–	0.66	> 0.5	> 0.5	> 0.75	1.34 (1.25)
Carswell	38	?	–	25	–	–	0.70	0.80	~ 1.5	
Rochechouart	–	19.5	24	16	0.69	0.84	0.59	0.59	~ 1 ?	1.25
Lappajärvi	–	17.0	21	14	0.78	~ 0.8	?	?	?	1.25

d_r = diameter of outer structural rim (apparent crater diameter)
d_{gr} = diameter of gravity anomaly at the present surface
d_{tc} = diameter of transient crater calculated according to d_{tc} = 0.65 d_r (Grieve et al. 1981)
d_m = diameter of impact melt sheet
d_{br} = diameter of breccia lens (allochthonous)
d_{sq} = diameter of zone of shocked quartz (> 7.5 GPa)
d_{sc} = diameter of zone of shatter cones (ca. > 5 - 2 GPa)
d_{db} = diameter of the zone of breccia dikes

the transient crater radius (Fig. 13). At Rochechouart the shatter coning extends to about 60%, at the larger Carswell and Charlevoix structures to about 80 and 90% of the transient crater radius (Lambert 1977; Robertson and Grieve 1977). The maximum extent of the allochthonous breccia layers approximates 70% (Ries), 85% (Rochechouart) and 80% (Lappajärvi) of the transient crater radius. Dike breccias are found even beyond this reference radius and are common within the peripheral ring fault system (e.g., Carswell; Wiest 1987), which itself is located beyond about 1.25 times the transient crater radius.

Any impact induced alteration of the basement rock must be in a certain relation to the geometry of the gravity anomaly and the region of reduced seismic velocity. The surface radius of the gravity anomaly at the Ries, Lappajärvi, and Rochechouart craters (Figs. 2, 4, and 5) corresponds obviously to a radius which is 1.2-1.3 times the transient crater radius. All the parameters discussed above and summarized in Table 3 and Fig. 13 can be used to estimate the apparent crater diameter d_r of eroded complex craters with central uplift. We believe that the characteristic parameters of Table 3 are applicable to craters in the 20 to 50 km diameter range.

Summary and Conclusions for the Formation of Complex Impact Craters

Following the basic concepts of cratering mechanics (Gault et al. 1968; Dence et al. 1977; Croft 1980, 1981, 1985; Melosh 1980; Kieffer and Simonds 1980; Grieve et al. 1981) the formation of an impact crater involves three main phases which are consecutive in principal but in reality are overlapping in time: (1) compression phase, (2) excavation phase, and (3) modification phase. These phases are connected with typical hypervelocity movements in the target area and are related to the half-spherical shock wave propagation (1), the material flow whose geometry is determined by the rarefaction waves (2), and the uplifting of the central crater combined with the subsidence of the rim as a consequence of the collapse of a transient crater cavity (3). We list below a few major theses which describe the processes and the permanent effects resulting from these processes:

1. The impact-induced shock wave compresses the target rocks in a half-spherical volume in which the material strength is exceeded. The rocks within this region which is far in excess of the radius of the transient crater cavity, are shock metamorphosed and brecciated. A first generation of dike breccias by frictional melting (pseudotachylite) is formed.

2. The target region shocked above about 30 GPa is completely excavated and leaves the crater as ejecta or resides as fragments within the allochthonous breccia lens of the final crater. The target region beyond the excavation cavity being in a cohesionless state of dilatation moves radially downward and outward. This material state allows for the formation of a second generation of dikes of complex geometry filled with polymict fragmental breccia (including fragments of pseudotachylite). The characteristic increase of the deformation intensity in the subcrater basement towards the crater rim region and the typical geometry of displaced megablocks (steeply dipping overturned blocks) in this area is the result of a specific flow geometry in this region which is transitional between downward and upward flow. Blocks with variable degree of brecciation and with two generations of dike breccias are ejected from the crater indicating the early time of brecciation and dike formation.

3. The uplifting of the central crater basement starts before the material flow and excavation is finished. It appears to be completed before a last generation of "intrusive"-like breccia dikes has been formed and the melt sheet has reached its equilibrium temperature (order of seconds to minutes). The dissection of the uppermost parautochthonous megablocks of the basement by intrusive clastic matrix or melt dikes belongs to the final stage of excavation and emplacement of polymict allochthonous breccia material inside the crater. The uplifting causes the compression of the shock pressure isobars of the basement into a flat bowl-shaped geometry which explains the

observed steep attentuation of shock pressure. The central uplift is combined with a simultaneous subsidence and deformation of the crater periphery causing the characteristic ring fault system. These processes take place as an early high-velocity crater modification induced by the collapse of the transient crater cavity.

4. A late modification of the final crater by gravity-induced vertical movements seems to be indicated in some craters like Carswell by the existence of nearly vertical fault systems at various radial ranges. These displacements lead to the formation of a final generation of breccia zones in the crater basement.

Acknowledgment. We would like to thank G.A. McCormack, F. Bartschat and C. Hilke for their help in preparing the manuscript. The research work was supported by the Deutsche Forschungsgemeinschaft. We are grateful for a special grant of the Barringer Crater Co. supporting the field work at the Rochechouart and Carswell craters.

References

Bischoff L and Oskierski W (1987) Fractures, pseudotachylite veins and breccia dikes in the crater floor of the Rochechouart impact structure, SW France, as indicator of crater forming process. In: Pohl J (ed), Research in terrestrial impact structures, Earth Evolution Sciences, Friedrich Vieweg & Sohn, Braunschweig, 5-29

Croft SK (1980) Cratering flow fields: Implications for the excavation and transient expansion stages of crater formation. Proc Lunar Planet Sci Conf 11, 2347-2378

Croft SK (1981) The excavation stage of basin formation: a qualitative model. Schultz PH and Merrill RB (eds). Multi-ring basins, Proc Lunar Planet Sci Conf 12A, 207-226

Croft SK (1985) The scaling of complex craters. Proc Lunar Planet Sci Conf 15th, J Geophys Res 90:C828-842

Dence MR, Grieve RAF and Robertson PB (1977) Terrestrial impact structures: Principle characteristics and energy considerations. In: Roddy DJ, Pepin RO and Merrill RB (eds). Impact and explosion cratering. Pergamon Press, New York, 247-275

Elo S (1976) A study of the gravity anomaly associated with Lake Lappajärvi, Finland. Geological Survey Finland Rep Q 20/21, 20 pp

Engelhardt W v and Graup G (1977) Stoßwellenmetamorphose im Kristallin der Forschungsbohrung Nördlingen 1973. Geological Bavarica 75:255-271

Gault DE, Quaide WL and Oberbeck VR (1968) Impact cratering mechanics and structures. In: French B and Short NM (eds): Shock Metamorphism of Natural Materials. Mono Book, Baltimore, 87-99

Geologica Bavarica (1974) Bd 72, Bayerisches Geologisches Landesamt, München, 98 pp

Geologica Bavarica (1977) Bd 75, Bayerisches Geologisches Landesamt, München, 470 pp

Grieve RAF (1987) Terrestrial impact structures. Ann Rev Earth Planet Sci 15:245-270

Grieve RAF and Floran RJ (1978) Manicouagan impact melt, Quebec. 2. Chemical interrelations with basement and formational processes. J Geophys Res 83:2761-2771

Grieve RAF, Dence MR, Robertson PB (1977) Cratering processes: As interpreted from the occurrence of impact melts. In: Roddy DJ, Pepin RO, Merrill RB (eds) Impact and explosion cratering. Pergamon Press, New York, 791-814

Grieve RAF, Robertson PB and Dence MR (1981) Constraints on the formation of ring impact structures, based on terrestrial data. Schultz PH and Merrill RB (eds). Multi-ring basins. Proc Lunar Planet Sci 12A, 37-57

Hüttner, R (1977) Makroskopische Beobachtungen zur Deformation des Kristallins in der Forschungsbohrung Nördlingen 1973. Geologica Bavarica 75:273-283

Jessberger EK und Reimold WU (1980) A late Cretaceous ^{40}Ar-^{39}Ar age of the Lappajärvi impact crater, Finland. J. Geophys. 48:57-59

Kieffer SW and Simonds CH (1980) The role of volatiles and lithology in the impact cratering process. Rev Geophys Space Phys 18:143-181

Lainé R, Alonso D and Svab M (eds) (1985) The Carswell structure uranium deposits, Saskatchewan. Geol Assoc of Canada Spec Paper 29

Lambert P (1977) The Rochechouart Crater: Shock zoning study. Earth Planet Sci Lett 35:258-268

Lambert P (1981) Breccia dikes: Geological constraints on the formation of complex craters. In: Schultz PH and Merrill RB (eds). Multi-ring Basins, Proc Lunar Planet Sci Conf 12A, 56-78

Lehtinen M (1976) Lake Lappajärvi, a meteorite impact site in western Finland. Dissertation, Geological Survey of Finland, Bull 282:92 pp

Melosh HJ (1980) Cratering mechanics observational, experimental, and theoretical. Ann Rev Earth Planet Sci 8:65-93

Oskierski W (1983) Geologisch-petrographische Untersuchungen im Zentralbereich der Impaktstruktur von Rochechouart, SW-Frankreich, unter besonderer Berücksichtigung der Petrographie und Geochemie von Brecciengängen des Krateruntergrundes. Diplomarbeit, Westfälische Wilhelms-Universität, Münster

Phinney WC and Simonds CH (1977) Dynamical implications of the petrology and distribution of impact melt rocks. In: Roddy DJ, Pepin RO and Merrill RB (eds) Impact and explosion cratering. Pergamon Press, New York, 771-790

Pohl J, Stöffler D, Gall H and Ernstson K (1977) The Ries impact crater. In: Roddy DJ, Pepin RO and Merrill RB (eds) Impact and explosion cratering. Pergamon Press, New York, 343-404

Pohl J, Ernstson K, and Lambert P (1978) Gravity measurements in the Rochechouart structure. Meteoritics 13:601-604

Rehfeldt A (1983) Petrographische und chemische Untersuchungen an Gangbreccien im kristallinen Krateruntergrund des Nördlinger Ries. Diplomarbeit, Westfälische Wilhelms-Universität, Münster

Reimold WU (1982) The Lappajärvi meteorite crater, Finland: petrology, Rb-Sr, major and trace element geochemistry of the impact melt and basement rocks. Geochim Cosmochim Acta 46:1203-1225

Reimold WU, Oskierski W, and Schmidt A (1983) Rb-Sr age dating of the Rochechouart impact event and geochemical implications for the formation of impact breccia dikes. Meteoritics 18:385-386

Robertson PB and Grieve RAF (1977) Shock attenuation at terrestrial impact structures. In: Roddy DJ, Pepin RO and Merrill RB (eds) Impact and explosion cratering. Pergamon Press, New York, 687-702

Staudacher et al. (1982) ^{40}Ar-^{39}Ar Ages of Rocks and Glasses from the Nördlinger Ries crater and the temperature history of impact breccias. J Geophys 51:1-11

Stöffler D (1977) Research drilling Nördlingen 1973: Polymict breccias, crater basement, and cratering model of the Ries impact structure. Geologica Bavarica 75:443-458

Stöffler D, Ewald U, Ostertag R, and Reimold WU (1977) Research drilling Nördlingen 1973 (Ries): Composition and texture of polymict impact breccias. Geologica Bavarica 75:163-189

Stöffler D, Knöll H-D, and Maerz U (1979) Terrestrial and lunar impact breccias and the classification of lunar highland rocks. Proc Lunar Planet Sci Conf 10th, 339-375

Wiest B (1987) Diskordante Gangbreccien und strukturelle Merkmale im Untergrund komplexer Impaktstrukturen und ihre Bedeutung für die Kraterbildungsprozesse. Dissertation, Westfälische Wilhelms-Universität, Münster

Possible Impact Structures in Sweden

F. E. WICKMAN[1]

Introduction

The concept of terrestrial impact structures is due to Barringer (1905), who suggested that the impact of a giant meteorite had formed what we now call Meteor Crater in Arizona. Amazingly, already in 1910 A.G. Högbom, Professor of Geology in the University of Uppsala, suggested that two structures in Sweden, Lake Mien and the Dellen Lakes could possibly be impact structures.

His suggestion was bold, since the lakes, regarded as Tertiary volcanoes, had little in common with Meteor Crater, which is surrounded by a strewn field of iron meteorites. Högbom's impact hypothesis, however, had no impact on the Swedish geologic community. To the best of my knowledge, for about 50 years it was not mentioned, quoted or discussed.

Impact structures came into focus again during the Sputnik era, and in 1963 Fredriksson and Wickman, in a review article about meteoritics, published a list of five possible impact sites in Sweden. At present the number of possible impact structures has grown to at least 15-20 (Tables 1, 2).

Table 1. Possible impact structures in Sweden discussed in this paper

No.	Name	No.	Name	No.	Name
1	Dobblon	6	Siljan Ring	11	Lake Mullsjön
2	Ledfat	7	Midsommar	12	Lake Hummeln
3	Bottenviken	8	Tvären Bay	13	Lake Mien
4	Lockne	9	Landsort Trench	14	Skedviken
5	Dellen Lakes	10	Granby	15	Hemmestorp

[1] Geologiska Institutionen, Stockholms Universitet
10691 Stockholm, Sweden

Fig. 1. Map of possible impact structures in Sweden. 1 Dobblon; 2 Ledfat; 3 Bottenviken; 4 Lockne; 5 Dellen Lakes; 6 Siljan Ring; 7 Midsommar; 8 Tvären Bay; 9 Landsort Trench; 10 Granby; 11 Lake Mullsjön; 12 Lake Hummeln; 13 Lake Mien; 14 Skedviken; 15 Hemmestorp

Table 2. Additional possible impact structures in Sweden according to Henkel (personal communication, 1987)

No.	Name[a]	No.	Name[a]
1	Ålaape, T,M,G	11	Loftahammar, T,E,B
2	Ålandshav, T	12	Nulanki, M
3	Alvasjön, T	13	Sautasjärvi, T,G,E
4	Andersviken, T	14	Skellefteå, T
5	Åvikebukten, T	15	Stensjön, T
6	Dala-Järna, T	16	Stordammen, T
7	Drängsjön, T	17	Styran, T
8	Hogane, T	18	Tjegelvas, M
9	Karlskoga, T	19	Unden, T
10	Kattfjärden, T	20	Västlandsjön, T

[a] Observed anomalies: T = topographic, M = magnetic, G = gravimetric, E = electric, B = breccia

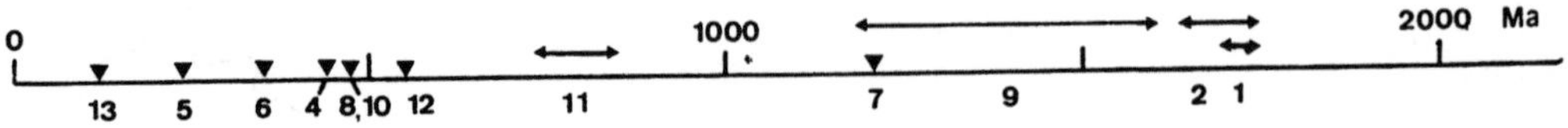

Fig. 2. Impact structure ages, if given in the text, Numbers see Fig. 1

All suggested Swedish impact structures are difficult to study for a variety of reasons, two of which may be mentioned: (1) most of them are depressions filled by unconsolidated sediments from the last glaciation and outcrops are usually rare; (2) only remnants of the structures still exist because of their great age.

Table 1 is the author's present tentative list of possible impact structures in Sweden, and Fig. 1 is a map of them. State geophysicist H. Henkel has kindly supplied an additional number of possible impact structures (Table 2).

The ages that are given in the descriptions of the impact structures are shown in Fig. 2. The number and ages of Proterozoic structures are poorly known, but if the Phanerozoic impact rate of Fig. 2 is used, i.e., 7 in 600 Ma, the expected number should be around 20-25. It would, however, be premature to discuss the likelihood of such a number.

The following brief descriptions of the individual structures only try to indicate our present knowledge and the author's tentative interpretations. The treatments are uneven, interest

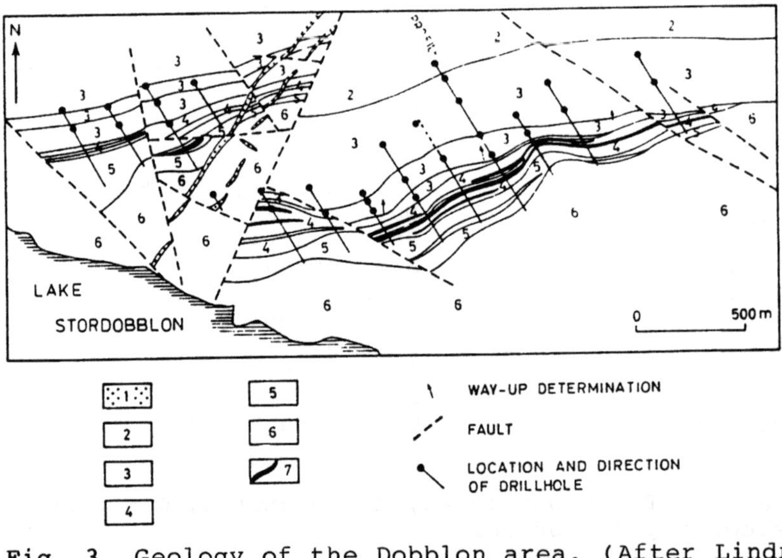

Fig. 3. Geology of the Dobblon area. (After Lindroos and Smellie 1979). 1 Dolerite dikes; 2 Gippervare porphyry; 3 conglomerates, tuffitic sandstones; 4 rhyolitic ignimbrites; 5 basal breccia; (3-5 Björnknösen Formation), 6 Older Granite; 7 U mineralizations

is focused on the geologically well-known structures which are here reinterpreted as impact structures. For space reasons only statements are given, discussions of the evidence must be given in other papers. The numbers given to a structure in Table 1 are used when convenient for identification.

Description of Possible Impact Structures

1. Dobblon (65°35'N 17°6'E). The structure is situated ca. 20 km W of Sorsele in Västerbotten County. Recognized by Högbom (1931) and discussed by Gavelin (1955) it was described in detail by Einarsson (1979). Uranium prospecting produced new research (Lindroos and Smellie 1979, Smellie 1982) based on many core drillings.

Table 3. The general stratigraphy of the Dobblon area (Lindroos and Smellie 1979) and ages (Lundqvist 1979)

Stratigraphy	Ages in Ma
Caledonides	
===============Major unconformity===============================	
Dolerite dikes	
===============Major unconformity===============================	
Sorsele granite	1590
===============Major unconformity===============================	
Dobblon Group	
Gippervare Formation	1690
Acid volcanics (1000+ m)	
Björnknösen Formation (0-430 m)	
Polymict conglomerate (0-200 m)	
Tuffitic sandstone (0-70 m)	
Rhyolitic ignimbrites (0-60 m)	
Basal breccia (0-100 m)	
===============Major unconformity===============================	
Older (Revsund?) Granite	1745-1750
===============Major unconformity===============================	
Paksjo Group	

Table 3 shows the stratigraphy and Fig. 3 a geologic map. The Paksjo Group consists of metasediments mainly of the greywacke type, which have been intruded by a massif called Older Granite, regarded as equivalent to the regionally common Revsund granite. The Björnknösen Formation of the Dobblon Group is the "impact formation", and the Gippervare Formation is an acid porphyry formation, distinctly younger than the Björnknösen Formation and also much more extensive. The Dobblon Group is intruded by a large massif of Sorsele Granite.

The stratigraphy of the Björnknösen Formation is similar to the one often found in impact structures:

Björnknösen Formation	**Impact Structure**
Polymict conglomerate	Sediment filling
Tuffitic sandstones	Tuffite and suevite
Rhyolitic ignimbrites	Impact melts etc
Basal breccia	Brecciated basement

The basal breccia reflects its basement, with angular rock clasts rounded upwards. Up to 5-cm-wide fractures filled with "mud and weathered residue" occur in the Older Granite. The rhyolitic ignimbrites, up to 60 m thick, first appear in the upper levels of the basal breccia. Both welded and unwelded parts occur with pumice and broken rock fragments. Matrix is a dense microcrystalline mass of angular crystals of quartz and feldspar together with altered glass (sericite-chlorite etc).

Table 4. Chemical analyses of older granite and ignimbrite from the Dobblon area

	Older Orig	Granite[a] Recalc	Ignimbrite[b]
SiO_2	75.0	76.1	76.7
TiO_2	0.19	0.19	0.22
Al_2O_3	12.9	13.1	13.3
Fe_2O_3[c]	-	1.7	1.6
(Fe_2O_3)	0.6		
(FeO)	1.0		
MnO	0.02	0.02	0.02
MgO	0.38	0.39	0.28
CaO	0.2	0.2	0.3
Na_2O	2.9	2.9	1.4
K_2O	5.4	5.5	6.3
P_2O_5	0.02	0.02	0.06
BaO	0.08	0.08	0.04
	98.69	100.10	100.16

[a] Older Granite from 1.5 km NE of eastern tip of Lake Stor-Dobblon (Einarsson 1979)
[b] Ignimbrite from bore hole 76001 (Lindroos and Smellie 1979)
[c] Fe_2O_3 = total iron calculated as Fe_2O_3

Lithophysae (0.5-15 cm) in a some meters thick zone are well-developed, as, e.g., in Mistastin Lake (Grieve 1975). The chemical compositions of Older Granite and ignimbrite are quite similar for Si, Ti, Al and total Fe (Table 4).

The well-stratified tuffitic sandstone includes beautiful ash textures, but also conglomeratic layers. Above it follows the polymict conglomerate, which in its lowest levels may contain pebbles of Older Granite and ignimbrite, but in general lacks local rock types.

The size of the original structure is unknown because of the intrusion of the Sorsele granites and the presence of Caledonian thrust rocks.

Some of the rocks have been studied by the Rb-Sr method (Welin 1970, Welin et al. 1971, 1977). The Gippervare Formation gave a good isochron, but the Older Granite gave no satisfactory age line. This is to be expected from the impact hypothesis, because exchange reactions start after the brecciation of the basement. The age of the structure is assumed to be somewhere between the age of the Gippervare Formation (1690 Ma) and 1745 Ma (if Older Granite = Revsund Granite = 1745 Ma).

Table 5. Stratigraphy of the Ledfat area according to Offerberg (1959) Rb-Sr ages according to Lundqvist (1979)

Dolerite (1215? Ma)			
Sorsele granite (1590 Ma), Ledfat granite (1645 Ma)			
Ledfat Series	Vargfors Conglomerates	Upper Vargfors conglomerate	Up to 30% granite pebbles
		Lower Vargfors conglomerate	No granite pebbles
	Lower Ledfat Formation	Tuffite Formation	Brownish violet sandstone Brick to violet red tuffite Brown violet sandstones
		Sandstone Formation	Arkose and feldspathic quartzites Oligomict conglomerates Brown violet sandstones
		Andesite Formation	Tuffs & tuffites Conglomerates Andesitic vesicular lavas
============Unconformity============			
Jörn granite (1850? Ma), Arvidsjaur granite (1740 Ma)			
Arvidsjaur Series		Liparites Volcanic breccias, agglomerates Dacites Andesites	

The structure shows great similarity to the Ledfat structure even in age. Both need careful study.

2. Ledfat (65°27'N 18°16'E). The structure is situated ca. 35 km E of Sorsele in Västerbotten County and is a 6-7-km-wide bowl-shaped structure of Early Proterozoic age. Recognized in 1921, it was described by Gavelin (1955) and in particular by Offerberg (1959). Current opinions regarding the structure are summarized by Lundqvist (1979). Here the Ledfat structure will be interpreted as an impact structure, particularly since this hypothesis seems to solve some of its stratigraphic problems.

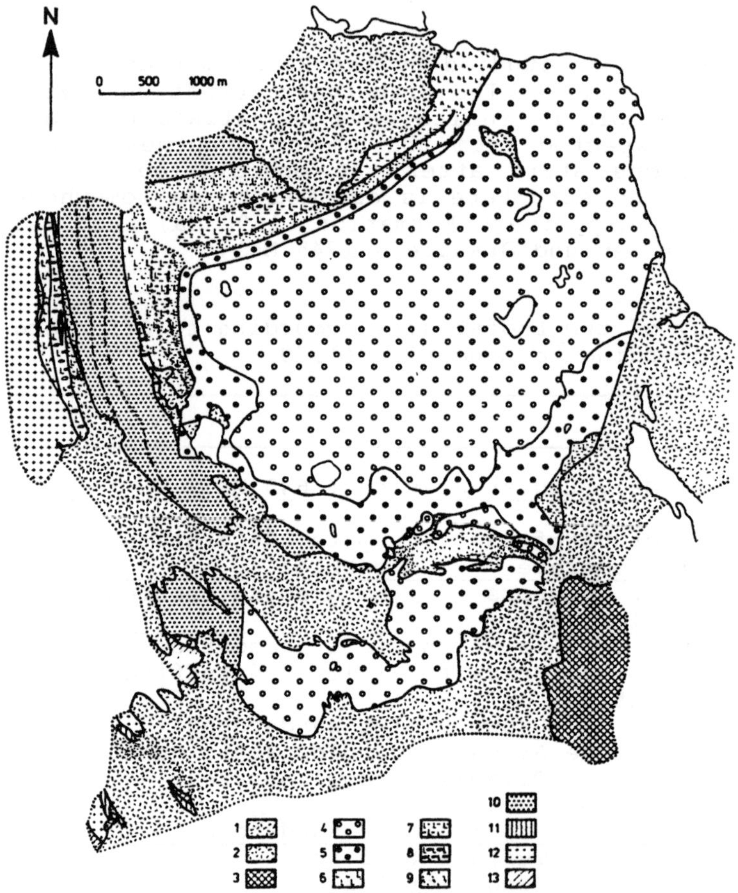

Fig. 4. Geology of the Ledfat area. (Simplified after Offerberg 1959). 1 Granite; 2 granite porphyry; 3 gabbro, Ledfat series: 4 Upper Vargfors congl.; 5 Lower Vargfors congl.; 6 sandstone; 7 tuffite; 8 conglomerate; 9 volcanics; 10 Sandstone formation. Arvidsjaur series: 11 acid volc.; 12 intermed. volc.; 13 basic volc

The stratigraphy given by Offerberg (1959) is shown in Table 5, and a simplified map is presented in Fig. 4. The two Vargfors conglomerates are regarded as sediments filling the original impact crater. The Upper Conglomerate is regarded as a fluvial sediment deposited when the structure had been eroded sufficiently tò receive more distant fluvial material. The Lower Conglomerate has a more complex origin with no granite pebbles, but local rocks of impact origin. The structure is not well preserved, particularly because Ledfat granite and perhaps also Sorsele granite have intruded it. Therefore the surviving parts of the structure have undergone thermal metamorphism (up to amphibolite mineral facies).

The impact occured in a continental area mainly covered by Arvidsjaur porphyries, arkose and feldspathic sandstones. The breccias present in the Arvidsjaur porphyries are partly interpreted as impact breccias, partly as preserved original brecciation. The andesitic vesicular lavas, conglomerates and tuffs and tuffites of the Andesite Formation are regarded as the normal sequence of impact sediments, while the rocks of the Tuffite Formation belong to the upper part of that sequence.

The Sandstone Formation is tentatively interpreted as partly consisting of remnants (mega-clasts) of pre-impact sandstones and arkoses arriving into the structure by big slides. The other units of the formation are regarded as mixtures of pre-impact rocks, Lower Vargfors conglomerates and tuffites.

There is a lack of outcrops in the critical areas and the contact relations between rocks are therefore obscure or unknown. The rock clasts present in all conglomerates in Ledfat seem to be consistent with the present hypothesis. Only new field work concentrated on critical evidence in critical areas can establish whether it is well founded or not.

The impact must be older than the Ledfat granite (1645 Ma). It must also be younger than the Arvidsjaur porphyries, which are considered to have a Revsund granite age (1745 Ma, Lundqvist 1979).

3. Bottenviken (64°10'N 22°21'E). This is an almost circular structure in the Bothnian Bay with a diameter of ca. 5 km. The depth to the basement is 500-600 m within the structure but only 80-130 m outside it. No aeromagnetic anomaly has been observed in the area of the structure.

Flodén (personal communication 1987) discovered the structure in 1978 during a cruise with R/S **Strombus**. A sampling of the structure is necessary for a better evaluation of its origin. It seems that its age is Phanerozoic, since the structure is not filled with sediments that have consolidated.

4. Lockne (62°59'N 14°40'E). The area is situated 20-30 km S of the town Östersund in the province of Jämtland; it is located just outside the eastern Caledonian margin and is characterized by Ordovician and Proterozoic rocks. The name refers only to the small "visible" part of the structure, its major part being covered by Caledonian nappes.

Wiman (1899) gave the first real description of the remarkable geology of the area, but the classic study is by Thorslund (1940). It has repeatedly been studied in the last 20 years, for example by Larsson (1973), Löfgren (1978), Lindström et al. (1983), and particularly by Simon (1984).

According the Thorslund (1940), in Middle and Upper Ordovician the Lockne area was a coastal landscape with Cambrian and Ordovician sediments and here and there monadnocks of denuded Proterozoic rocks, mainly granite and dolerite. This means that the Ordovician limestones formed in very shallow water, and this picture has been accepted by most stratigraphers. Lindström and his co-workers, however, have argued for a deposition in relatively deep water with slumping processes. The main argument is the difficulty of explaining, from Thorslund's picture, the presence of anomalous chaotic breccias with clasts exceeding 1000 m^3 in size as well as graded polymict arenites. These "anomalous" sediments occur only in Caradoc, otherwise the sedimentary rocks of the Lockne area belong to the normal Baltoscandian Ordovician sequence. The monadnocks are also unusual because they consist of "arkose-breccia". Attention will therefore be focused on the anomalous rocks.

The arkose-breccia is a monomict breccia of the bedrock, mainly granite but also other Proterozoic rocks, and there is no sharp boundary to the ordinary bedrock. The arkose-breccia forms a NW-SE zone in the Lockne area, and is often impregnated by bitumen, a feature only observed in the arkose-breccia. The new studies (Simon 1984) have shown that the monadnocks are allochthonous, for example a ca. 50-m-high granite hill in a rock quarry is underlain by Cambrian alum shale. It has been shown that the movements occurred in Caradoc.

The Lockne breccia is a polymict breccia with clasts from any of the older formations, from Proterozoic up to Caradoc. In fact Simon (1984) has shown that the Lockne breccia is both underlain and overlain by limestones belonging to the Caradocian conodont subzone **Prioniodus gerdae**. The orientation of the clasts seems to be random, the sorting is poor, the matrix always contains limestone as an important constituent. Its thickness may exceed 20 m.

The loftarsten (sometimes translated as loftarstone) is a tuffaceous, lithic greywacke, which essentially occurs together with the Lockne breccia. The two rocks can be followed for ca. 55 km roughly N-S along the strip of autochthonous Lower Paleozoic at the Caledonian front. The loftarsten is up to 5 m thick, often graded, and its clasts are essentially similar to those of the Lockne breccia. New investigations have shown (Simon 1984) that the loftarsten contains up to 20% clasts of pyroclastic, not epiclastic, rocks, originally glass, now chert. The Lockne breccia may also contain such clasts. The result is unexpected, since no volcanism is known in the area.

The evidence found by Lindström and co-workers is important and must be explained, but also the evidence for a shallow sea scenario is impossible to neglect. N.B. Svensson and the present author therefore suggest a big Caradocian impact west of Östersund in the Lake Storsjön area. The Lockne area is regarded as a part of the crater wall around the central depression. The paleogeography of Thorslund (1940) is accepted, except for the existence of monadnocks, which may be totally or partially absent as the result of later impact movements. The megaclasts of Lindström and co-workers are regarded as impact sediments. There is no difficulty in finding examples with similar large transports in impact structures.

The arkose-breccia is regarded as shocked and moved basement rocks, the Lockne breccia as roughly equivalent to the Bunte Breccia of the German Ries structure, and the pyroclastic rocks formed at the impact. There is one fundamental difference between the two structures: Ries formed within a continental area, Lockne in a shallow sea. The geologic processes are therefore difficult to discuss at present.

The impact crater was so large that a central uplift must have formed. After an interval of ca. 50-100 Ma, Caledonian nappes

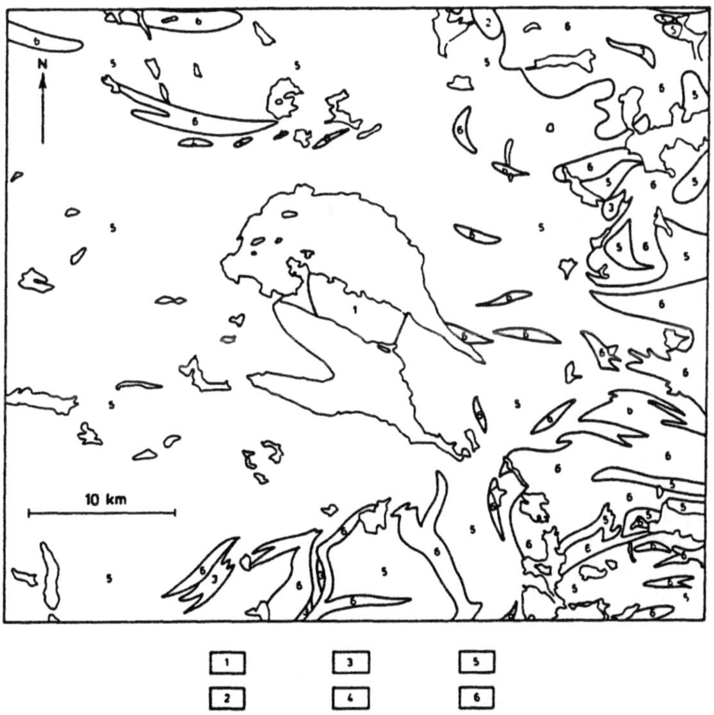

Fig. 5. Geologic map of the Dellen Lakes area. (After Lundegårdh 1967). 1 Dellen rocks; 2 Jotnian dolerites; 3 younger granite; 4 amphibolite; 5 older granites to quartz diorites ("Ljusnedal granite"); 6 sedimentary gneiss

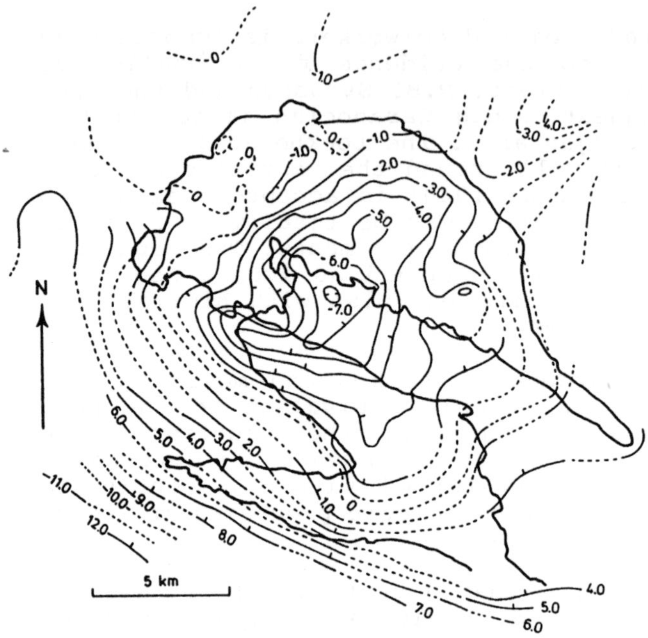

Fig. 6. Map of Bouguer anomalies in mgal over the Dellen Lakes area. (Courtesy Svensson personal communication 1987)

moved over the impact crater area but stopped before the "wall area" around Lockne. The consequences of the present hypothesis for the geologic development of the Lake Storsjön area are too many and too complex to be mentioned and discussed here.

The age of the structure is, as mentioned, well defined: the conodont subzone **Priniodus gerdae** in the Upper Ordovician.

5. Dellen Lakes (61°50'N 16°45'E). The two lakes are divided by an isthmus, and are situated ca. 25 km WNW of the town Hudiksvall on the Bothnian Sea coast. The diameter of the structure is somewhere between 10-15 km. (Fig. 5).

The structure was discovered by Svenonius (1888), who described it as the remnants of a Tertiary andesite volcano. Brögger (1895) suggested that the hypersthene-andesite (better dacite-liparite) should have a special name, dellenite. As mentioned, Högbom (1910) and Fredriksson and Wickman (1963) suggested that the structure was an impact. Redaelli (1957) concluded that the structure was a complex volcano, and this interpretation was accepted by Lundegårdh (1967).

Svensson (1968) found planar elements in quartz grains from rock fragments in dellenite and pointed out the similarity between Dellen rocks and those from the Ries structure in Germany. He also made a gravity survey of the lakes, and found a Bouguer anomaly of 5-10 mgal (Fig. 6). The uncertainty depends on the limited extension of the area measured. At present the bottoms of the lakes are being studied by seismic methods

(Flodén personal communication 1987) and the whole area by magnetic, gravimetric and other geophysical methods (Henkel personal communication 1987).

The basement of the structure mainly consists of Early Proterozoic granodiorites with an age around 1800-1950 Ma (Lundqvist 1979). Due to the lack of outcrops, the stratigraphy of the Dellen rocks is poorly known, but glassy and devitrified dellenites, tuff-breccias and tuffs dominate. The isthmus may be a part of an uplift and/or showing the remnants of the impact melt layer. If so, its chemical composition should be similar to that of the basement rocks. A preliminary test is shown in Table 6 between analyses of dellenite and the "Ljusnedal granite". The similarity is evident.

The age of the structure is still uncertain, but it is geologically young. Bylund (1974) seemingly confirmed the suggestion by Svenonius and suggested a Lower Tertiary age from a paleomagnetic study. However, an Ar-Ar study by Bottomley et al. (1977) gave an unexpected result: an unvitrified dellenite glass 235 Ma, a devitrified one 100 Ma, both with excellently defined plateaus. Åberg (personal communication 1987) has made a preliminary K-Ar dating of two dellenite glasses, and obtained 225 and 252 Ma respectively. It seems that the age is Triassic or Permian; further studies are necessary and are in progress.

6. Siljan Ring (61°5'N 15°00'E). The structure is the place of this conference. Therefore only a very brief summary is given of earlier views of the structure. Also a list of references and short notes of more recent papers not connected with the gas project of Vattenfall are given.

Table 6. Chemical analyses of dellenite and "Ljusdal granite" Analyses recalculated to the sum of 100

	Dellenite[a]	"Ljusdal granite"[b]
SiO_2	70.8	70.5
Al_2O_3	14.2	14.3
Fe_2O_3 c	4.3	4.1
MgO	0.9	0.9
CaO	2.3	2.4
Na_2O	2.9	2.9
K_2O	4.5	4.8
	99.9	99.9

[a] Average of the six analyses of dellenite given by Svenonius (1888)
[b] Average of the four analyses of the "Ljusdal granite" given by Lundegardh (1967)
[c] Fe_2O_3 = total iron calculated as Fe_2O_3

Hjelmqvist (1966) wrote the standard geologic description of the region, but already in 1740 a geologic paper about the Boda area was published by D. Tilas. W. Hisinger, around 1830, drew the first geologic map where the ring can be seen. A. G. Nathorst realized the importance of fault tectonics in the ring.

Due to the lack of outcrops, both in the ring and the uplift progress, understanding of the structure has been slow and difficult. Nevertheless many formation hypotheses have been published. Most popular have been various variations of volcanism plus some sort of caldera formation. Typical examples are Asklund (1931 and later) and Rutten (1966). Cloos (1933) suggested a special tectonic origin with the help of a regional tension, which should produce a central disc surrounded by an annular depression. The impact hypothesis was first published by Fredriksson and Wickman (1963) and Wickman et al. (1963).

The structure consists of an approximately circular central uplift, ca. 30 km in diameter and made up of Proterozoic rocks, mainly granites but also many other kinds of rocks. The uplift is surrounded by an annular depression with an outer diameter of ca. 44 km. The depression (the ring) is known for its occurence of fossil-rich Ordovician and Silurian sedimentary rocks, sometimes occurring as slab-shaped mega-clasts with sizes over 1 km and often chaotically mixed. Proterozoic rocks are also important constituents in the ring.

Svensson (1967) found planar elements in quartz from the uplift area, and Tamminen and Wickman (1980) studied cleavage fractures and planar elements in pegmatitic quartz from the same area. Svensson (1973) found the first examples of shatter cones, and Wickman (1980) tried to determined the impact center using shatter cone orientations (Fig. 7).

Hedström (1894) and Hjelmqvist (1966) described breccias, Stam (1967) studied the tectonics in the ring, Thorslund and Auton (1975) discussed the finding of deformed fossils in the Paleozoic limestones, and Hjelmqvist (1977) a contact between granite and sandstone on the uplift. Thorslund and Collini (1980) reported a probably Eocambrian sandstone found at Djupgrav. Nyström (1983) observed that in Vamhus, inside the ring, Dala porphyrites belong to the prehnite-pumpellyite mineral facies, while in their source area, just W of the ring they belong to a higher (actinolite-pumpellyite) mineral facies.

Lindblom and Wickman (1985) studied the fluid inclusions in quartz from a post-impact uplift breccia. The vein has been cemented by meteoric water (1.0 eq wt% NaCl) initially over 300°. Wickman (1981) discussed the displacements in the structure and Wickman and Nyström (1985) used Ordovician basal conglomerate pebbles as an example.

The Pb-Zn mineralizations in the ring have been studied by Welin (1959), the Pb isotopes by Wickman et al. (1963) and Johansson (1984), who also studied the S isotopes. He describes the deposits as carbonate-hosted calcite-fluorite-barite-sphalerite ("schalenblende")-galenapyrite-marcasite ores, enriched in radiogenic lead and heavy sulfide sulfur.

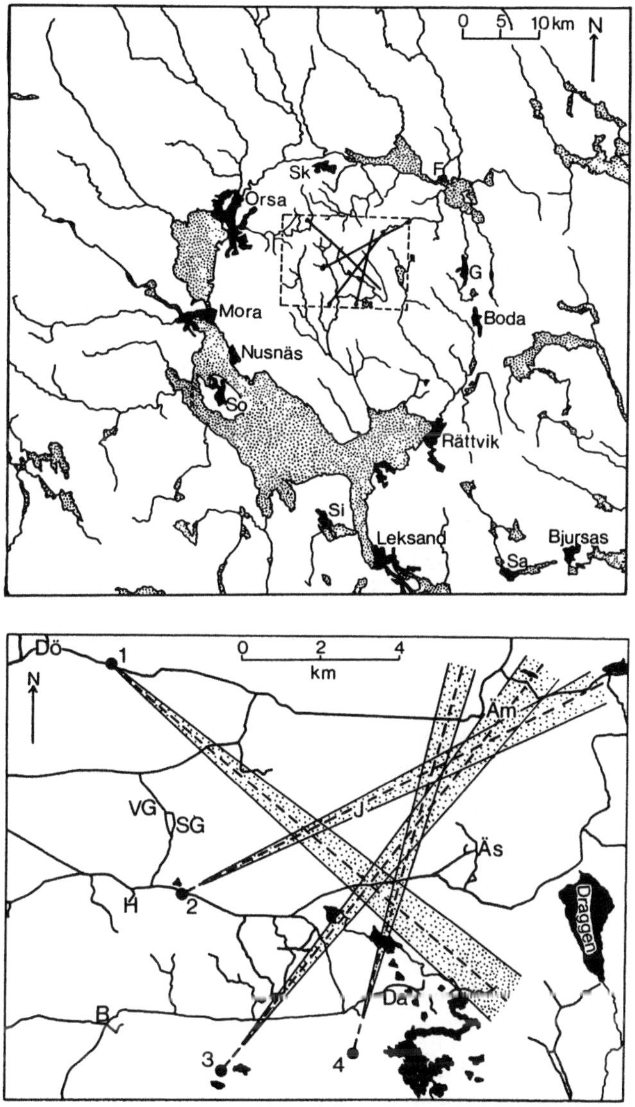

Fig. 7. Impact center of the Siljan Ring structure from directions of shatter cone axes. (Wickman 1980)

Geologic evidence points to a Postsilurian age. Bottomley et al. (1978) studied the age of shock melt from the central uplift by the Ar-Ar method and obtained 362 Ma. Åberg (personal communication 1987) has dated shock melt from another locality on the uplift by the K-Ar method and got the value 349 Ma. Åberg and Bollmark have studied the effects, if any, of the impact on the basement rocks. in 1983 they dated a composite dike using the Rb-Sr method, and in 1985 they studied the retention of U and Pb in zircons from the central uplift granite. In both cases the shock effect was found to be small.

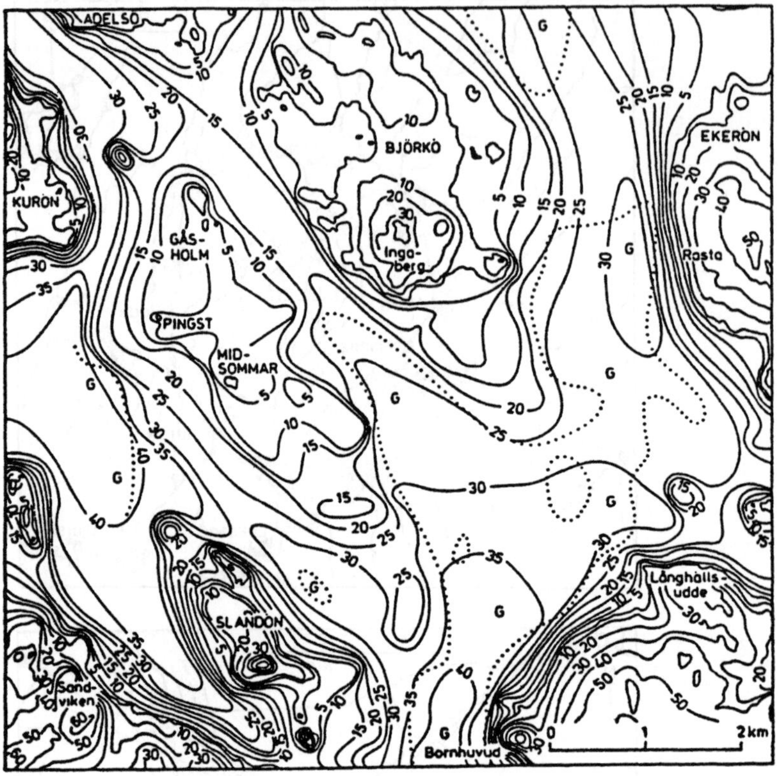

Fig. 8. Map of the Midsommar structure area. (Flodén et al. 1987)

7. Midsommar (59°18'N 17°35'E). This structure is centered in the Södra Björkfjärden Bay in Lake Mälaren, ca. 30 km W of the center of Stockholm City. The name refers to an islet in the bay (Fig. 8).

Törnebohm (1862) described a red sandstone series from the island Ekerön, later known as the Mälar Sandstones. Studied by many geologists, the standard description is the paper by Gorbatschev and Kint (1961). According to them, the presumably Jotnian sandstone series was formed in an arid area with considerable morphological relief, since some beds indicate a talus origin. The fluvial sediments were deposited in a continuously filling basin. The small outcrop areas are restricted to Rasta on the island Eekrön and the islets Midsommar and Pingst in the bay. The stratigraphy, as given by Gorbatschev and Kint (1961), is shown in Table 7.

Outcrops of the basement granite close to the structure are distinguished by an intensely red color, then follows a maroon, thoroughly silicified, ferruginous, basal claystone, which also fills the cracks present in the contacting basement granite. In the claystone, angular fragments of wholly unweathered microcline-perthite occur, as well as small amygdules filled with

Table 7. Stratigraphy of the sandstone series in the Midsommar structure according to Gorbatschev and Kint (1961)

		Thickness (m)
Sandstone unit	Sandstones, arkoses, polymict rounded conglomerates	More than 300
Conglomerate unit	Sharpstone conglomerates of local rocks	More than 40
Claystone unit	Dense rock with some local angular fragments	ca. 1.7
Major unconformity		
Svecokarelian basement	Granites, gneisses etc	

cryptocrystalline quartz partly showing relict banding. The claystone gradually passes into fine-grained sandstone.

On the claystone follows a coarse sharpstone conglomerate (=breccia), almost totally consisting of strictly local poorly sorted granite. Upwards the breccia becomes increasingly better-rounded and bedded with interrupting layers of arkosic sandstone. Above this unit follows the sandstone formation with increasingly rounded boulders.

According to Gorbatschev and Kint (1961), quartz ofter contains Böhm striation and it also shows "lamellar deformation lineation". The sandstones on the island Ekerön contain secondary minerals (Gorbatschev 1962), indicating the circulation of hot pore-solutions forming minerals like sphene, epidote, and laumontite cement.

It is evident that the observations by Gorbatschev and Kint (1961) can be interpreted as indicating a Proterozoic impact structure, preliminary defined by the Rasta area and the islet Pingst, that is having a diameter of ca. 6-8 km. Flodén et al. (1987) are therefore presently studying the structure with seismic and other methods.

The age of the structure is important, and the claystone, which is interpreted as a ahock melt, has therefore been dated by the K-Ar method. The two samples give the ages 1216 and 1203 Ma, or an average of 1210 Ma. This can be regarded as a confirmation of the supposed Jotnian age (see Lundqvist 1979).

8. Tvären Bay (58°46'N 17°25'E). The bay is situated ca. 70 km SSW of Stockholm. In the center of the bay a 2-km-wide, 80-m-deep circular depression occurs, and a boulder fan of Paleozoic sedimentary rocks is present S of the bay (Asklund

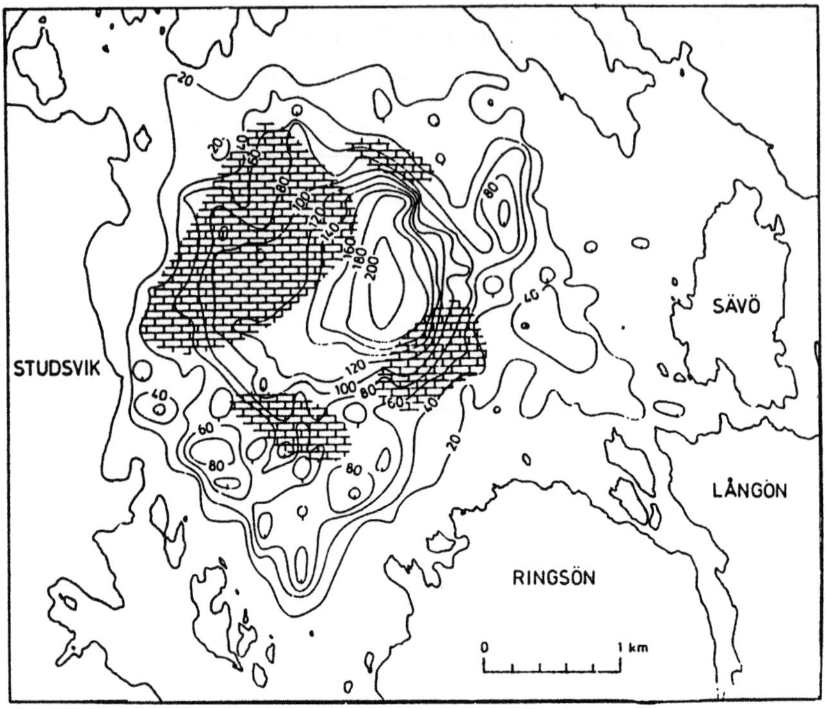

Fig. 9. Map of the Tvären Bay area. Tentative Paleozoic sediments indicated. (Flodén et al. 1986)

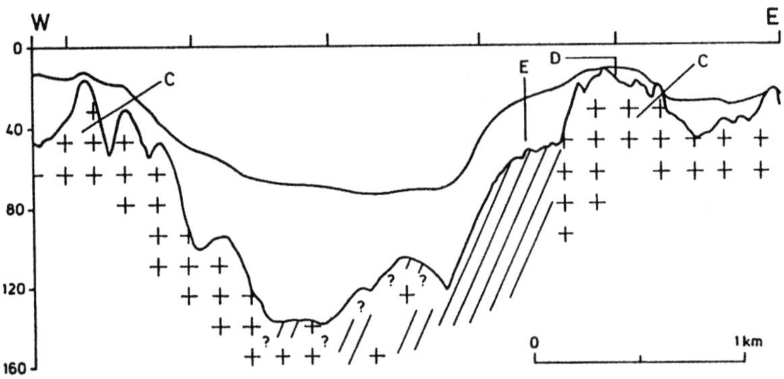

Fig. 10. Profile over the Tvären Bay structure with Paleozoic sediments indicated. (Flodén et al. 1986)

1931). Thorslund (1940) showed that the main component of the fan is Middle Ordovician Dalby limestone containing sharp-edged fragments of crystalline bedrock, mainly migmatites. A conglomerate also contains Orthoceratite limestone pebbles with a well-preserved conodont fauna (Bergström 1962), indicating an age around Lanna-Volkhov or lower Kunda. The outcrops around the bay sometimes contain breccias.

A seismic investigation (Flodén et al. 1986) resulted in a tentative geologic map (Figs. 9 and 10) and showed that the depth to the bedrock in the structure is somewhat more than 200 m. The original depth is estimated to be ca. 250-300 m.

The age seems to be close to that of the boundary between Lanna-Volkhov and Kunda.

9. **Landsort Trench** (58°36'N 18°30'E). The structure is a hook-shaped depression ca. 75 km SSE of Stockholm. The trench (max. depth 459 m) is located in the NNW part of the structure, which has a diameter of ca. 38 km (Fig. 11). Regionally the structure is situated in a drenched cuesta landscape with Cambrian to Silurian sedimentary rocks. Wickman (1973) suggested that the trench may be part of the annular ring of an impact structure.

Flodén (1980, personal communication 1987) studied the structure by geophysical methods, magnetically it is "smoothed-out"

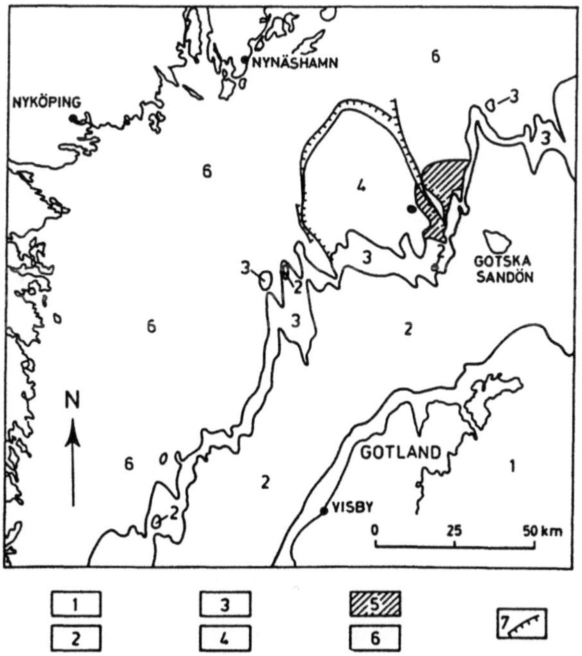

Fig. 11. The Landsort Trench area. (After Flodén 1980). 1 Silurian; 2 Ordovician; 3 Cambrian; 4 Jotnian sediments; 5 Subjotnian rocks; 6 crystalline basement; 7 boundary of the Landsort Trench

and the depth to the bedrock in the trench is over 500 m. The geology shown in the map (Fig. 11) is based on the variations of the acustic velocities in the rocks. According to Flodén (1980) the central part of the structure in the main consists of Jotnian (ca. 1200 Ma) sediments while Subjotnian (ca. 1600 Ma) rocks occur in SE.

If the shaky assumption is made that the Jotnian sediments represent fill in the structure, the age is probably in the age bracket 1200-1600 Ma.

10. Granby (58°25.5'N 14°56'E). The structure, close to the small town of Vadstena, is oval-shaped (3.5 x 2.9 km) with a depth of over 377 m. No outcrops exist, as was discovered by totally ten core drillings (Bruun and Dahlman 1980, 1982). A geologic map and a vertical section are shown in Figs. 12 and 13.

The Proterozoic crystalline rocks below the Granby sediments are monomict-oligomict breccias, followed by ca. 70 m angular clasts of Proterozoic rocks and Cambrian and Lower Ordovician sedimentary rocks, in a matrix of almost black, clayey sediments. Above this layer follows 20 m of fragment-free matrix ("shale"). Then comes a distinct sequence of mudstones and limestones belonging to the Kunda etage of Lower Ordovician. This sequence is 170 m thick against normally 20 m in this

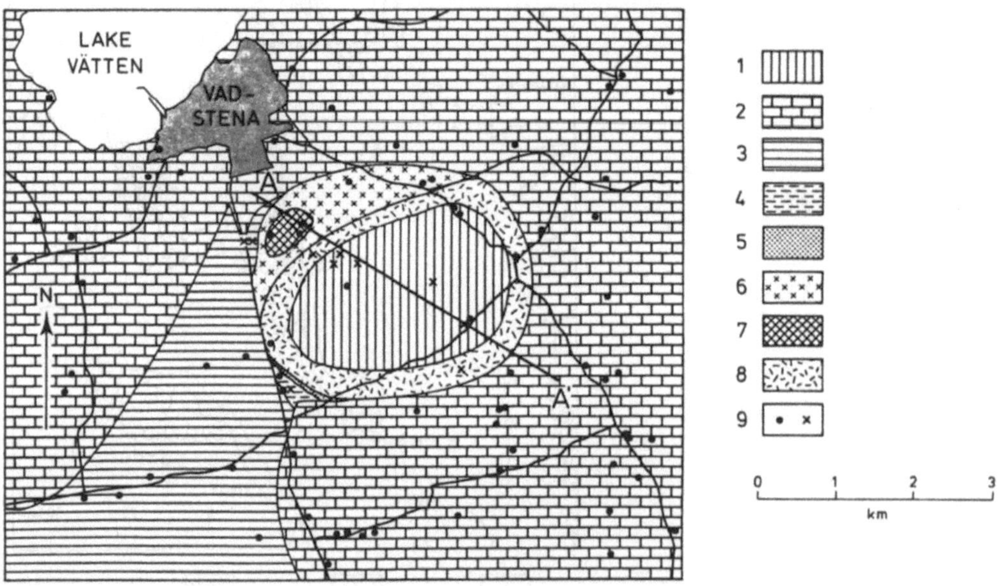

Fig. 12. Geologic map of the Granby structure. (After Bruun and Dahlman 1980, 1982). 1 Granby sediments; 2 Ordovician limestones; 3 alum shales; 4 Cambrian shale; 5 Cambrian sandstone; 6 Proterozoic granodiorite; 7 Proterozoic metabasite; 8 Proterozoic metavolcanite; 9 drillings, left churn, right core

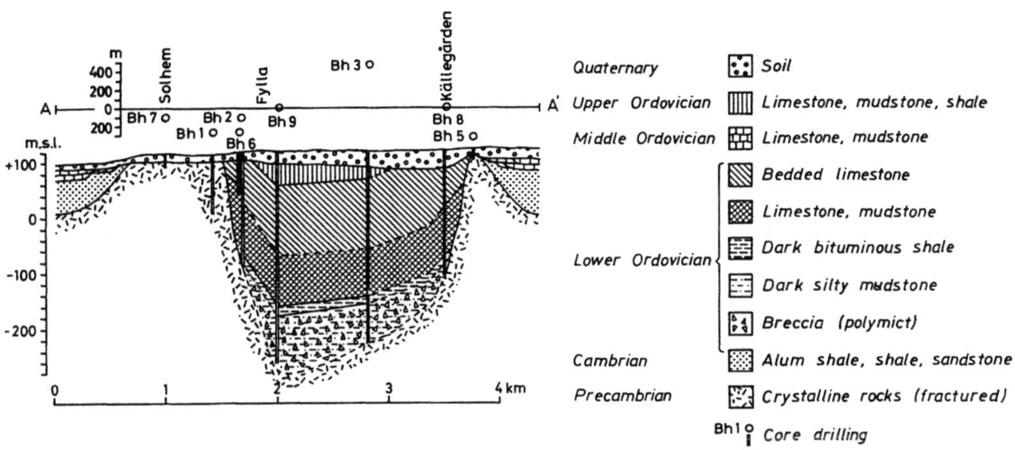

Fig. 13. Profile A-A'of the Granby structure. (After Bruun and Dahlman 1982)

area. Middle Ordovician rocks are ca. 135 m against normally 65. Youngest filling of the depression is mainly Upper Ordovician and probably some Lower Silurian limestone.

The impact seems to have happened in a shelf area during the youngest part of the Lanna-Volkhov or perhaps the oldest part of the Kunda etage. Åberg (personal communication 1987) has made a preliminary K-Ar dating of a "shale" sample and obtained 472 Ma, indicating a Lower Ordovician age.

The Granby and Tvären Bay structures are not far from each other, and seemingly formed at about the same time. Evidently only core drilling in the Tvären Bay can give further clues.

11. Lake Mullsjön (58°19'N 14°56'E). This structure is situated to the small town of Hjo around the round, shallow, 2-2.5-km-wide Lake Mullsjön. A 4.5-km-wide circle can be drawn around the lake, where all known outcrops (Blomberg 1906) except one lie outside the circle or its periphery. There is also a definitely larger depression defined by the surrounding hills. Lind (1972) made a regional gravimetric study, which included this area, and on his map of Bouguer anomalies a minimum can be seen some kilometers N-NE of Lake Mullsjön. Good indications for an impact structure are lacking.

It is generally assumed that the depression is filled by sandstones belonging to the Visingsö group, which are Late Riphean to Vendian (750-850 Ma). The structure needs investigations.

12. Lake Hummeln (57°22'N 16°15'E). This structure is situated about 20 km NW of the town of Oskarshamn. The presence of Paleozoic sedimentary rock boulders south of the lake has been known since long. A granite breccia has also been noted on the southern shore of the lake. Nordenskjöld (1944) found a round

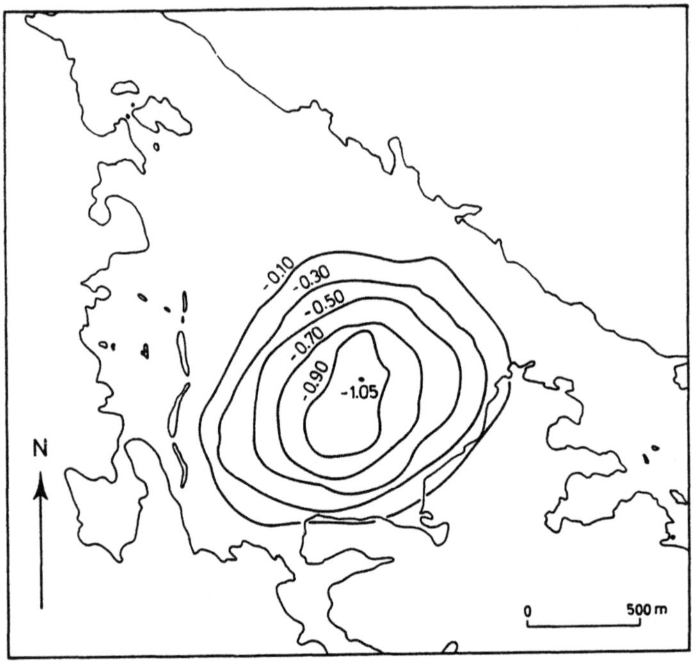

Fig. 14. Map of the Bouguer anomalies in mgal over depression in Lake Hummeln. (After Svensson 1966)

1.2-km-wide, 61-m-deep depression in the lake. Asklund (in Nordenskjöld 1944) suggested that the depression is a cryptovolcanic structure, Fredrikson and Wickman (1963) an impact structure. Svensson (1966) made a gravimetric survey of the area and studied the breccia. The Bouguer anomalies (Fig. 14) show a small deficit which is consistent with results from similar impact structures (Table 8).

The fan boulders consist of sedimentary rocks from Middle Ordovician limestones down to Lower Cambrian sandstone. The age of the structure is therefore assumed to be Lower Cambrian, since

Table 8. Bouguer anomaly deficit for some structures in mgal

Structure	Diameter (m)	Deficit
Meteor Crater	1175	-0.75
Lake Hummeln	1200	-1.0
Holleford	2350	-2.2

no Late Precambrian sediments have been reported among the boulders found.

13. **Lake Mien (56°25'N 14°52'E).** The structure is situated ca. 30 km N of the town of Karlshamn. Holst (1890) reported the occurrence of rhyolitic rocks in the lake and on the islet Ramsö, and suggested that Lake Mien was the remnants of a Tertiary volcano (Fig. 15). Högbom (1910) and Fredriksson and Wickman (1963) pointed out that Lake Mien was possibly an impact structure.

Svensson and Wickman (1965) and Svensson (1969) identified coesite in its rocks, and von Engelhardt and Stöffler (1965) found planar elements in quartz from the structure. Stanfors (1969, 1973) described briefly cores drilled on the islet Ramsö, and concluded that it was unlikely but not disproved that the Lake Mien was an impact structure. Vogel (1973) come to same conclusion from a gravimetric study. Henkel (1982), on the other hand, from a more detailed study, found a residual gravity anomaly of ca. -6 mgal, and that the structure has a highly magnetic impactite sheet 2.2 km wide and 250 m thick underlain by least 1.8 km fractured rocks. Ekelund (1986) found from a very detailed geophysical study of Ramsö that the data indicated the presence of several small bodies with properties similar to native iron and at only moderate depth.

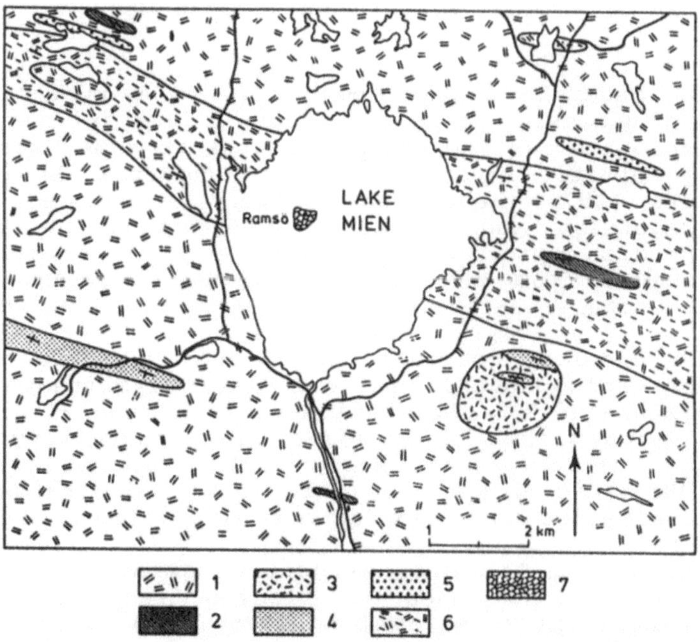

Fig. 15. Geology of the Lake Mien area. (After Stanfors 1973) 1 Gneissic granite; 2 amphibolite; 3 younger granite ("Spinkamåla"); 4 "coastal gneiss"; 5 porphyry; 6 gneissic granite with supracrustal rocks; 7 rhyolitic rocks

Table 9. Chemical analyses of Lake Mien rocks. Analyses recalculated to the sum of 100

	1	2	3	4
SiO_2	71.9	71.9	72.33	67.41
TiO_2	0.4	0.4	0.35	0.65
AL_2O_3	14.1	14.1	14.00	15.18
$Fe_2O_3{}^t$	2.95	2.95	2.76	4.77
MnO	0.05	0.1	0.08	0.10
MgO	0.9	0.6	0.57	1.21
CaO	2.2	1.8	1.65	3.32
Na_2O	3.05	3.2	3.18	3.57
K_2O	4.5	5.0	5.09	3.79
	100.05	100.05	100.01	100.00

1 Average of the so-called A-type rhyolite (Holst 1890, Stanfors 1973)
2 Mixture of 91% of analysis No. 3 and 9% of No. 4
3 Average of two "Spinkamåla granites", 2 km from Lake Mien (Stanfors 1973)
4 Average of two grey gneisses, at most 2.5 km from Lake Mien (Stanfors 1973).
^{t} Fe_2O_3 = total iron calculated as Fe_2O_3

The basement consists of Early to Middle Proterozoic granitic and gneissic rocks (Fig. 15). The drilled cores from Ramsö exhibit black and grey rhyolites to some extent interstratified by tuffitic layers similar to suevites. Besides coesite, Carstens (1975) described "ballen" structures after quartz, and Fregerslev and Carstens (1976) show the presence of small chromian spinels with variable Cr_2O_3 content, which probably derived from the biotite in the original rocks. A positive Ir-anomaly has also been found in the Mien rocks.

The chemical compositions of the rhyolitic rocks and the basement rocks reveal that the black rhyolite can be regarded as a mixture of 91% "Spinkamåla granite" and 9% older gneissic rocks ("Tving granite") (based on the "refractory" elements Si, Ti, Al, Fe) (Table 9). Welin (1975) determined the initial $^{87}Sr/^{86}Sr$ ratio of the rhyolitic rocks and obtained the value 0.741, a value not obtained for ordinary volcanic rhyolites.

The age of the Mien structure has been determined repeatedly. Storzer (1971) used the fission track method and obtained the minimum age 92 Ma. Welin (1975, 1979) used the K-Ar method and obtained the value 115 Ma, but he regarded a higher age, like 120 Ma, as more likely. Bottomley et al. (1978), using the Ar-Ar method, found the age to be 119 Ma. Thus the structure is Cretaceous.

Finally it may be pointed out that the result that the "Spinkamåla granite" is the dominating ingredient of the rhyolitic rock mixture and not the "Tving granite" can be strengthened from the study by Welin (1975). If the Sr age is computed from the Rb, Sr, and the Sr-isotope ratio values the result is 1271 Ma referred to the time of the impact or 1390 Ma. This value is consistent with age values found for the "Spinkamåla granite".

14. Skedviken (59°46'N 18°16'E). The structure is situated ca. 6 km WNW of the small town of Rimbo. It was discovered as an anomaly on an airborne VLF map over the area, and was studied more closely by Claesson (1985) using several geophysical methods.

Lake Skedviken is shaped roughly like a triangle, with a maximum depth of 6 m. According to Claesson, the structure forms a nearly circular topographic depression in the regional Subcambrian peneplain, with a diameter of 5.5 km. The depth of the depression measured from the peneplain to the bottom of the lake is ca. 30-40 m.

According to the aeromagnetic survey the magnetic field is uniform and featureless. A Bouguer anomaly map based on a special gravity survey shows an almost circular gravity minimum of 4 mgal over the topographic depression (Fig. 16) and with a diameter of ca. 7 km.

According to Claesson, a local pattern of radial and tangential lineaments and fracture zones can be seen around the topographic depression. The bedrock in the area consists of Early Proterozoic granites and gneisses. Most likely this structure is Proterozoic.

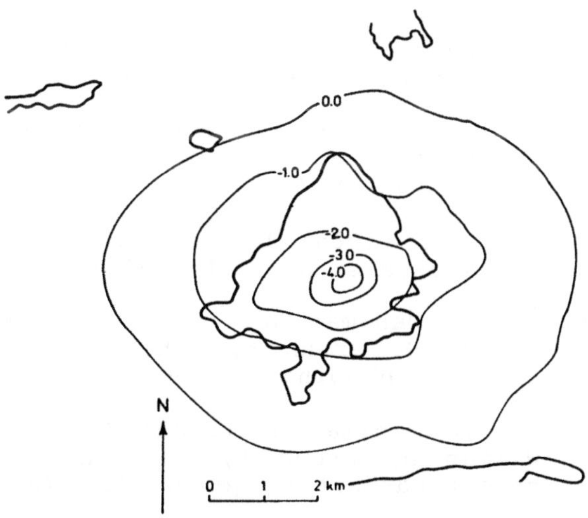

Fig. 16. Map of Bouguer anomalies in mgal over depression around the Skedviken structure. (After Claesson 1985)

15. **Hemmestorp** (56°18'N 14°3'E). The structure is situated ca. 30 km NNW of Kristianstad in the province of Scania, and was recognized by Henkel (1983) as a magnetic anomaly at Hemmestorp. The size is 6 x 9 km. The magnetic anomaly coincides with a minor topographic depression, ca. 60 m against ca. 75 m a.s.l. Henkel also made a gravimetric survey of the area, but could not observe any anomaly. He noted that the structure can be deeply eroded impact structure. If it is, it must most probably be Proterozoic.

References

Åberg G, Bollmark B (1983) Rb-Sr dating of a postorogenic Svecokarelian composite dike within the Siljan ring structure, central Sweden. Geol Foeren Stochlm Foerh 107:75-77

Åberg G, Bollmark B (1985) Retention of U and Pb in zircons from shocked granite in the Siljan impact structure, Sweden. Earth Planet Sci Lett 74:347-349

Asklund B (1931) Fennoskandias ålder, kontinentrörelser och relief. In: Ramsay W (ed) Geologines grunder, del II. Schildt, Stockholm, pp 355-381

Barringer DM (1905) Coon Mountain and its crater. Proc Natl Acad Sci Philadelphia 57:861-886

Bergström SM (1962) Conodonts from the Ludibundus Limestone (Middle Ordovician) of the Tvären area (S.E.Sweden). Ark Miner Geol 3:1-61

Blomberg A (1906) Beskrifning till kartbladet Hjo. Sver Geol Unders Aa 132

Bottomley RJ, York D, Grieve RAF (1977) ^{40}Ar-^{38}Ar dating of Scandinavian impact craters. Meteoritics 12: 182-183

Bottomley RJ, York D, Grieve RAF (1978) ^{40}Ar-^{38}Ar ages of Scandinavian impact structures: I Mien and Siljan. Contrib Mineral Petrol 68:79-84

Brögger WC (1985) Die Eruptivgesteine des Kristianiagebietes II. Vid-Selsk Skr I Math-Naturv Kl 1895:No 7, 59

Bruun Å, Dahlman B (1980) Den paleozoiska berggrunden. In: Wikman H, Bruun Å, Dahlman B Beskrivning till berggrundskartan Linköping NV. Sver Geol Unders Af 119:68

Bruun Å, Dahlman B (1982) Den paleozoiska berggrunden. In: Wikman H, Bruun Å, Dahlman B, Vidal G, Beskrivning till berggrundskartan Hjo NO. Sver Geol Unders Af 120:76

Bylund G (1974) Palaeomagnetism of a probable meteorite impact, the Dellen structure. Geol Foeren Stochlm Foerh 96:275-278

Carstens H (1975) Thermal history of impact melt rocks in the Fennoscandian Shield. Contrib Mineral Petrol 50:145-155

Claesson O (1985) The Skedviken structure. Geofysikrapport BRAP 85 405. Sver Geol Unders Berggrundsbyrån

Cloos H (1933) Über Biegungsbrücke und selektive Zerlegung. Geol Rundsch 24:203-219

Einarsson Ö (1979) Stratigrafisk och petrografisk undersökning av den prekambriska berggrunden i Dobblonområdet, Västerbottens län. Sver Geol Unders C 748

Ekelund A (1986) Geophysical results from Ramsö Island in Lake Mien, an impact site in southern Sweden. Lunars Planet Sci 17. Abstr Pap (17th Lunar Planet Sci Conf, Houston)

Engelhardt W v, Stöffler D (1965) Spaltflächen im Quartz als Anzeichen für Einschläge grosser Meteoriten. Naturwissenschaften 52: 489-490

Flodén T (1980) Seismic stratigraphy and bedrock geology of the central Baltic. Stochlm Contrib Geol 35:1-240

Flodén T, Tunander P, Wickman FE (1986) The Tvären Bay structure, an astrobleme in southeastern Sweden. Geol Foeren Stochlm Foerh 108: 225-234

Flodén T, Söderberg P, Wickman FE (1987) Midsommar, a possible Middle Proterozoic impact structure west of Stocholm, Sweden. (in prep.)

Fredriksson K, Wickman FE (1963) Meteoriter. Naturv Forsk, Årsbok, pp 121-157

Fregerslev S, Carstens H (1976) Chromian spinels in impact melt rocks of Lake Mien, Sweden. Geol Foeren Stockhlm Foerh 98:316-321

Gavelin S (1955) Beskrivning till berggrundskarta över Västerbottens län. Sver Geol Unders Ca 37

Gorbatschev R (1962) Secondary sphene in the Mälar Sandstone. Geol Foeren Stochlm Foerh 84:32-37

Gorbatschev R, Kint O (1961) The Jotnian Mälar Sandstone of the Stocholm region, Sweden. Bull Geol Inst Univ Uppsala 40:51-68

Grieve RAF (1975) Petrology and chemistry of the impact melt at Mistastin Lake crater, Labrador. Bull Geol Soc Am 86:1617-1629

Hedström H (1894) Geologiska notiser fran Dalarne. I Breccior fran Siljanstrakten. Geol Foeren Stochlm Foerh 16:585-591

Henkel H (1982) The Lake Mien structure. Sver Geol Unders Geofys Rap 8221

Henkel H (1983) Geofysiska Undersökningar. In: Wikman H, Bergström J, Lidmar-Bergström K (eds) Beskrivning till berggrundskarten Kristianstad NO. Sver Geol Unders Af 127:137-165

Hjelmqvist S (1966) Beskrivning till berggrundskarta över Kopparbergs län. Sver Geol Unders Ca 40

Hjelmqvist S (1977) An astroblematic contact granite/sandstone. Bull Geol Inst Univ Uppsala NS 7:67-70

Högbom A (1931) Nya iakttagelser inom Norr- och Västerbottens urberg. Geol Foeren Stochlm Foerh 53:415-438

Högbom AG (1910) Minutes of Geol Soc Stockholm meeting March 3rd, 1910. Geol Foeren Stockhlm Foerh 32:482-483

Holst NO (1890) Ryoliten vid sjön Mien. Sver Geol Unders C 110

Johansson A (1984) Geochemical studies on the Boda Pb-Zn deposit in the Siljan astrobleme, central Sweden. Geol Foeren Stochlm Foerh 106:15-25

Larsson K (1973) The Lower Viruan in the autochthonous Ordovician sequence of Jämtland. Sver Geol Unders C 683

Lind G (1972) The gravity and geology of the Vättern area. Geol Foeren Stochlm Foerh 94:245-258

Lindblom S, Wickman FE (1985) Fluid inclusions in quartz from a quartz breccia in the Siljan Ring structure, central Sweden. Geol Foeren Stochlm Foerh 107:53-58

Lindroos H, Smellie J (1979) A stratabound uranium occurrence within Middle Precambrian ignimbrites at Duobblon, Northern Sweden. Econ Geol 74:1118-1130

Lindström M, Simon S, Paul B, Kessler K (1983) The Ordovician and its mass movements in the Lockne area near the Caledonian margin, central Sweden. Geolog Palaentol 17:17-27

Löfgren A (1978) Arenigian and Llanvirnian conodonts from Jämtland, northern Sweden. Fossils Strata 13:1-129

Lundegårdh PH (1967) Berggrunden i Gävleborgs län. Sver Geol Unders Ba 22

Lundqvist T (1979) The Precambrian of Sweden. Sver Geol Unders C 768

Nordenskjöld CE (1944) Morfologiska studier inom övergångsområdet mellan kalmarslätten och Tjust. Doct Thes, Dep Geogr, Univ Lund

Nyström JO (1983) Pumpellyite-bearing rocks in Central Sweden and extent of host rock alterations as control of pumpellyite composition. Contrib Mineral Petrol 83:159-168

Offerberg J (1959) Rocks and stratigraphy of the Ledfat area. Västerbotten County, Northern Sweden. Sver Geol Unders C 564

Redaelli LL (1957) A petrological investigation in Lake Norra Dellen by means of frog-man equipment. Sver Geol Unders C 548

Rutten GM (1966) The Siljan Ring of Paleozoic, Central Sweden: A posthumous ring-complex of a late Precambrian Dala porphyries caldera. Geol Mijnb 45:125-136

Simon S (1984) Stratigraphie, Petrographie und Entstehungsbedingungen von Grobklastika in der autochthonen, ordovizischen Schichtenfolge Jämtlands (Schweden). Doct Thes, Univ Marburg/Lahn

Smellie JAT (1982) The mineralogy and genesis of uranium in rhyolitic ignimbrites of Precambrian age from Duobblon, Sweden. Mineral Mag 46:189-201

Stam JC (1967) On the geology and tectonics of the Lake Siljan area, Central Sweden. Geol Mijnb 46:467-481

Stanfors R (1969) Lake Mien - an astrobleme or a volcanotectonic structure. Geol Foeren Stochlm Foerh 91:73-86

Stanfors R (1973) Mien-strukturen - en kryptoexplosive bildning i Fennoskandias urberg. Doct Thes, Univ Lund

Storzer D (1971) Fission-track dating of some impact craters in the age range between 6000 y. and 300 m.y. Meteoritics 6:319

Svenonius F (1888) Andesit fran Norra Dellen i Helsingland. Geol Foeren Stochlm Foerh 10:262-285

Svensson NB (1966) En kvalitativ tolkning av en gravimetermätning över sjön Hummeln. Thes, Dep, Geol, Univ Stockholm

Svensson NB (1967) Probable meteorite impact crater in Central Sweden. Nature (London) Phys Sci 229:90-92

Svensson NB (1968) The Dellen Lakes, a probable meteorite impact in Central Sweden. Geol Foeren Stochlm Foerh 90:314-316

Svensson NB (1969) Lake Mien, southern Sweden - a possible astrobleme. Geol Foeren Stochlm Foerh 91:101-110

Svensson NB (1973) Shatter cones from the Siljan structure, Central Sweden. Geol Foeren Stochlm Foerh 95:139-143

Svensson NB, Wickman FE (1965) Coesite from Lake Mien, southern Sweden. Nature (London) 205:1202-1203

Tamminen J, Wickman FE (1980) Shock effects in pegmatitic quartz from the Siljan ring structure, central Sweden. Geol Foeren Stochlm Foerh 102:275-278

Thorslund P (1940) On the Chasmops series of Jemtland and Södermanland (Tvären). Sver Geol Unders C 436

Thorslund P, Auton C (1975) Evidence of meteorite impact in the Siljan structure, Central Sweden. Bull Geol Inst Univ Uppsala NS 6:69-72

Thorslund P, Collini B (1980) An Eocambrian sequence in the Siljan ring structure. Geol Foeren Stochlm Foerh 102:188

Törnebohm AE (1862) Några ord till upplysning om bladet "Södertelge". Sver Geol Unders Aa 4

Vidal G (1982) Den prepaleozoiska sedimentära berggrunden. In: Wikman H, Bruun Å, Dahlman B, Vidal G, Beskrivning till berggrundskartan Hjo NO. Sver Geol Unders Af 120 p 52

Vogel A (1973) A study on gravity of the Lake Mien structure. Rep 11, Dep Solid Earth Phys, Univ Uppsala

Welin E (1959) Till kännedomen om sulfidmalmerna och de zonerade zinkbländena i Boda och Solleröns kommuner, Kopparbergs län. Geol Foeren Stochlm Foerh 81:495-513

Welin E (1970) Den svekofenniska orogena zonen i norra Sverige. En preliminär diskussion. Geol Foeren Stochlm Foerh 92:433-451

Welin E (1975) K-Ar dating and Sr-isotope composition of rhyolitic rocks from Lake Mien in southern Sweden. Geol Foeren Stochlm Foerh 97:307-311

Welin E (1979) Tabulation of recalculated radiometric ages published 1960-1979 for rocks and minerals in Sweden. Geol Foeren Stochlm Foerh 101:309-320

Welin E, Christiansson K, Nilsson Ö (1971) Rb-Sr radiometric ages of extrusive and intrusive rocks in northern Sweden I. Sver Geol Unders C 666

Welin E, Einarsson Ö, Gustafsson B, Lindberg R, Christiansson K, Johansson G, Nilsson Ö (1977) Radiometric ages of intrusive rocks in northern Sweden II. Sver Geol Unders C 731

Wickman FE (1973) The origin of the Landsort Trench in the Baltic. Geol Foeren Stochlm Foerh 95:339-341

Wickman FE (1980) The impact center of the Siljan structure, central Sweden. Geol Foeren Stochlm Foerh 102:105-109

Wickman FE (1981) Displacements and the Siljan Ring structure. Geogr Ann 63A:313-317

Wickman FE, Nyström JO (1985) The Proterozoic rock pebbles of the basal Ordovician conglomerate in the Siljan Ring structure and their significance. Geol Foeren Stochom Foerh 106:215-218

Wickman FE, Blomqvist NG, Geijer P, Parwel A, Ubisch H v, Welin E (1963) Isotopic constitution of ore lead in Sweden. Ark Mineral Geol 3:193-257

Wiman C (1899) Eine untersilurische Litoralfacies bei Locknesjön in Jemtland. Bull Geol Inst Univ Upsala 4:133-151

The Formation of Large Impact Structures and Constraints on the Nature of Siljan

R. A. F. GRIEVE[1]

Introduction

The Siljan ring structure has the form of a 32-km diameter uplifted central core of Dala granites of Sub-Jotnian age surrounded by a 45-km diameter annular depression, which is partially filled with lakes and contains down-faulted Ordovician and Silurian sediments (Fig. 1). The characteristic ring form, the presence of shatter cones, allochthonous breccias, breccia and impact melt dikes and microscopic, shock-produced planar features in the central granites provide abundant evidence that the Siljan structure is the remnant of a large impact structure (Fredriksson and Wickman 1963; Rondot 1975; Svensson 1971, 1973). Erosion has removed most of the original morphology and impact lithologies, exposing the impact structure at a level below its original floor. Isotopic dating of pseudotachylite from a dike (Fig. 1) place the age of the impact event at 326 ± 1 Ma (Bottomley et al. 1978) and the original, pre-erosional diameter of the structure is estimated to have been approximately 52 km (Grieve 1982).

Current understanding of the formational processes in impact cratering events comes from the geologic and geophysical character of terrestrial impact structures, studies of the morphometric parameters of fresh lunar and planetary craters, small-scale experimental and high-energy (conventional and nuclear) explosion craters, numerical simulations using continuum-mechanical cratering codes and theoretical considerations. These studies, particularly those at terrestrial craters, provide a considerable base of knowledge from which to constrain the original nature of the Siljan structure. Following a general treatment of cratering processes, this contribution considers the formation and some of the characteristics of an impact structure, such as Siljan, and discusses the spatial relationship of the Gravberg deep drill hole with respect to the original structure.

---

[1] Geophysics Division, Geological Survey of Canada, 1 Observatory Crescent, Ottawa, Ontario, Canada, K1A 0Y3

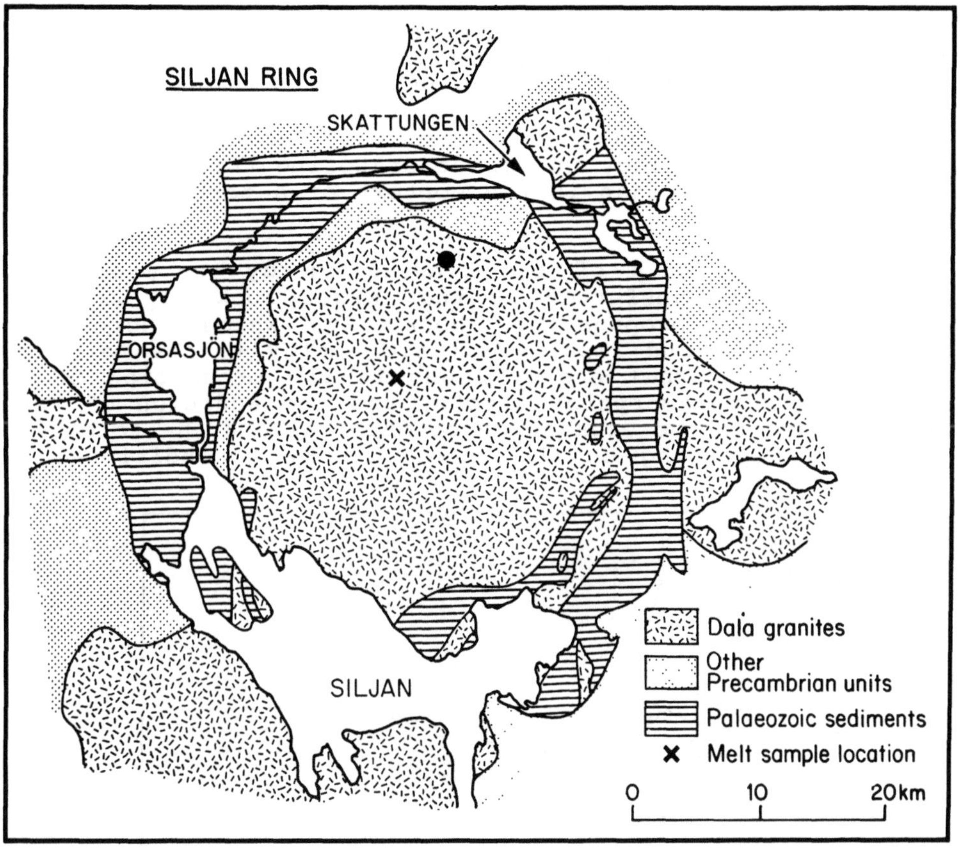

Fig. 1. Geologic sketch map of Siljan ring structure (Bottomley et al. 1978). Location of Gravberg drill hole is indicated by black dot

The Impact Process

Approximately 120 terrestrial impact structures ranging in size from a few tens of meters to 140 km are currently known (Grieve and Robertson 1987). Observational data on terrestrial impact structures can be found in papers dealing with individual structures and in such summaries and compilations as French and Short (1968), Masaitis et al. (1980), Roddy et al. (1977) and Grieve (1987). The terrestrial population of impact structures follows the characteristic morphological progression with size that is exhibited on other planetary bodies.

Smaller craters are essentially a bowl-shaped depression partially filled by breccias and surrounded by an uplifted rim. These are known as simple craters. Larger structures have an uplifted central area, which is morphologically expressed as a central peak and/or ring, surrounded by a down-faulted annulus or trough and a normally faulted rim area. These structures, known collectively as complex structures, have a relatively shallower depth/diameter ratio than simple craters and, on earth, occur at diameters greater than 4 km in crystalline rocks (Dence et al. 1977). Detailed discussion of the stages of the cratering process can be found in such papers as Gault et al. (1968), Dence et al. (1977), Croft (1980), Grieve et al. (1977, 1981) and others. Outlined here is some general background and terminology on impact cratering that will allow the reader to place Siljan in the context of a terrestrial complex impact structure.

On impact, the bulk of the kinetic energy contained in the projectile is transferred to the target rocks by means of a hemispherically propagating shock wave which increases the internal energy and imparts kinetic energy to the target. The increase in internal energy leads to the production of diagnostic shock metamorphic effects, and the addition of kinetic energy leads to the excavation of material and the formation of a crater. Depending on the specific impact conditions, i.e.,

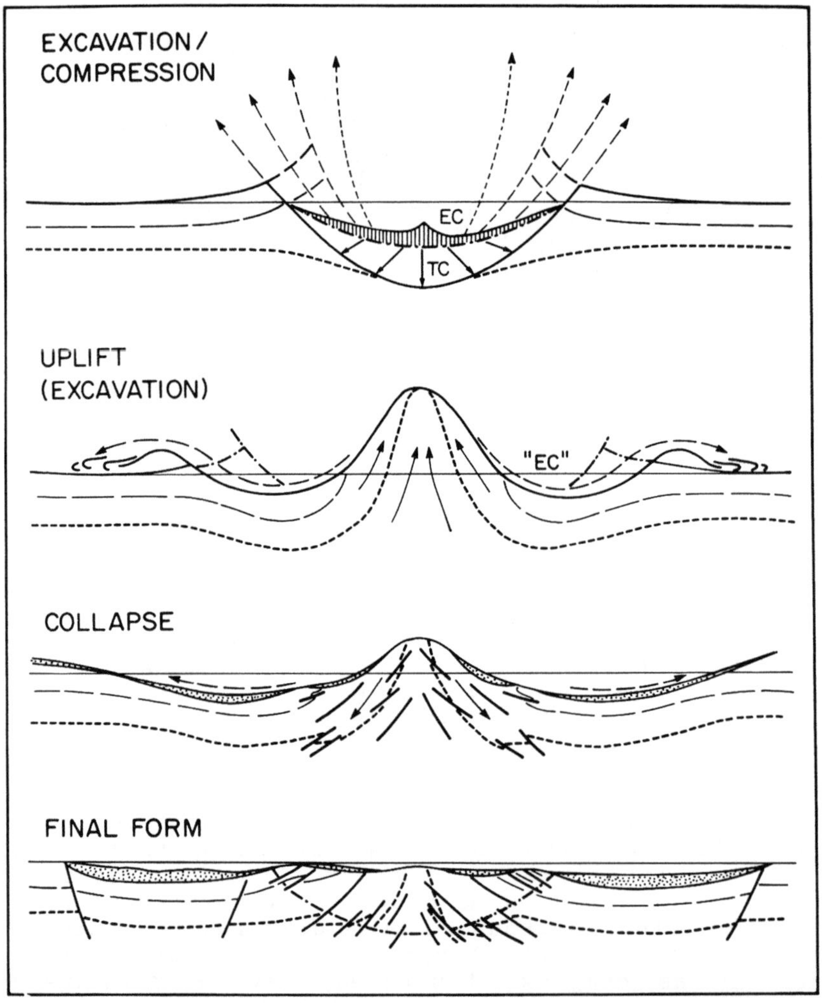

impact velocity, and physical properties of projectile and target, peak pressures at the point of impact are in the megabar to tens of megabar range, and associated post-shock temperatures are above the vaporization temperature of silicates. Particle motions induced by the shock and subsequent release or rarefaction waves, which bring the target material back to ambient pressure, produce a so-called cratering flow-field in the target. This flow-field has two complimentary spatial components. Outwards and upwards particle motions in the upper component of the flow-field excavate target material by ballistic ejection, resulting in an excavated cavity of maximum depth d_e and diameter D_e. As the near-surface portion of the cavity grows by excavation, the second, lower component of the flow-field, which involves the volume immediately under the impact point, leads to the displacement of deeper target material, thereby enlarging the cavity downward and outward. This combination of excavation and displacement defines a transient cavity, which corresponds to the maximum growth of the cavity and is approximately three times deeper than the excavated cavity (Dence et al. 1977). As the name indicates, this transient cavity does not survive throughout the entire cratering process. Modification processes operate subsequent to, or possibly even during transient cavity growth, and serve to shallow the transient cavity and bring it to some equilibrium form. In simple craters, the displacements induced in forming the transient cavity floor are almost entirely "locked-in" and modification is by wall failure. Wall failure serves to slightly enlarge the cavity diameter, with the inward collapse of the cavity walls leading to the formation of an interior allochthonous breccia lens. This is not the case for larger, complex structures. In complex structures, the transient cavity floor undergoes considerable uplift and the rim area collapses down-

---

Fig. 2. Schematic diagram illustrating the formation of a complex structure such as Siljan (Grieve et al. 1981). **Excavation/compression** stage - maximum growth of the excavated (EC) and transient cavities (TC). **Shaded region** depicts flow-field configuration of melted and brecciated material beneath maximum penetration of meteorite. **Dashed envelopes** illustrate ejection trajectories from a series of earlier, transitional stages in cavity growth and are a first-order approximation of cavity shape at these times. **Uplift (excavation)** stage - maximum rebound of transient cavity floor above regional surface. Excavation may continue with material driven outward over cavity rim in turbulent flow producing secondary removal of surface material beyond the diameter of major excavation. **Collapse** stage - overheightened, physically unstable uplift collapses with overthrusting and stacking of strata. Excess volume is accommodated in topographic ring peaks on flanks at approximately the diameter of major excavation. Breccia (and impact melt) accumulates on flanks of uplift and in peripheral depression. **Final form** - central uplift collapses to below original surface. Outer rim is produced by downfaulting, further deepening the peripheral depression. Limit of intense brecciation and faulting shown by **dashed envelope**

ward (Fig. 2). Although not fully understood, this extensive modification appears to be initiated under plastic failure conditions, during which the rocks have little or no yield strength and behave hydrodynamically. Various suggestions for producing the physical conditions required for this hydrodynamic response of the target rocks include: crushing, shearing, frictional melting between blocks and acoustic fluidization. The driving mechanism for uplift is generally considered to be elastic rebound of the transient cavity floor, in combination with the release of gravitational potential energy from the original overheightened transient cavity rim area. Further details of the formation of complex crater-forms can be found in papers in Schultz and Merrill (1981).

It is sufficient to note that in the case of a structure such as Siljan, the central Dala granites represent the eroded remnant of the uplifted transient cavity floor and that the position of the Paleozoic sediments was outside the original area of excavation, i.e., the transient cavity, and they owe their preservation to down-faulting accompanying the collapse of the transient cavity rim area during modification. The geologic and structural characteristics of Siljan are consistent with those of other large impact structures, and constraints on the Siljan impact event and the original parameters of the structure can be drawn from our present understanding of large-scale terrestrial impact events. A number of such inferences are given below.

Energy Estimate and Extent of Shock Heating at Siljan

Kinetic energy-diameter scaling relationships for the transient cavity at simple craters are the subject of a recent review by Holsapple and Schmidt (1982). Their scaling relationships, however, require information on impact conditions. As with most structures, the impact parameters are virtually unknown for the Siljan event. The problems with determining precise scaling relations are compounded considerably when considering complex structures in the size range of tens of kilometers, which have undergone a good deal of post-excavation modification. One type of scaling relation for complex structures has the form:

$$D = F\, E^{1/\alpha}, \qquad (1)$$

where D is final diameter, F is a function of parameters such as impact velocity, target and projectile density, gravity, and target strength. E is kinetic energy and α is an exponent with a range of from 3.4 to 4.0.

Based on a number of observations, Dence et al. (1977) determined an empirical scaling relation for complex terrestrial craters in crystalline targets where

$$D = 1.96 \times 10^{-5}\, E^{1/3.4}, \qquad (2)$$

and D is in km and E in Joules. For a diameter of 52 km, this relation gives an estimated kinetic energy of 1.4×10^{22}J for

Siljan, approximately an order of magnitude greater energy than the annual output of internal energy of the earth. This should be regarded as a minimum estimate, as the Dence at al. (1977) relation does not fully account for energy losses accompanying the formation of superheated vapor and melt (Grieve and Cintala 1981).

Although it is theoretically possible to differentiate the velocity and mass components of such an energy estimate (Grieve and Cintala 1981), the basic observational data are lacking for Siljan. As impact velocity, projectile type and projectile size control peak pressures, temperatures and the radial distance to a given shock isobar or post-shock isotherm, only general statements can be made about the post-impact pressure-temperature distribution at Siljan. For illustration, it has been assumed that the Siljan event was the result of the impact of a chondritic body impacting granite at 25 km/s, the r.m.s. impact velocity of earth-crossing Apollo bodies (Shoemaker 1977). For a spherical body with a density of 3000 kg/m^3, the diameter of the projectile would be on the order of 2-3 km. Energy partitioning calculations for a chondritic projectile impacting granite at 25 km/s indicate that peak pressures at the point of impact will be on the order of 8 megabars, and 65% of the projectile kinetic energy will be partitioned into waste heat in the target (Cintala and Grieve 1984).

Most of this waste heat is concentrated close to the point of impact, where model calculations for a 2-3 km body suggest that it resulted in the melting and vaporization of approximately 250 km^3 of the target rocks at Siljan. Although some of the melted and vaporized material will be ejected from the growing cavity by the cratering flow-field, some will remain to form an annular melt sheet in the center of the final crater (Grieve et al. 1977). This sheet, with an initial temperature on the order of 2200°C, may have originally been ~200 m thick at Siljan and essentially covered the central area of the final structure and extended out to the area of the Paleozoic sediments. There is insufficient data, however, to determine whether it originally covered some of the Paleozoic sediments and produced local thermal metamorphic effects. Such a thickness of impact melt would require on the order of 1000-2000 years to crystallize completely (Onorato et al. 1978) and perhaps 10,000 years to cool completely to ambient temperature.

Post-shock temperatures associated with release from shock compression attenuate relatively rapidly from the point of impact (Grieve and Cintala 1981; Cintala 1984). For a granitic target, material shocked above approximately 50 GPa will have sufficient post-shock heat to be melted and/or vaporized. All this material, as well as some slightly lower shocked material, is set in motion during crater formation and ultimately constitutes allochtonous deposits in and around the crater. Peak pressures recorded in rocks in the autochthonous target rocks at the center of central uplifts at terrestrial complex craters are generally in the 25-30 GPa range, corresponding to post-shock temperatures of only 100-200°C. At the transient cavity rim, now corresponding approximately to the area of Paleozoic sediments at Siljan, pressures had attenuated to the 0.5-1.0 GPa range, with little or no associated post-shock heating.

Estimates of Extent of Excavation and Uplift

The depth of excavation at complex structures is the subject of some debate (see papers in Schultz and Merrill 1981). There is, however, a general consensus that it is not proportionally greater than that experienced in smaller simple craters. Data from code-calculations of nuclear explosions and the results from experimental impacts place the maximum depth of excavation, d_e, for simple craters at:

$$d_e = 0.1\ D_e, \tag{3}$$

where D_e is the diameter of the excavated cavity and is equivalent to the diameter of the transient cavity at the original

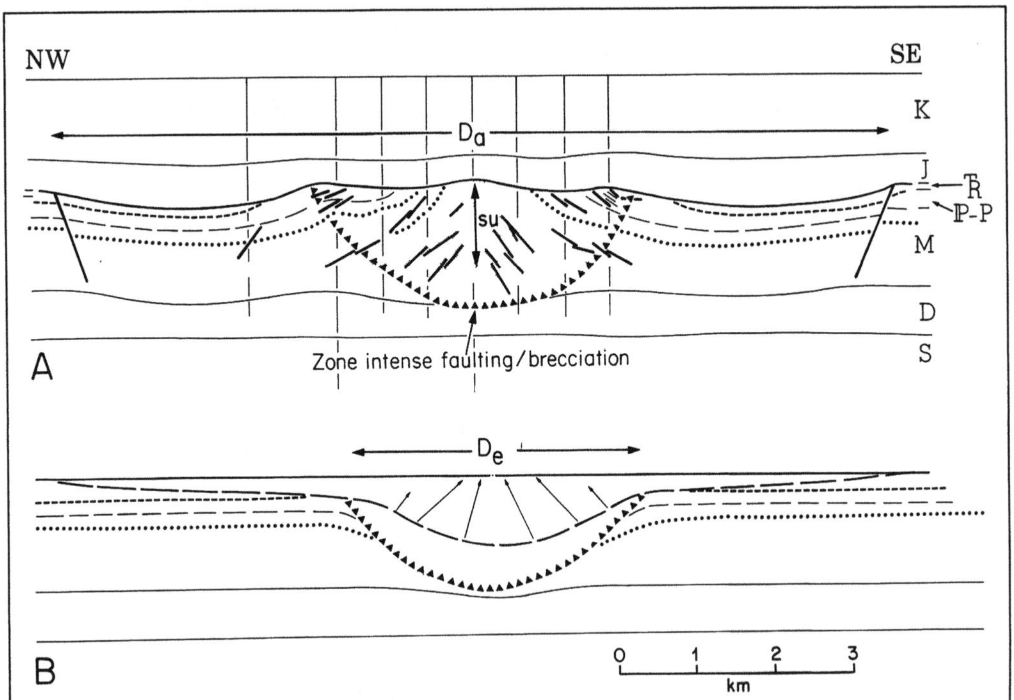

Fig. 3. Red Wing Creek, N. Dakota - an example of a complex impact structure in sedimentary rocks. A cross-section of present morphology and structure from drilling and seismic data, illustrates stratigraphic uplift (SU) of Mississippian and disturbance as deep as the Devonian (D). The margin of the final crater (D_a) is marked by steep, normal faults. Structure is largely preserved under Jurassic (J) and Cretaceous (K) post-crater sediments. Reconstruction of the excavated cavity indicates deep central excavation within D_e and shallow removal of strata to D_a. (Grieve et al. 1981)

ground surface (Croft 1980, Grieve and Garvin 1984). Due to the considerable amount of post-excavation modification, however, some uncertainty is involved in estimating D_e at complex structures. From stratigraphic data at complex structures in sedimentary targets (Fig. 3), the final spatial distribution of near-surface sediments (such as the Paleozoic sediments at Siljan) and the limit of diagnostic shock effects, Grieve et al. (1981) estimated that:

$$D_e = 0.5 - 0.65\ D, \qquad (4)$$

where D is the final rim diameter at complex structures. A more recent compilation of data by Croft (1985), which includes morphometric data from lunar craters, suggests the following empirical relationship for terrestrial complex structures formed in crystalline targets:

$$D_e = 1.23\ D^{0.85}. \qquad (5)$$

For Siljan with D = 52 km, relation (4) results in a D_e estimate of 26-34 km, whereas relation (5) results in 35 km. On the basis of Eq. (3), this gives an estimated maximum depth of excavation for Siljan of 2.6-3.5 km.

An alternative measure of the depth of excavation is the amount of stratigraphic uplift undergone by the central rocks (Fig. 3). This is based on the argument that, at least in fresh craters, higher stratigraphic units now absent from the central uplift were removed largely by excavation. Observational data on the amount of stratigraphic uplift, which is defined as the observed amount of uplift undergone by the deepest marker horizon now exposed at the center of a complex structure, at a number of terrestrial structures results in the relationship:

$$SU = 0.06D^{1.1}, \qquad (6)$$

where SU is stratigraphic uplift and D is final rim diameter (Fig. 4; Grieve et al. 1981). The calculated value for SU at Siljan is, therefore, 4.6 km. As a depth of excavation estimate, this is somewhat in excess of that determined above. Given the uncertainty in the data used to define the above empirical relations, however, this is not considered to be a major discrepancy.

There are few large impact structures which have sufficient stratigraphy and minimal erosion to provide a direct measure of maximum excavation depths. These figures are, however, consistent with observations at the 22 km diameter Haughton impact structure. This 21.5 Ma-old structure is formed in 1.6 km of Paleozoic sediments overlying Precambrian crystalline basement. The ejecta is, in part, preserved and contains clasts of Precambrian rocks, indicating excavation to at least 1.6 km. Scaling this excavation diameter to depth ratio to 52 km gives an estimated excavation depth of 3.8 km at Siljan.

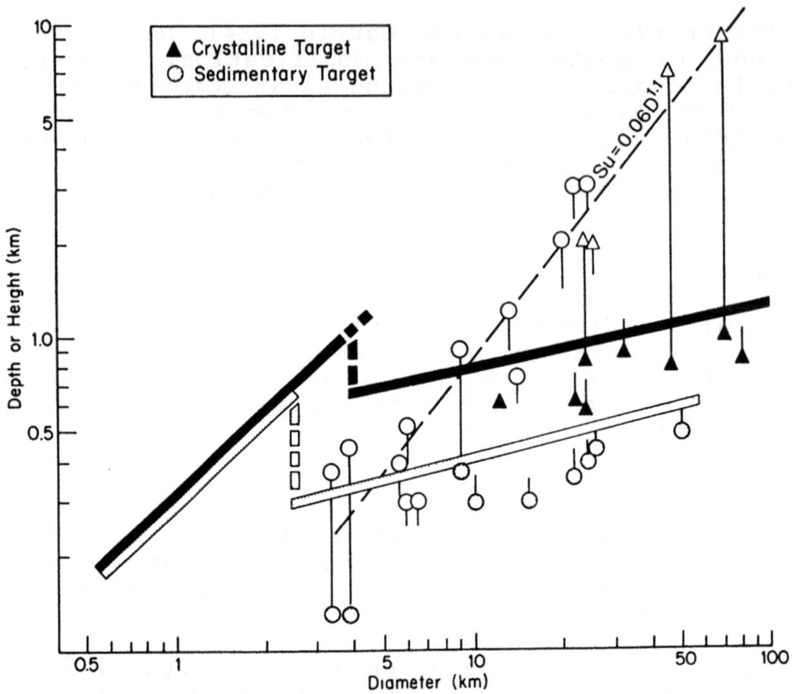

Fig. 4. Depth/diameter relationships (**solid lines**) and structural uplift (**broken line**) at terrestrial craters. Structures in sedimentary targets develop complex form at smaller diameters than those in crystalline targets, reflecting the physical properties of target. Physical heights of central uplifts or rings (**closed symbols**) never exceed final crater depth, indicating final height is an equilibrium feature. Amount of structural uplift (**open symbols**) increases with diameter

Extent of Fracturing

Considerable brecciation and fracturing accompany crater formation and are generally reflected in the occurrence of a gravity low over impact structures. Although often complicated by regional gradients and local density contrasts, the common gravity signature of both terrestrial and lunar complex impact structures is an outer negative ring, beginning in the general area of the original rim area, which then grades upwards in value towards the center (Fig. 5; Dvorak and Phillips 1977; Sweeney 1978; Barlow 1979). For example, the peripheral gravity low at the Manicouagan structure can be modeled with a density contrast of -130 kg/m^3, which corresponds to a fracture porosity of 5-10%, depending on whether the fractures are water-filled or not. By comparison, the central uplift area has little or no density contrast, with a residual anomaly of 0 mGal (Fig. 5; Sweeney 1978).

This outer negative ring can be attributed to increased fracture porosity in the area of downfaulting in the rim area. The central portion of the residual gravity field over complex

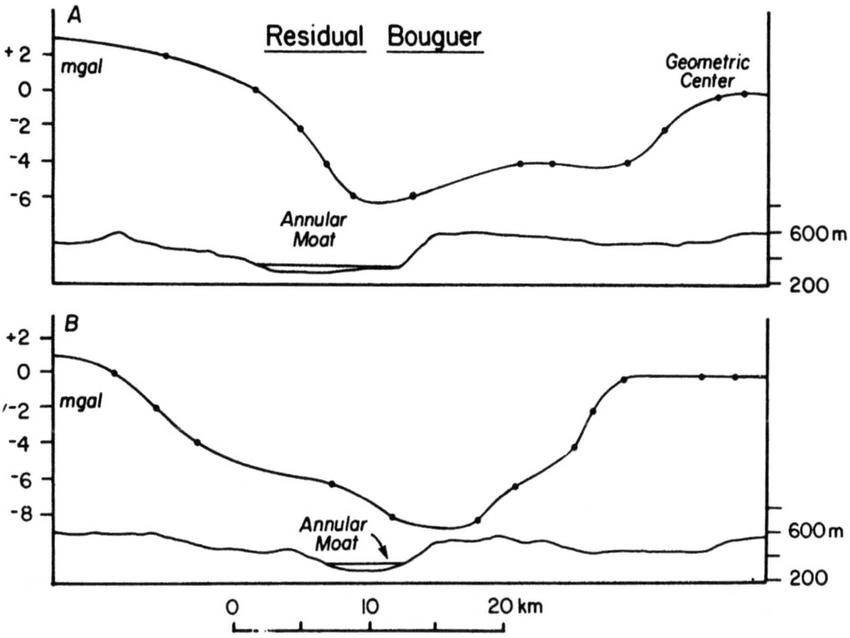

Fig. 5. Residual Bouguer gravity profiles, after removal of regional field, associated with the Manicouagan structure. Profiles constructed from Fig. 5 in Sweeney (1978). The minimum gravity values are in the area of the annular moat, which is structurally equivalent to the area of Paleozoic sediments at Siljan

structures is still generally negative with respect to the regional field, also indicating increased fracturing. The fact that it is higher than the outer negative ring can be best explained by a relative reduction in fracture porosity due to the nature of the uplift process, where the trajectories are upwards and inwards. This results in a relatively compressive regime, with closing of some of the fractures, in the final central uplifted volume.

The residual gravity signature of Siljan (Vattenfall 1985) is apparently more complex than that at many other complex structures. The presence of a gravity low in the east and south, in the area between Boda and Rattvik, appears consistent, however, with the above generalizations. The concentric -20 mGal low in the north is centered over Skattungen in the area of the downdropped sediments (Fig. 1). Its concentric nature, however, suggests that its ultimate cause goes beyond simple impact-induced fracture porosity and may be related to subsurface rock-type contrasts. It may well reflect the occurrence of the northerly Siljan granite, which is relatively less dense, 2604 kg/m^3, compared to other granites in the area, 2668 kg/m^3.

A zone or halo of increased fracturing can be observed surrounding some terrestrial craters. The fracture zone can be

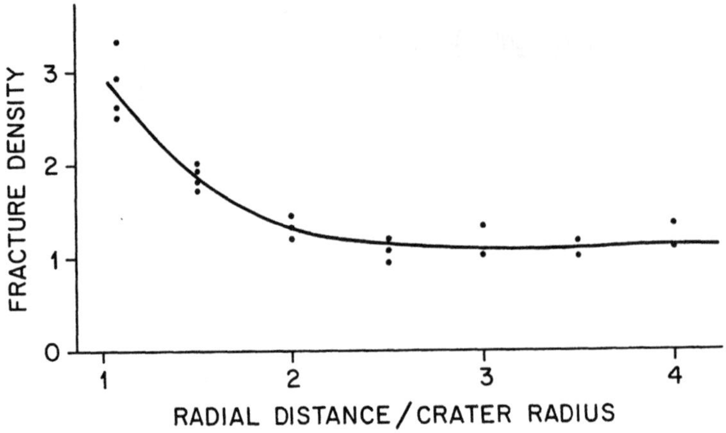

Fig. 6. Change in relative fracture density beyond the crater rim against normalized radial distance at the Elgygytgyn structure. (After Gurov and Gurova 1982)

considered to be that volume beyond the actual crater that was stressed above the inherent dynamic strength of the target rocks. One of the few quantitative analyses of fracture density around craters has been undertaken by Gurov and Gurova (1982). They indicate that increased fracturing extends for approximately one crater diameter beyond the relatively fresh 23 km diameter, 3.5 ± 0.5-Ma-old Elgygytgyn structure in Siberia. They suggest that fracturing decays as approximately R^{-3}, where R is radial distance from the crater (Fig. 6). Similar-sized fracture haloes can be observed on LANDSAT images of more deeply eroded structures such as Manicouagan (Dence 1977; Grieve and Head 1983) and Clearwater (Dence et al. 1977) in the Canadian Shield. In the case of Siljan, therefore, increased fracturing is expected to have occurred for a distance of approximately 50 km beyond the original rim. Such fracturing is limited to the near-surface and is generally insufficient to produce a recognizable gravity signature.

Drilling and reflection seismic profiles indicate that major structural disturbance beneath complex impact structures extends in the center to a maximum depth of approximately 1.5SU (Fig. 3; Brennan et al. 1975; Brown 1973). It involves a roughly paraboloid volume extending upwards, as expected, to the surface limit of obvious structural uplift and modification. At Siljan, therefore, the depth of major structural disturbance is expected to have originally extended to a depth of approximately 7 km. Gravity modeling at Siljan suggests a maximum depth for major disturbance and fracturing of approximately 5 km (Vattenfall 1985). This depth, however, is highly dependent on the chosen density contrast in the model; in this case, -120 kg/m^3, a value similar to that for Manicouagan. A lower density contrast would increase this depth. It is important to note that gravity models have sharp boundaries separating volumes of contrasting density. In the case of density contrasts due largely to porosity changes from impact-induced fracturing, this

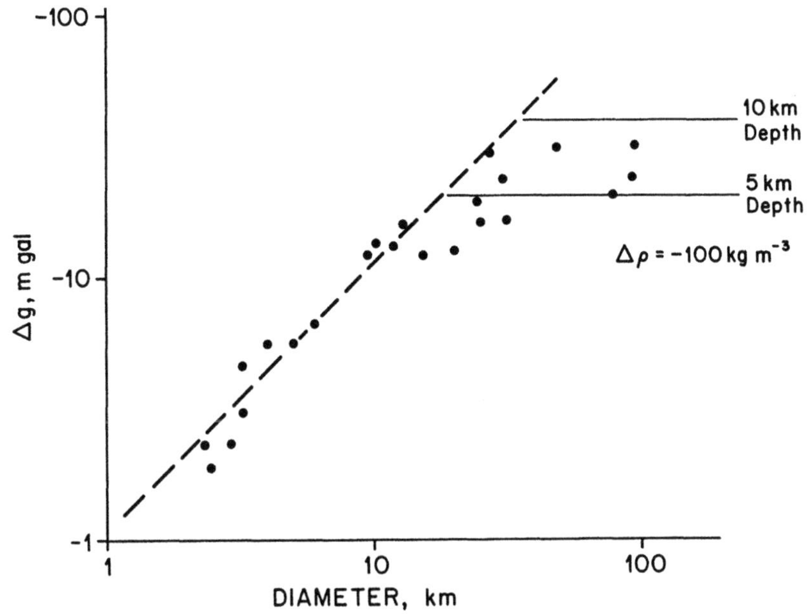

Fig. 7. Variation in residual Bouguer gravity anomaly at terrestrial impact structures with diameter. Note scales are logarithmic and $-\Delta g$ increases with diameter until ~20-30 km, where it stablises. For a fracture-induced density contrast of -100 kg/m^3, maximum gravity anomalies correspond to fracturing to 5-10 km depth

is unlikely to be the case. Boundaries will be gradual, with fracture density decreasing gradually with distance.

A plot of the gravity anomaly against diameter of impact structures shows a decrease in gravity, $-\Delta g$, with diameter (Fig. 7). The anomaly, however, does not increase beyond certain limits and stablilizes at -20 to -30 mGal (Dabizha and Fedynsky 1977). These values correspond to model depths of 5-10 km for major structural disturbance and fracturing, using a density contrast of approximately -100 kg/m^3. At these depths, the lithostatic load is a few kilobars, sufficient to close open fractures.

During the impact event itself, however, tensile failure may theoretically extend down to the depth where the lithostatic pressure exceeds the strength of the rarefaction wave (Grieve and Robertson 1984). The dynamic tensile strength of crustal rocks is low and rarely exceeds 0.04 GPa (Grady and Hollenbach 1979). A first-order calculation, based on a 2-km chondritic projectile impacting granite at 25 km/s and the shock wave attenuation model of Cintala (1984), indicates that impact-induced fracturing in the center of Siljan could occur to depths on the order of 40 km (Fig. 8). In the calculation, it was assumed that the tensile stress associated with the rare-

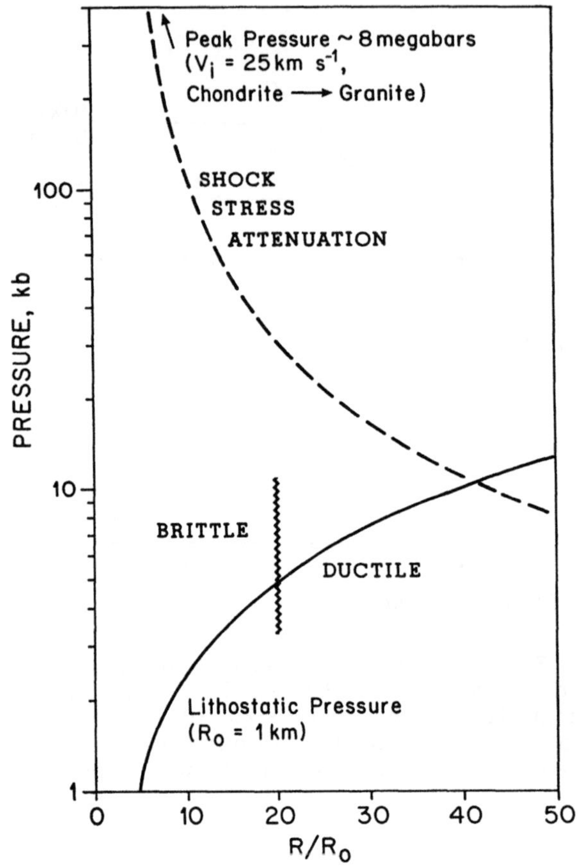

Fig. 8. Pressure against normalized radial distance, where R_o is projectile radius. Shock attenuation stress is calculated for a chondritic body impacting granite at 25 km/s (V_i) and corresponds to the pressure values axially below the impact. The lithostatic pressure was calculated and scaled for R_o = 1 km. Note that the intersection of the shock stress and lithostatic pressure curves occurs at depths where ductile behavior is expected under normal tectonic strain rates

faction wave was equal in magnitude to that of the shock wave at any given radial distance. Thus, the Siljan impact event had the potential to initially fracture, at least in the center, the entire crustal column.

There are two possible problems with this hypothesis. Firstly, there is considerable geophysical evidence that the lower continental crust behaves as ductile material, unlike the brittle upper crust (Fig. 8; Kusnir and Park 1984; Meissner and Strehlau 1982). The gradual build-up of strain in tectonism is released by creep in the lower crust and by brittle failure in the upper crust. Material behavior, however, is dependent on strain rate, and impact stresses are essentially instantaneous, being imposed 10^{20} times faster than in tectonic processes. Thus, under the very rapid strain rates associated with impact, it is possible that the lower crust may behave in a brittle fashion.

Secondly, the gravity data over craters indicate that many of the initial fractures are closed by compaction and may be healed due to later mineral precipitation. This is consistent with the assumption that pore space is virtually eliminated at depths greater than 8 km due to lithostatic pressure (Perrier and Quiblier 1974). The recent results of the Kola drill hole, however, indicate that circulating fluids in fracture zones exist at greater depths (Koglovsky 1982). Therefore, whether or not some of the initial, deep crustal fracturing beneath Siljan could remain open due to circulating fluids is an unanswered question. Deep faulting by transient high stress and circulating fluids has been suggested as a possible hypothesis to explain vertical geomagnetic induction vectors (Bailey et al. 1974) at the 46-km Charlevoix impact structure (Jones 1987).

Erosion Estimate

The original topographic relief of the Siljan structure has been removed by erosion and the sub-structure of the crater floor exposed. Although there is considerable scatter in the depth-diameter relationship for terrestrial complex structures in crystalline targets, it appears to conform to:

$$d = 0.5D^{0.3}, \tag{7}$$

where d is original, pre-erosional depth to the autochthonous rocks of the crater floor (Grieve 1984). Thus, approximately 1.5 km of erosion is required to remove the topographic rim and expose the crater floor at Siljan. It does not necessarily mean, however, that 1.5 km of erosion has occurred throughout the entire Siljan area, e.g., in the area of the down-dropped Paleozoic sediments. The absence of any expression of a central positive topographic feature such as a peak and/or rings, as well as the removal of any annular melt sheet, suggest at least 1 km of erosion in the center (Fig. 4). It should be noted, however, that the original physical heights of central features (Fig. 4) are even less well constrained than depth estimates. This estimate of 1 km is a minimum, as the peak recorded pressures in the autochthonous rocks exposed in the center of Siljan are on the order of 12 GPa, based on the occurrence of $(10\bar{1}3)$ planar features in quartz grains (Svensson 1973). This is lower than the expected 25-30 GPa at the center of terrestrial complex structures, and is indicative of erosion at Siljan to below the original autochthonous crater floor.

Spatial Relation of Gravberg Drill Hole to the Structure

The Gravberg drill hole is located ~14 km from the geometric center of the Siljan structure. Experience with recorded shock distribution and shatter cone orientations suggests that the center of the Siljan structure is expected to be at or near the geometric center of the annulus formed by the Paleozoic sediments. The suggestion by Wickman (1980) that the center is located 4-9 km north of the geometric center is based on a data set of only four shatter cone localities, with three of the localities lying within 90°C of arc. In addition, shatter cone orientations at locality 2 of Wickman (1980) differ from those reported from the same locality by Svensson and Robertson (pers. commun. in Wickman 1980). If the Svensson and Robertson orientations for locality 2 are used in combination with Wickman's other data, the shatter cone "center" occurs ~5 km south of that suggested by Wickman (1980) and is close to the geometric center. Given the limited data set and the complex nature of block movements during modification in the center of complex structures, there is insufficient evidence to call for an exceptionally off-center location in the case of Siljan.

A number of observations suggest that the Gravberg drill hole is located near or outside the structural trace of the original zone of excavation. Firstly, empirical relations presented earlier suggest that the pre-erosional diameter of the excavated cavity at Siljan was in the range 26-34 km. Erosion will reduce the surface expression of the originally excavated zone, which is now filled and expressed as uplifted and structurally disturbed material. Erosion estimates of 1.5 km (this work) to 2 km (B. Collini pers. commun. 1986) suggest that the present expression of the excavated diameter would be reduced by ~4 km, i.e., will be 22-30 km. Taking the geometric center as the impact point places Gravberg at a diameter of 28 km, which is greater than the lower estimate and close to the higher estimate for the present expression of the diameter of the excavated cavity. It is important to note that these figures are from empirical relations, based on observations at a number of impact structures, and depend upon knowledge or an estimate of the original diameter. Thus, specific values calculated for a particular structure may be in slight error. In the case of Siljan, however, there are additional observational data that can be interpreted in favor of Gravberg lying exterior to the original excavated cavity.

Near-surface rocks are removed by ejection from the excavated cavity in an impact event. In the case of Siljan, the target at the time of impact consisted of Precambrian granites, gneisses and other rocks overlain by a relatively thin sequence of Ordovician and Silurian sediments. These sediments are now missing from the central area of Siljan but are preserved as a downfaulted annulus within the peripheral valley. They also occur as isolated outcrops closer to the center. The closest outcrops are ~13 km from the center. Thus, they confine the limit of near-surface excavation to a diameter of ~26 km, i.e., inside that of the Gravberg drill hole. Some caution, however, must be exercised, as the closest outcrops of sediments may have been structurally displaced inwards during cavity modification and uplift (Bischoff and Oskierski 1985). The figure, however, is consistent with that derived from dimensional relationships.

Seismic investigations along a N-S line through Gravberg indicate a series of horizontal reflectors. Their presence, absence and geometry provide information on the subsurface nature of the Siljan structure. The major reflector closest to the surface disappears just south of Gravberg and, with increasing depth, the other reflectors disappear at progressively greater distances to the south. The southerly termination of the reflectors thus defines an arc with increasing depth to the south (Juhlin and Pedersen 1987). The geometry is very similar to that shown in Fig. 3 for the limit of major structural disturbance associated with modification of the excavated and transient cavities during central uplift formation. If radially symmetric, this geometry defines the present expression of the excavated cavity as 22 km in diameter. Very deep reflectors continue south with less disruption and rise in the center of the structure. These may define the trace of uplift in the center. It is estimated at ~3 km, which is close to that based on previous empirical relations, 4.6 km, after correction for erosion. If these reflectors are pre-impact, their presence at Gravberg, but absence to the south, suggests that Gravberg lies outside the area of major modification, i.e., outside the excavated cavity.

Based on the above, it is believed that the Gravberg drill hole is located near or, more likely, just outside the limit of the excavated cavity and zone of major uplift and structural disturbance at Siljan. It would, therefore, lie in the inner portion of the so-called peripheral trough. Here the rocks have suffered some block displacements but have not undergone the major disturbance associated with central uplift formation.

Summary of Constraints

The preceding discussion has presented a number of observations regarding the original and present nature of the Siljan structure. Some of these observations are summarized below.

1. Initial impact conditions are unknown. A reasonable assumption, however, is that the structure was the result of the impact of an asteroidal-sized body in the size-range of a few kilometers. Impact energy was in range of 10^{22}J or greater and accompanying peak pressures were in the megabar range with temperatures close to the impact point in excess of the vaporization temperature of silicates.

2. Excavation was to a depth of several kilometers (~2.5-4.5 km), sufficient to penetrate through the Paleozoic sedimentary cover into the underlying Precambrian granites. Deep excavation, however, was limited to the central area, approximately 25-35 km in diameter, and now occupied by uplifted Dala granites.

3. Following excavation and large-scale modification associated with the impact, the final Siljan structure is estimated to have had a diameter of 52 km. It had a complex form with a central topographic peak and/or rings and a complexly faulted rim area. The preserved Paleozoic sediments were outside the area of deep excavation and represent down-faulted material from the interior of the final rim-area. The post-modification depth of the original structure may have been on the order of 1.5 km and the central structure(s) may have formed topographic peaks on the order of 1 km in relief. These would have been surrounded by an annular sheet of impact melt rocks and allochthonous breccias, possibly up to 200 m thick, extending out to the area of the Paleozoic sediments. Shock pressures recorded in the uplifted Dala granites were 25-30 GPa.

Post-impact shock heating in the uplifted center was 100-200°C. This was in addition to the geothermal temperature of 75°C associated with the uplift of the central granites from an original, pre-impact depth of 4-5 km. The post-shock temperature attenuated rapidly away from the center and was probably insignificant in the area of the area of the Paleozoic sediments. Post-shock temperatures and those associated with the virtually instantaneous uplift of deep-seated rocks were supplemented by a thermal event associated with conductive cooling of the overlying melt. This thermal metamorphism was relatively local and confined to the volume of rock essentially in contact with the melt sheet.

5. Fracturing associated with the impact event was extensive, extending out to possibly 50 km beyond the original rim of Siljan. Major structural disturbances in the center are likely to have extended to depths of 7 km. If the lower crust behaves in a brittle fashion at very high strain rates, there is the potential for increased, initial fracturing to lower crustal depths of 40 km. Many of these initial fractures may have been closed almost immediately by normal lithostatic pressure. In addition, many of the upper crustal fractures in the central area would have been closed by the inwards and upwards movement of material during the formation of the central uplift. This would have been supplemented by fracturing-healing due to hydrothermal deposition of minerals in the elevated temperature regime of the central uplift. Fractures and faults in the rim-area were not as deep-seated as in the center but are likely to have remained relatively open due to their location in an area to tension and low post-shock heating.

6. Erosion has since removed much of the topographic expression of Siljan, exposing and removing some of the original crater floor. Erosion in the area of the original rim and central structure(s) is estimated to have been on the order of 1.5 and 1.0 km, respectively. The extent of erosion of the Paleozoic sediments may have been less, as they were initially in a relatively low topographic area.

7. Based on various lines of evidence, it is believed that the Gravberg drill hole is located slightly outside the present expression of the excavated and transient cavities and the zone of major structural disturbance and uplift. This corresponds to the inner margin of the area which underwent largely block displacements during cavity modification. If this relative location is correct, sub-surface geometry dictates that the Gravberg hole would not intersect the highly disturbed zone at depth.

Acknowledgments. I would like to thank T. Lindbo of Vattenfall for the opportunity to visit Siljan and discuss its nature with those involved in the Deep Gas Project. Special thanks are extended to B. Collini for sharing his knowledge of Siljan.
Geological Survey of Canada Contribution 27887

References

Bailey RC, Edwards RN, Garland GD, Kurtz R, Pitcher DH (1974) Electrical conductivity studies over a tectonically active area in eastern Canada. J Geomagn Geoelectr 26:125-146

Barlow BC (1979) Gravity investigations at the Gosses Bluff impact structure, Central Australia. Bur Min Res Geol Geophys 4:323-339

Bischoff L, Oskierski W (1985) First results of a multidisciplinary analysis of the Haughton crater, Devon Island, Canada. Lunar Planet Sci XVI:65-66

Bottomley RJ, York D, Grieve RAF (1978) ^{40}Ar - ^{39}Ar ages of Scandinavian impact structures: I Mien and Siljan. Contrib Mineral Petrol 68:79-84

Brennan RL, Peterson BL, Smith HJ (1975) The origin of Red Wing Creek structure, McKenzie County, N.Dakota. Wyo Geol Assoc Earth Sci Bull 8:1-41

Brown AR (1973) A detailed seismic study of Gosses Bluff, Northern Territory. Bur Min Res Geol Geophys Rep 163:1-43

Cintala MJ (1984) A method for evaluating shock propagation and its thermal effects during impact events. Lunar Planet Sci XV:154-155

Cintala MJ, Grieve RAF (1984) Energy partitioning during terrestrial impact events: Melt production and scaling laws. Lunar Planet Sci XV:156-157

Croft SK (1980) Cratering flow fields: Implications for the excavation and transient expansion stages of crater formation. Proc Lunar Planet Sci Conf 11:2347-2378

Croft SK (1985) The scaling of complex craters. J Geophys Res 90:C828-C842

Dabizha AI, Fedynsky VV (1977) The special features of the gravitational filed of astroblemes. Meteoritika 36:113-119 (in Russian)

Dence MR (1977) The Manicouagan impact structure, Quebec, from Skylab. In: Skylab explores the Earth. NASA SP-380, Washington, p 175

Dence MR, Grieve RAF, Robertson PB (1977) Terrestrial impact structures: Principal characteristics and energy considerations. In: Roddy DJ, Pepin RO, Merrill RB (eds) Impact and explosion cratering. Pergamon New York, p 247

Dvorak J, Phillips RJ (1977) The nature of gravity anomalies associated with young large lunar craters. Geophys Res Lett 4:380-382

Fredriksson K, Wickman FE (1963) Meteoriter. Sven Nat 16:121-157

French BM, Short NM (eds) (1968) Shock metamorphism of natural materials. Mono, Baltimore, p 644

Gault DE, Quaide WL, Oberbeck VR (1968) Impact cratering mechanics and structures. In: French BM, Short NM (eds) Shock metamorphism of natural materials. Mono, Baltimore, p 87

Grady DE, Hollenbach RE (1979) Dynamic fracture strength of rock. Geophys Res Lett 6:73-76

Grieve RAF (1982) The record of impact on earth: Implications for a major Cretaceous/Tertiary event. Geol Soc Am Spec Pap 190:25-38

Grieve RAF (1984) The impact cratering rate in recent times. J Geophys Res 89:B403-B408

Grieve RAF (1987) Terrestrial impact structures. Annu Rev Earth Planet Sci 15:245-270

Grieve RAF, Cintala MJ (1981) A method for estimating the initial impact conditions of terrestrial cratering events, exemplified by its application to Brent crater, Ontario. Proc Lunar Planet Sci Conf 12B:1607-1621

Grieve RAF, Garvin JB (1984) A geometric model for excavation and modification at terrestrial simple impact craters. J Geophys Res 89:11561-11572

Grieve RAF, Head JW (1983) The Manicouagan impact stucture: An analysis of its original dimensions and form. J Geophys Res 88:A807-A818

Grieve RAF, Robertson PB (1984) The potential for the disturbance of a buried nuclear waste vault by large-scale meteorite impact. Proc workshop on transitional processes. At Energy Can Ltd AECL Rep 7822:231-260

Grieve RAF, Robertson PB (1987) Terrestrial impact structures. Geol Surv Can Map 1658A

Grieve RAF, Dence MR, Robertson PB (1977) Cratering processes as interpreted from the occurence of impact melts. In: Roddy DJ, Repin RO, Merrill RB (eds) Impact and explosion cratering. Pergamon, New York, p 791

Grieve RAF, Robertson PB, Dence MR (1981) Constraints on the formation of ring impact structures, based on terrestrial data. In: Schultz PH, Merrill RB (eds) Proc Conf Multi-Ring Basins. Pergamon, New York, p 37

Gurov EP, Gurova EP (1982) Some regularities of the areal spreading of fractures around Elgygytgyn crater. Lunar Planet Sci XIII:291-292

Holsapple KA, Schmidt RA (1982) On the scaling of crater dimensions 2, impact processes. J Geophys Res 87:1849-1870

Jones AG (1987) Are impact-generated lower-crustal faults observable? Earth Planet Sci Lett 85:248-252

Juhlin C, Pedersen LB (1988) Reflection seismic investigations of the Siljan impact structure. J Geophys Res (in press)

Koglovsky YA (1982) Kola super-deep: Interim results and prospects. Episodes 4:9-12

Kusnir NJ, Park NG (1984) Intraplate lithosphere deformation and the strength of the lithosphere. Geophys J R Astron Soc 79:513-538

Masaitis VL, Danilin AN, Mashak MS, Raichlin AI, Selivanvskaya TV, Shadyenkov YM (1980) The Geology of Astroblemes. Nedra, Moscow, p 331 (in Russian)

Meissner R, Strehlau J (1982) Limits of stresses in continental crusts and their relation to the depth-frequency distribution of shallow earthquakes. Tectonics 1:73-89

Onorato PIK, Uhlmann DR, Simmonds CH (1978) The thermal history of the Manicouagan impact melt sheet, Quebec. J Geophys Res 83:2789-2798

Perrier R, Quiblier J (1974) Thickness changes in sedimentary layers during compaction history: Methods for quantative evaluation. Am Assoc Petrol Geol Bull 58:507-517

Roddy DJ, Pepin RO, Merrill RB (eds) (1977) Impact and explosion cratering. Pergamon, New York, p 1301

Rondot J (1975) Comparaison entre les astroblèmes de Siljan, Suède, et de Charlevoix, Québec. Bull Geol Inst Univ Uppsala 6:85-92

Schultz PH, Merrill RB (eds) (1981) Proc Conf Multi-Ring Basins. Pergamon, New York, p 295

Shoemaker EM (1977) Astronomically observable crater-forming projectiles. In: Roddy DJ, Pepin RO, Merrill RB (eds) Impact and explosion cratering. Pergamon, New York, p 617

Svensson NB (1971) Probable meteorite impact crater in central Sweden. Nature (London) 229:90-92

Svensson NB (1973) Shatter cones from the Siljan structure, central Sweden. Geol Foeren Stockholm Foerh 95:137-143

Sweeney JF (1978) Gravity study of great impact. J Geophys Res 83:2809-2815

Vattenfall (1985) The Siljan ring structure investigations and evaluations. Gas Project G 2, Rep No 4

Wickman FE (1980) The impact center of the Siljan structure, central Sweden. Geol Foeren Stockholm Foerh 102:105-109

Geological Setting of the Siljan Ring Structure

B. Collini[1]

Origin

The suggestion that the Siljan Ring structure was caused by a meteorite impact was first launched by Wickman et al. (1963) and Fredriksson and Wickman (1963). Pioneer works also include the findings of shocked quartz by Svensson (1971) and shatter cones by Svensson, Robertson and Chao (Svensson 1973).

The impact conception is now generally accepted. The interpretation of the structure as due to "normal" faulting activity has not found much support. The high-pressure features of the area are too difficult to explain along such lines.

An alternative approach of current interest is a cryptovolcanic origin of the Siljan Ring structure being debated for other ring-shaped structures of the world. Earlier, variants of this conception were discussed by Asklund (1936), Thorslund (1961) and Rutten (1966). In the Abstracts volume of this Symposium, a number of crucial points in the choice between impact and cryptovolcanic origin of the Siljan Ring structure are pointed out by Nicolaysen and Duane (1987). In the author's opinion, a decisive observation is obtained from the seismic north-south profile through the area. The central part of the structure is closed downward by a nearly unbroken subhorizontal reflector at 13-16 km depth. This fact is not consistent with a cryptovolcanic origin.

Bedrock

The impact area is situated on a very complicated bedrock, in a boundary zone between old bedrock (oldest supracrustals, prim- and ser-orogenic intrusions) in the east and younger bedrock (porphyries, Digerberg beds) in the west (Table 1). For detailed information the reader is referred to Hjelmqvist (1966).

The central part of the structure, surrounded by Palaeozoic sedimentary rocks, consists of Dala granites, the felsic Siljan

---

[1] Prof. em., Department of Mineralogy and Petrology, Uppsala University, Box 555, S-751 22 Uppsala, Sweden

Table 1. Chronology of bedrock units of the Siljan Ring structure and surrounding areas

Age, Ma	Bedrock
Postorogenic and anorogenic complexes and events (as regards the Svecokarelian orogeny)	
361.9 ± 1.1 (349 ± 2; 342 ± 3)	Siljan meteorite impact
410 - 440	Silurian
440 - 500	Ordovician
(590 - 670)	Vendian (?) sandstone
926 ± 13 (surface); 906 ± 6 (1500 m)	Subhorizontal dolerites (Gravberg)
900 - 970	Dolerites (NNW-SSE)
ca. 1220 (1215 - 1290)	Åsby dolerites
	Dala sandstone
1370	Tuna dolerites (NNE-SSW)
1635	Dala porphyries
1646 - 1670	"Porphyrite" (W of Gulleråsen)
1675 - 1735	Dala granites (Siljan and Järna granite; diorite)
	Digerberg beds. Red porphyrite
Serorogenic intrusions	
1750 - 1800	Fellingsbro granite. Migmatites
Primorogenic intrusions	
>1845	Gneiss-granites
Oldest supracrustals	
>1880	Leksand quartzite. Leptites (acid and basic metavolcanics)

Compiled after Åberg and Bollmark 1983, Åberg and Collini 1987, Åberg and Fredriksson 1984, Åberg (personal communication), Bottomley et al. 1978, Hjelmqvist 1966, Kresten 1986, Lundqvist 1979, Patchett 1978.

granite in the northern half and the more basic Järna granite in the south. Older rocks, partly of minor extension, are migmatite, quartzite, gabbro, porphyries, porphyrite and sediments of the Digerberg beds (Table 1). The main units are shown in Fig. 1.

The central, crystalline part of the structure has been cut by at least four generations of basic dikes (porphyrites and dolerites; Table 1); at least two of the dolerite groups continue beyond the Palaeozoic ring. As described elsewhere (Juhlin 1987, Karlsson 1987), a number of subhorizontal dolerite sheets, some of them being parts of seismic reflectors, have been pene-

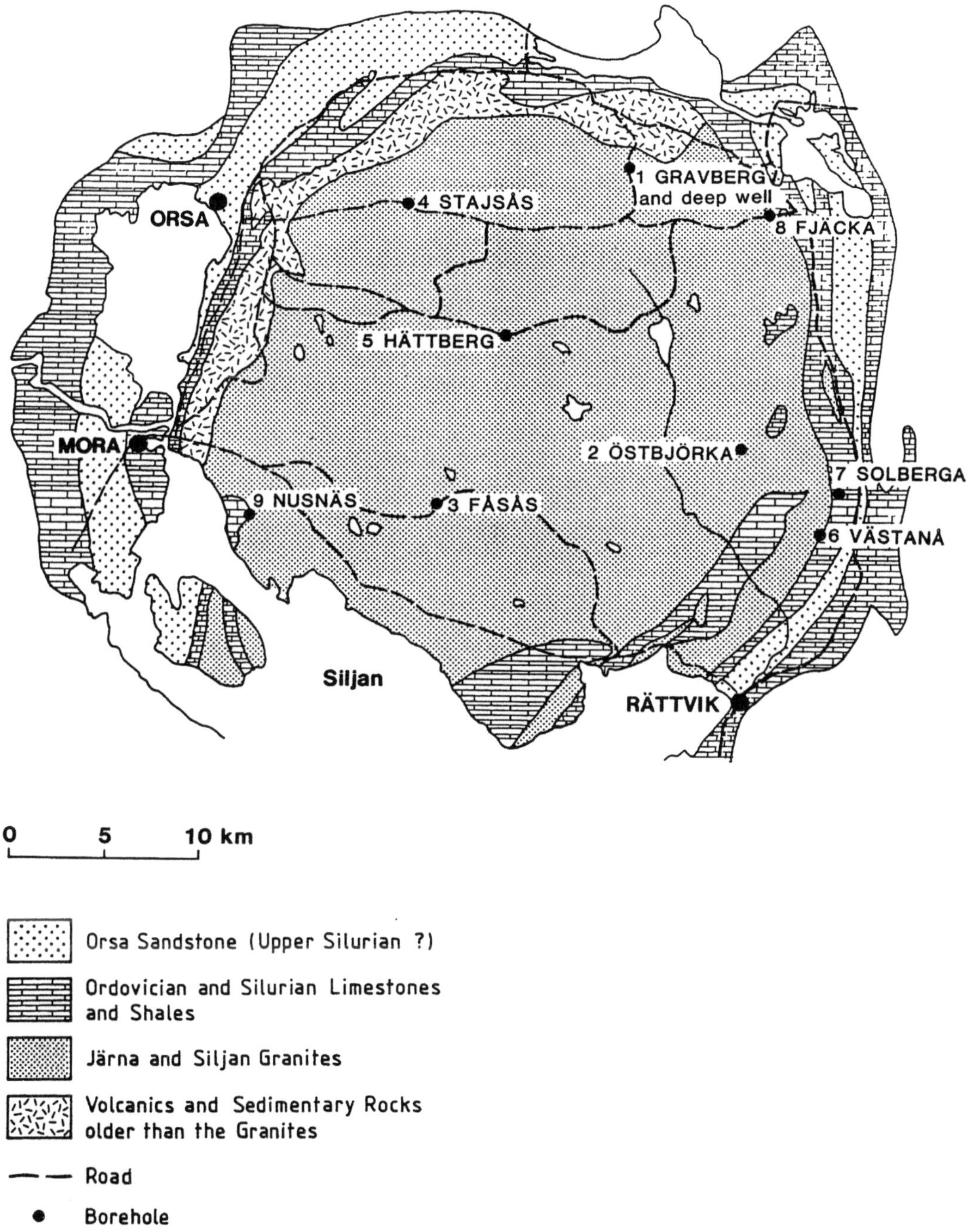

Fig. 1. Bedrock map of the central part of the Siljan Ring structure. (Simplified after Hjelmquist 1966)

trated in the deep well at Gravberg. Considering similarities in age (Table 1) and composition, the vertical NNW-SSE-striking dolerites are most likely feeders to the subhorizontal system.

It has been rumoured that some dolerite dikes, especially the NNW-SSE-striking ones, may be impact-generated or at least post-Ordovician. There are no facts available to support such an idea.

The ring of Palaeozoic sedimentary bedrock round the central cupola is largely a megabreccia built up of kilometre- to metre-sized blocks of Ordovician and Silurian sedimentary rocks. Exceptions may be the Orsa area and some other, small areas, such as the Solberga mound area, probably in original position or only slightly down-faulted. In the inner part of the breccia ring, especially in its eastern part, the abundance of granite blocks increases. In the outer part, crystalline blocks still occur, though rarely. These originate from the walls or rim of stages in the late evolution of the crater. In a couple of cases, the source area of the blocks can be traced. The thickness of the Palaeozoic ring is up to 350 m, according to seismic data.

Several shallow core borings sunk for prospecting purposes in the Palaeozoic ring show reversed stratigraphical sequences, with blocks of Ordovician rocks resting on upper Silurian Orsa sandstone. In most cases, the sandstone belongs to blocks in the megabreccia and not to undisturbed bedrock; it may, in turn, rest on Ordovician rocks and/or strongly brecciated Dala granite.

Impact Features

Shatter cones have been registered all over the central part of the structure. For some peripheral localities the term conoid structure is used until its true character has been revealed - if at all possible in these areas of low shock pressure. Systematic measurements of the orientation of the cone axes are being made. They are, however, often difficult to accomplish due to incomplete shape of the shatter cones. Preliminarily, the measurements point at a central target area of the meteorite.

Microscopic shock-metamorphic features are systematically registered in rocks from outcrops, from the core drillings and from the deep well at Gravberg. Shock pressure from the area (Svensson 1973, Tamminen and Wickman 1980, Grieve personal communication 1984; Åberg and Bollmark 1985) are in the interval 12-17 GPa. New measurements, so far, are of the same size. The fairly low values fit well with the deep erosion level of the impact area.

The melts of the area, now appearing as devitrified glasses, are interpreted as pseudotachylites formed through early modification movements. Along a broad dike of dolerite of Åsby type (Table 1) pseudotachylites of doleritic composition occur. One of these was used by Bottomley et al. (1978) for dating the impact using the ^{40}Ar-^{39}Ar method (Table 1). Later datings of similar pseudotachylites confirm the age found.

True impact melts have not, so far, been found. Melts from two localities in the northern periphery of the central area, with a more acid composition than that of all its crystalline rocks, are suspected to be true impact melts. Their geochemistry is, however, ambiguous, and their dating so far not definite.

Tectonics

Geological mapping and seismic and VLF profiles have revealed a system of concentric faulting of the central crystalline area certainly caused by the meteorite impact and the following modification movements. Due to the location of the geophysical profiles and the distribution of outcrops, the eastern and northernmost parts are best known. The Granån-Styggforsen fault zone cuts the northeastern part of the crystalline area in NNW-SSE direction close to the Gravberg deep well. The first movements in this zone are older than the concentric fault system. The age relation between these movements and the subhorizontal fracturing intruded by dolerites is not yet known. The zone was reactivated by the impact: faulting took place during the modification stage.

References

Åberg G, Bollmark B (1983) Rb-Sr-dating of a postorogenic Svecokarelian composite dike within the Siljan ring structure, central Sweden. Geol Foeren Stockholm Foerh 105:75-77

Åberg G, Bollmark B (1985) Retention of U and Pb in zircons from shocked granite in the Siljan impact structure, Sweden. Earth Planet Sci Lett 74:347-349

Åberg G, Collini B (1987) K-Ar analyses from the Deep Earth Gas Project at Siljan, southern central Sweden. A preliminary report (manuscript)

Åberg G, Fredriksson G (1984) Radiometric dating of the postorogenic Järna granite, south central Sweden. Geol Foeren Stockholm Foerh 106:171-174

Asklund B (1936) Discussion in: Thorslund P (1936) Några stratigrafiska och tektoniska rön från Siljansområdet. Geol Foeren Stockholm Foerh 58:380-387

Bottomley RJ, York D, Grieve RAF (1978) ^{40}Ar-^{39}Ar ages of Scandinavian impact structures: I Mien and Siljan. Contrib Mineral Petrol 68:79-84

Fredriksson K, Wickman FE (1963) Meteoriter (Meteorites). Svensk Naturvet 1963

Hjelmqvist S (1966) Beskrivning till berggrundskarta över Kopparbergs län. Summary: Description to map of the pre-Quaternary rocks of the Kopparberg County, central Sweden. Geol Surv Sweden Ca 40

Juhlin C (1987) Interpretation of the seismic reflectors in the Gravberg-1 well. 3rd Int Symp Obs Cont Crust Drilling. Mora and Orsa, Sweden, Sept 1987

Karlsson D, (1987) Investigations of rocks from the Gravberg-1 well: petrological and geochemical results. 3rd Int Symp Obs Cont Crust Drilling. Mora and Orsa, Sweden, Sept 1987

Kresten P (1986) Geochemistry and tectonic setting of metavolcanics and granitoids from the Falun area, south central Sweden. Geol Foeren Stockholm Foerh 107:275-285

Lundqvist Th (1979) The Precambrian of Sweden. Geol Surv Sweden C 768

Nicolaysen LO, Duane MJ (1987) The Siljan structure interpreted as an internally generated feature and not a relict impact crater. Abstr. 3rd Int Symp Obs Cont Crust Drilling. Mora and Orsa, Sweden, Sept 1987, pp 79-80

Patchett PJ (1978) Rb/Sr ages of Precambrian dolerites and syenites in southern and central Sweden. Geol Surv Sweden C 747

Rutten MG (1966) The Siljan Ring of Paleozoic, central Sweden: a posthumous ringcomplex of a late Precambrian Dala porphyries caldera. Geol Mijnbouw 45:125-136

Svensson NB (1971) Probable meteorite-impact crater in central Sweden. Nature (London) Phys Sci 229:90-92

Svensson NB (1973) Shatter cones from the Siljan structure, central Sweden. Geol Foeren Stockholm Foerh 95:139-143

Tamminen J, Wickman FE (1980) Shock effects in pegmatitic quartz from the Siljan ring structure, central Sweden. Geol Foeren Stockholm Foerh 102:275-278

Thorslund P (1961) Rättvikstraktens geologiska historia. Rättvik

Wickman FE, Blomqvist NG, Geijer P, Parwel A, v. Ubisch H, Welin E (1963) Isotopic constitution of ore lead in Sweden. Ark Mineral Geol 3:11

Deep Crustal Structure in the Vicinity of the Siljan Ring

C. E. LUND, R. G. ROBERTS, T. DAHL-JENSEN and J. LINDGREN[1]

Introduction

The Siljan Ring Structure is generally believed to be an impact crater caused by a meteor impact about 362 million years ago (Bottomley et al. 1978). Some of the most convincing evidence that the structure is an impact crater is provided by the characteristic ring structure and the presence of shatter cones and impact melts (Collini this Vol.; Fredriksson and Wickman 1963; Wickman et al. 1963; Svensson 1971, 1973; Stöffler 1971; Rondot 1975; Thorslund and Auton 1975; Tamminen and Wickman 1980). While there is still a certain amount of evidence against the ring structure being due to such an impact (Wickman et al. 1963; Hjelmqvist 1966; Svensson 1971), the above evidence is generally considered convincing. According to Grieve (this Vol.) the Siljan Ring Structure belongs to the group of complex terrestrial impact craters with a developed central topographic peak and a rim area, complexly faulted. A complex impact structure is the final result of very complicated processes which at present are not fully understood. When a projectile hits the target rock, an excavated crater is formed by the shock wave. This excavated crater develops into the transient crater during the rarefraction part of the shockwave. The transient crater exists as an unstable phase and the whole cratering process is developed into its final structure through the formation of a megablock zone in the rim area outside the transient crater which has now collapsed. The Siljan structure has been greatly eroded since the impact. Its pre-erosional diameter has been estimated to about 52 km. The erosion in the center of the structure is estimated to at least 1 km depth. The excavation of material caused by the impact was most probably confined to depths less than about 3-4 km and the probable diameter of the excavated crater has been estimated to 26-34 km. The cavity depth, that is the depth of transient crater, may be two to three times deeper than that for the excavated crater. The diameter of the transient crater might be slightly larger then that for the excavated crater (Grieve this Vol.; Croft 1985).

Gravity data shows a Bouguer gravity anomaly with a relative amplitude of about -20 mgal, lying in the northeastern part of

---

1 University of Uppsala, Section of Solid Earth Physics, Box 556, 751 22 Uppsala, Sweden

the structure (Dyrelius 1987). Three-dimensional modeling shows that this gravity low may be explained by a mass deficit extending down to a depth of about 7 km. Thus at its present stage of interpretation, the gravity data is unable to resolve any deviation from lateral homogeneity at a depth of over a few kilometres.

A dense grid of magnetotelluric measurements have been made to map the electrical conductivity structure of the impact area (Zhang et al. 1987). The resistivity is generally about 10^4 ohm/m from the surface down to the "lower crust" at about 20 km. In the seismically defined central uplift area, two-dimensional modeling has shown enhanced electrical conductivity from a depth of about 4 km. One possible explanation for this can be the existence of fluids migrated through fractures caused by the impact (Juhlin and Pedersen 1987). Interpretation of the magnetotelluric data relating to the lower crust is difficult, presumably because of complex crustal structure. The data at periods corresponding to lower crustal depths shows a highly anisotropic impedance tensor. Such anisotropy is observed at many locations in SW Sweden (Rasmussen 1987). The estimated lower crustal resistivity is 50-500 ohm/m, which may be slightly more conductive than can be considered "typical".

As part of the Swedish State Power Board's extensive studies prior to the commencement of the deep drilling at Gravberg, approximately 80 km of reflection seismic data was shot during 1985 by the University of Uppsala (Fig. 1). Explosives of 5 and 10 kg were used as a source. The shot and geophone spacings were 150 to 200 m and 50 m respectively. The recording and shooting parameters used for the profiles 5 and 7 are seen in Table 1. After trial studies using geophone strings, single 28 Hz geophones were used on most of the profiles. These were carefully mounted - for the profiles shot in winter the geophones were mounted in holes drilled in the frozen topsoil or ice. Various different shot-receiver configurations were used, with a maximum offset of 6 km for the data presented here.

Figure 2 shows a CDP-stack of data from profiles 5 and 7 (see Table 2 for a summary of processing steps). In this paper we restrict ourselves to a discussion of the deeper structure. The more shallow structures are discussed in some detail by Juhlin and Pedersen (1987). The seismic reflection measurements revealed a highly reflective upper crust. Several unusually strong reflectors were recorded down to about 3 s TWT. Later drillings down to about 6 km depth identified these reflectors as being caused by dolerites or a series of dolerites and neighboring fracturing. In the interpretation by Juhlin and Pedersen (1987), it is concluded that the diameter of the transient crater developed by the impact most probably is about 22 km. This is considerably smaller than was earlier estimated from surface geology alone. The uplift at a depth of 5 to 8 km was estimated to be 3 km.

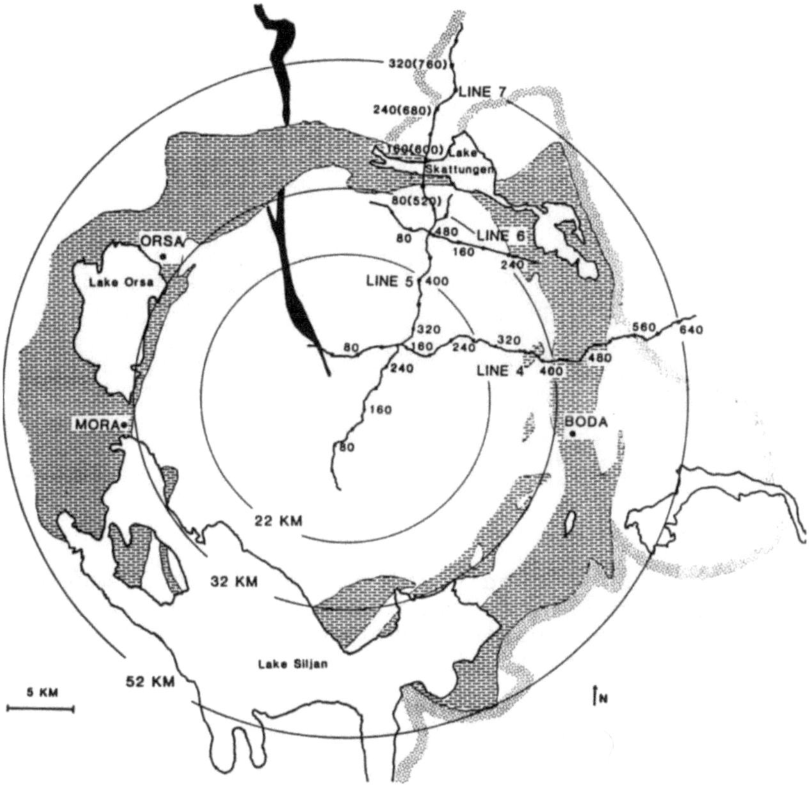

Fig. 1. Map of the Siljan ring area, showing the position of the four main seismic reflection profiles

Table 1. Recording and shooting parameters

	Profile 5	Profile 7
Channels	96	120
Geophones	28 Hz single	28 Hz single
Sampling rate	1 ms	2 ms
Charges	5 kg TNT	10 kg TNT
Fold	12	20
Geophone group distance	50 m	50 m
Shot point distance	200 m	150 m
Shooting configurance	split spread	end on
Recording length	20 s	20 s

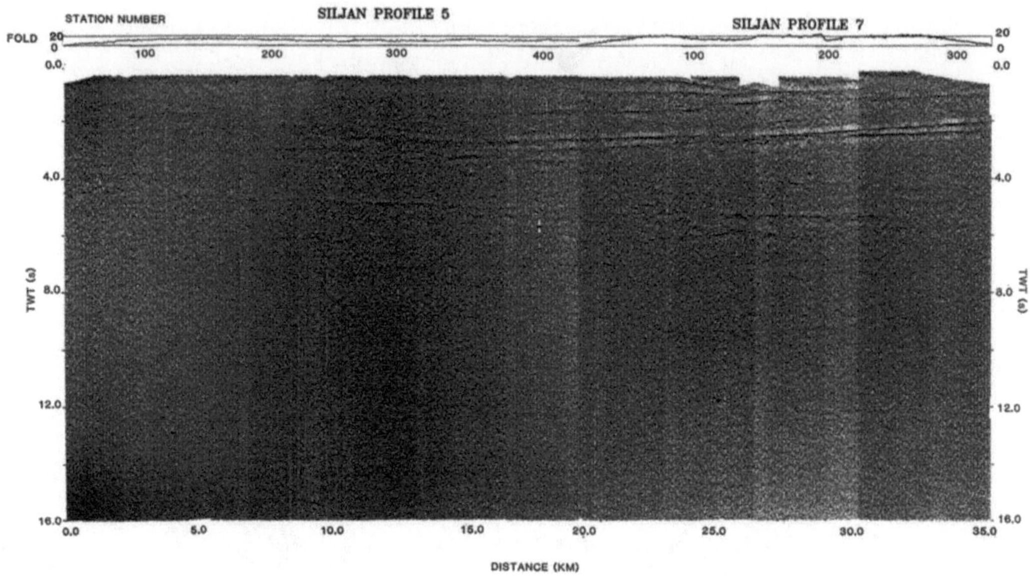

Fig. 2. A CDP Stacked section of profiles 5 and 7 (Fig. 1). The processing steps are described in Table 2. The stack is not true scale, the section covering 35 km and the total depth 16 s TWT corresponding to ~50 km

Deep Reflections

Between 3 and 13 s TWT several bands of reflected energy are observed. The first of these lies at 3-4 s TWT, which corresponds to a depth of about 9 to 12 km. Much of this energy can be piecewise correlated indicating a horizontally stratified structure. In the time window 4.2 to 6.3 s TWT (13-19 km) there appear to be several reflectors. The uppermost of these reflectors is dipping smoothly northwards from a depth corresponding to 4.2 s TWT (about 13 km) in the south to a depth corresponding to 5.3 s TWT (about 16 km) in the north. Considering the results from the profile 6 (perpendicular to profiles 5 and 7)

Table 2. Processing steps

Demultiplex	
Gain restoration	
Trace and record edit	
CDP sorting	
Static corrections	Elevation and residual (pr.7)
Trace balancing	Equal RMS
Digital bandpass filter	Lowcut 20 Hz, highcut 80 H
Trace mute	
NMO-removal	
Stack	
Trace scaling for display	400 ms AGC window

the actual slope of this reflector seems to be to the northwest. This reflector seems to be less well developed in the southernmost and central part of the profile. However, this might be an effect of different acquisition parameters on profiles 5 and 7, and the lack of residual statics on profile 5. Within 1 s below the above mentioned reflector other apparent reflectors are clearly recorded. In the time window 8 to 10 s (about 25 to 31 km depth) weak energy has been recorded giving the impression of a southwards dipping structure. At 12 to 14 s TWT (about 38 to 44 km depth) a band of reflectors are recorded.

The charges and experimental configuration used were designed to be optimal in resolving the relatively shallow structure (down to about 15 km). Thus the signal strength and signal-to-noise ratio are rather low at travel times corresponding to middle and lower crustal depths. We can enhance the signal-to-noise ration and obtain a more objective analysis of the lower crust by examining power and coherence measures stacked over a group of CDP's. An example of such results is presented in Fig. 3. These results can be regarded as representative for data from the Siljan area. We present (A) Uncorrelated power - After normal moveout the power within a moving time window is calculated for each individual trace. The power at corresponding TWTs is then summed for each CDP, and these powers are then added. (B) Correlated power at each CDP - The data is CDP stacked and the power within moving time windows is calculated. The power from all CDP's is summed. (C) Summed semblance - The semblance (Hatton et al. 1986) is calculated at each individual CDP within the same time windows as above, and the semblance from all CDP's are added. (D) Total stack power - A normal moveout is applied and all traces are summed. The power of the stacked trace within each window is calculated. (E) Total semblance - The semblance associated with the total stack power (E). The time window used was 200 ms and the time step 66 ms.

The features interpreted from the CDP stacked sections are now clearly seen, and can be more clearly resolved. The apparent reflectors at 5-6 and 11-14 s TWT can be clearly seen in the uncorrelated power plot. The feature at about 9 s is not apparent there, but can bee clearly seen on the semblance plot.

Comparison with Seismic Results from Other Areas

It is interesting to attempt to compare our results about the middle and lower crust in the Siljan area with results from other areas on the Baltic Shield, in order to try to answer two questions; (1) Is there any evidence that the deeper part of the crust has been affected by the supposed meteor impact? (2) Apart from the ring structure itself, is this area typical for the Baltic Shield? If not, this might suggest that the more shallow results even from outside the supposed impact area are atypical for the shield as a whole. Comparable results for this part of the Baltic shield are rather sparse, and we are forced to consider data collected at some distance from Siljan, and

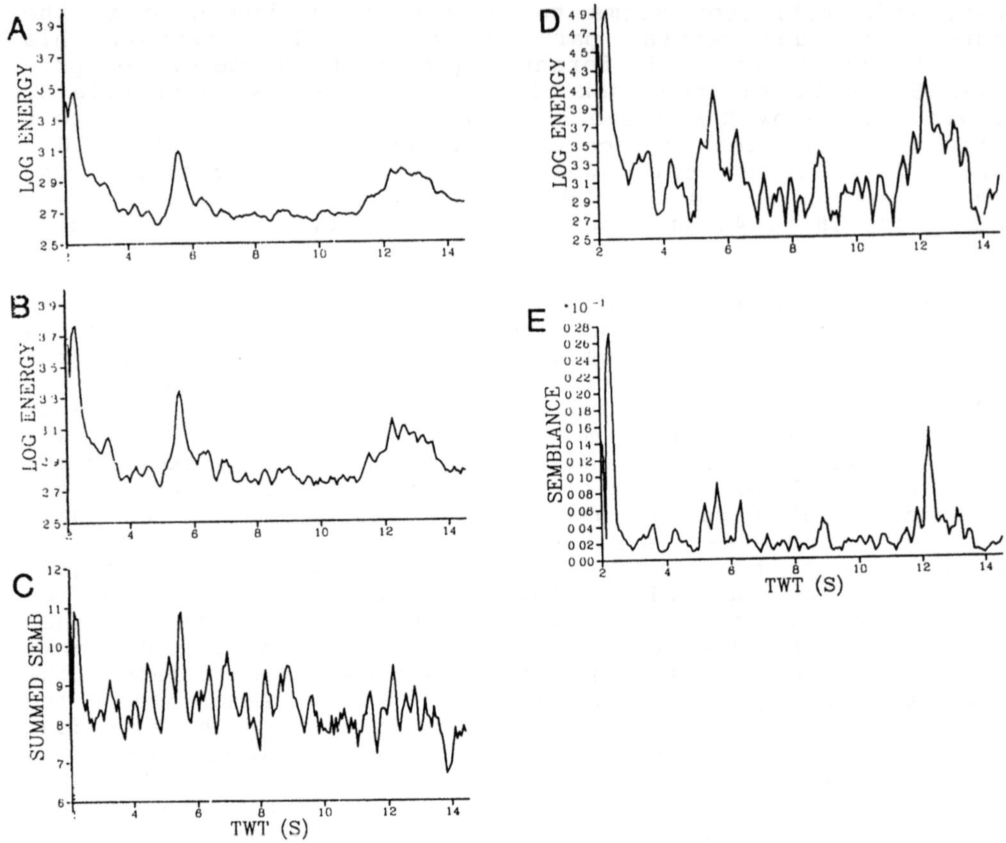

Fig. 3 a-e. a Uncorrelated power. b Correlated power. c Stacked semblance. d Total stack power. e Total semblance for 50 CDP gathers starting at station 290 on profile 7 (for a full explanation see text). The power is scaled to allow for the decay of amplitude with time due to spherical spreading. A linear noise trend has been removed

even lying on other structural units. However, as discussed below, certain common features are found in several of the data sets.

One parameter which has been estimated by various methods at a large number of locations is depth to Moho. The earthquake studies of Slunga (1985) indicate a Moho depth in the Siljan area of about 45 km, although resolution appears to be rather poor in this area (depending on earthquake and receiver locations). This can be considered to be a depth fairly typical for the Baltic Shield for results from this technique.

Spectral ratio results from Hagfors seismological observatory which is believed to lie on the same structural unit as the Siljan Ring Structure, about 100 km to the SW, give a Moho depth of 45 km. Similar analyses of data from Uppsala (200 km SE of Siljan) give 40 km (Bungum et al. 1980).

The FENNOLORA long-range refraction profile ran some 100 km to the east of Siljan. Models from what we consider to be the most relevant part of FENNOLORA for comparison indicate velocity boundaries at about 7, 20 and 44-48 km depth (Guggisberg 1986). The 20-km boundary is normally regarded as marking the transition from the upper to the lower crust, with an associated velocity increase from about 6.4 to 6.6 km/s. The Moho boundary varies between 44 and 48 km depth with a velocity increase from about 7.3 to 8.1 km/s. The latter two boundaries presumably correspond to the bands of energy at 5-6 and 12-13 s TWT recorded in the Siljan Ring at about 16-19 and 38-43 km depth respectively (Fig. 2).

Seismic reflection results from Värmlandsnäs, some 200 km to the South-west of Siljan show a rather reflective lower crust, with distinct bands of energy at 4, 9, 10 and 13 s TWT corresponding to about 12, 28, 32 and 43 km depth respectively (Lund et al. 1987). The reflectors recorded at 4 s TWT (12 km depth) are interpreted as the upper/lower crust transition. The Moho is considered to be marked by the lower boundary of the reflective band at 13 s TWT (43 km). The Värmlandsnäs results can be considered representative of results from that part of the EUGENO refraction-reflection profiles which was on the part of the shield closest to Siljan (see Lund et al. 1987). However, all the EUGENO profiles, including the Vämlandsnäs reflection profile, are situated on a geologically younger crustal unit than all the above discussed results.

Conclusions

Comparison of the data sets indicates that a "typical" section through the crust on this part of the Baltic Shield has a gradual, possibly discontinuous, increase in velocity down to depth varying between 12 and 20 km, where a velocity increase of about 0.2-0.3 km/s occurs. This increase in velocity is regarded as the boundary between the upper and lower crust. In some areas, this velocity increase is most probably discontinuous. In the Siljan Ring area the upper/lower crust transition may be marked by the dipping reflector at 13-16 km or by that at 19-km depth. It is not possible at this stage to judge whether or not the reflector recorded at 13-16 km depth is of the same origin as the shallower reflectors, which have been shown to be made up of dolerites. This reflector is not well developed in the area of Gravberg drill hole. If this reflector is caused by dolerite intrusion, comparison with the FENNOLORA data suggested that there might be a discontinuous velocity increase associated with the 19 km reflector.

The comparison also shows a great variation in the Moho depth between 38 and 46 km in the area which we have considered. The transition between the lower crust and the upper mantle is often observed as a strong band of reflectors spanning about 1 to 2 s TWT. In reflection seismics the Moho is sometimes defined where the number of reflections decreases (Gibbs 1986). In the Siljan Ring area the Moho seems to appear at a depth of about 45 km. At this stage it is not possible to resolve unambiguously whether the Moho appears layered due to real layering, a rough (laterally inhomogeneous) Moho, or lateral velocity inhomogeneites above the Moho, although an interpretation of similar data in terms of true layering is preferred by many authors (Braile and Chiang 1986; Behrens et al. 1987; Deichman and Ansorge 1983; Sandmeier and Wenzel 1986).

Some of the weak reflectors above the Moho in the Siljan data are dipping. Possible explanations for these include compositional changes, mylonitic zones and fault planes (Smithson et al. 1986; Jones and Nur 1984). These and other possible explanations for lower crustal reflectors do not suggest that dipping structures are implausible.

We can conclude that the presently available data do not indicate that the lower crust in the Siljan Ring area is anomalous for the region, and no influence from the impact can be resolved. The only possible exception to this is the magnetotelluric data, but it is not clear whether the deviation of lower crustal resistivity from "typical" values can be resolved.

References

Behrens K, Goldflam S, Heikkinen P, Hirschleber H, Lindqvist G, Lund C-E (1987) Reflection seismic measurements along the granulite belt of the Polar profile in the northern Baltic Shield, northern Finland. Tectonophysics (accepted)

Bottomley RJ, York D, Grieve RAF (1978) 40Ar-39Ar ages of Scandinavian impact structures: Mien and Siljan. Contrib Mineral Petrol 68:79-84

Braile LW, Chiang CS (1986) The continental Mohorovicic discontinuity: Results from near-vertical and wide angle reflection studies. In: Barazangi M, Brown L (eds) Reflection seismology: The continental crust. AGU Geodyn Ser 14. AGU, Washington DC, pp 257-272

Bungum H, Pirhonen SE, Husebye ES (1980) Crustal thichness in Fennoscandia. Geophys J R Astron Soc 63:759-774

Croft SK (1985) The scaling of complex craters. J Geophys Res 90:C828-C842

Deichman N, Ansorge J (1983) Evidence for lamination in the lower continental crust beneath the Black Forest (Soutwestern Germany). J Geophys 52:109-118

Dyrelius D (1987) The gravity field of the Siljan Ring structure Presented at:3rd Int Symp Observ Cont Crust through Drilling, Mora, Sweden, Sept 7-10

Fredriksson K, Wickman FE (1963) Meteoriter. Sven Nat 16:121-157

Gibbs AK (1986) Seismic reflection profiles of Precambrian crust: A qualitative assessment. In: Barazangi M, Brown L (eds) Reflection seismology: The continental crust. AGU Geodyn Ser 14. AGU, Washington DC, Ser 14 pp 95-106

Guggisberg BC (1986) Eine zweidimensionale Refraktionsseismische Interpretation der Geschwindigkeits-Tiefen-Struktur des Oberen Erdmantels unter dem Fennoscandischen Schild (Projekt Fennolora). Diss ETH N 7945, Zurich, 199 pp

Hatton L, Worthington MH, Makin J (1986) Seismic Data Processing. Blackwell, Oxford, 177 pp

Hjelmqvist S (1966) Beskrivning till berggrundskarta över Kopparbergs län. Swed Geol Surv, Uppsala, Ca 40

Jones TD, Nur A (1984) The nature of seismic reflections from deep crustal fault zones. J Geophys Res 89:B5:3153-3171

Juhlin C, Pederson LB (1987) Reflection seismic investigations of the Siljan impact structure, Sweden. J Geophys Res 92:813:14113-14122

Lund C-E, Roberts RG, Juhlin C (1987) The reflectivity of the lower crust in SW Sweden. Ann Geophys 5B:(4):375-380

Rasmussen T (1987) Magnetotellurics in southwestern Sweden; Evidence for anisotropy in the lower crust. J Geophys Res (in press)

Rondot J (1975) Comparaison entre les astroblemes de Siljan, Suede, et de Charlevoix, Quebec. Bull Geol Inst Univ Uppsala 6:85-92

Sandmeier K-J, Wenzel F (1986) Synthetic seismograms for a complex crustal model. Geophys Res Lett 13:1:22-25

Slunga R (1985) The seismicity of Southern Sweden, 1979-1984, Final Rep. FOA Rep C 20572-T1

Smithson SB, Johnson RA, Hurich A (1986) Crustal reflection and crustal structure. In: Barazangi M, Brown L (eds) Reflection seismology in the continental crust. AGU Geodyn Ser 14: AGU, Washington DC pp 21-32

Stöffler D (1971) Progressive metamorphism and classification of shocked and brecciated crystalline rocks at impact centers. J Geophys Res 76:5541-5551

Svensson N-B (1971) Probable meteorite impact crater in central Sweden. Nat Phys Sci 229:3:90-92

Svensson N-B (1973) Shatter cones from the Siljan structure, central Sweden. Geol Foeren Stockholm Foerh 95:137-143

Tamminen J, Wickman FE (1980) Shock effects in pegmatic quartz from the Siljan ring structure, central Sweden. Geol Foeren Stockholm Foerh 102:275-278

Thorslund, P. and Auton, C., (1975). Evidence of meteorite impact in the Siljan structure, central Sweden. Bull Geol Inst Univ Uppsala NS 6:69-72

Wickman FE, Blomqvist NG, Geijer P, Parwel A, Ubisch H v, Welin E (1963) Isotopic constitution of ore lead in Sweden. Ark Mineral Geol 3:193-257

Zhang P, Rasmussen T, Pedersen LB (1987) Electrical resistivity structure of the Siljan impact region. J Geophys Res (in press)

MIX
Papier aus verantwortungsvollen Quellen
Paper from responsible sources
FSC® C105338

If you have any concerns about our products,
you can contact us on
ProductSafety@springernature.com

In case Publisher is established outside the EU,
the EU authorized representative is:
Springer Nature Customer Service Center GmbH
Europaplatz 3, 69115 Heidelberg, Germany

Printed by Libri Plureos GmbH
in Hamburg, Germany